Materials for High Temperature
Power Generation and Process Plant Applications

Also from IOM Communications

B726 Cyclic Oxidation of High Temperature Materials

B708 Advanced Heat Resistant Steel for Power Generation

B689 Advances in Turbine Materials, Design and Manufacturing

B723 Microstructural Stability of Creep Resistant Alloys for
High Temperature Plant Applications

B667 Microstructural Development and Stability in
High Chromium Ferritic Power Plant Steels

Forthcoming

B693 High Temperature Surface Engineering

B736 Parsons 2000

Materials for High Temperature Power Generation and Process Plant Applications

Edited by
A. Strang

Proceedings of the session on High Temperature
Power Plant and Process Plant Applications
from the Institute of Materials
Materials Congress '98 –
Frontiers in Materials Science and Technology

Book 728
Published in 2000 by
IOM Communications Ltd
1 Carlton House Terrace
London SW1Y 5DB

© IOM Communications Ltd 2000

IOM Communications Ltd
is a wholly-owned subsidiary of
The Institute of Materials

ISBN 1 86125 103 3

Typeset in the UK by
Alden Bookset Ltd, Oxford

Printed and bound in the UK at
The University Press, Cambridge

Contents

Foreword

Materials Congress '98 – Frontiers in Materials Science and Technology, organised by the Institute of Materials, was held from the 6–8 April 1998 at the Royal Agricultural College, Cirencester. This event, which was attended by more than 450 delegates from academia and industry, consisted of technical sessions dealing with materials issues involving rubbers, polymers, composites, ceramics, ferrous and nonferrous metals and alloys in applications ranging from medical prosthesis to advanced power plant engineering.

These proceedings contain the ten most important papers presented in the session of the Congress concerned with materials for high temperature power plant and process plant applications organised by the Institute of Materials High Temperature Materials Performance Committee. The selected papers are largely in the form of critical reviews which not only highlight the development of materials to meet specific applications but also comment on solutions used for both current and future applications. The papers presented by Thornton and Fleming *et al.* are concerned with and highlight the problems associated with the selection of materials and manufacturing processes for critical components for large steam turbines and boiler plant where design lives in excess of 200 000 hours are now required at stresses of up to 300 bar and temperatures approaching 600°C. Starr's paper critically reviews materials for advanced heat exchanger applications such as those used in indirectly fired and recuperative gas turbines, fluidised bed combustion, coal gasification and waste incineration systems, where the effects of corrosion often severely life-limits components. The paper by Barnes *et al.* considers the key issues concerned with the welding and fabrication of key components for advanced power plant particularly in terms of their properties and expected service performance. Material integrity is of paramount importance in critical components such as welded steam chests, pipework and large rotors for large turbine generator applications and the material considerations concerned with defect tolerance for such applications is critically reviewed in the paper by Holdsworth. A review of typical problems encountered during the operation of high temperature plant is presented in the paper by Townsend while the extensive review by Bhadeshia *et al.* considers present assessment procedures for the life extension of power plant components. The important role of surface engineering in the reliable operation of all high temperature power and process plant is reviewed in the paper by Nicholls and Rickerby. Significant developments are still continuing in this field with 'designed surfaces' likely to become a routine part of design procedures for many future component and plant applications. Surface engineering already plays a significant part in ensuring that many gas turbine components such as blades, combustion cans and other hot gas parts of the system achieve their planned design lives. The paper by Piearcey considers these problems as well as others associated with the use of poor quality fuels and marine environments with the effects on the performance of high temperature alloys in industrial gas turbine plant. Finally Dominy *et al.* present an engineer's view in a review of the application and role of ceramics and CMCs in modern aero gas turbines. These reviews not only discuss the evolution of advanced materials for high temperature power and process plant applications but also indicate the developments in materials which must take place to achieve the higher efficiency and environmentally acceptable plant for the 21st Century and beyond.

As this book goes to print Materials Congress 2000 is already planned for 12–14 April 2000 and addressing the theme of Materials for the 21st Century.

Andrew Strang
Chairman of the Institute of Materials
High Temperature Materials Performance Committee

Progress in the Manufacture of Materials for Advanced High Efficiency Steam Turbines

D. V. THORNTON
ALSTOM Energy Ltd, Steam Turbines, Newbold Road, Rugby CV21 2NH, UK

ABSTRACT

Marked improvements in both the unit size and the thermal efficiency of steam turbines have occurred over the last forty years. To a large degree this has been the direct result of advances in the available size, quality and mechanical properties of the forgings and castings from which the most critical components are manufactured. The typical unit size in the 1950s was about 100 MW whereas fossil-fired units currently being tendered can develop as much as 1000 MW and nuclear powered units up to 1500 MW from one single line machine. Over this period the thermal efficiency of fossil-fired units has been improved from 30% to 49%. This increase in the efficiency of fuel conversion not only provides economic benefit but also serves to reduce power station emissions, a 1% efficiency increase of a 680 MW unit resulting in a lifetime reduction of some 0.8 million tonnes of carbon dioxide. The improvements in unit size and efficiency have been achieved by advances in the steam conditions from 100 bar/500°C to 300 bar/600°C, made possible by the improved high temperature strength of available alloy steels, and technological progress in the manufacture of large forgings and castings.

The metallurgical developments underpinning these advances have been in a variety of disciplines which include steelmaking, forging, heat treatment, foundry technology, alloy design and fabrication processes.

1 INTRODUCTION

It is now well over 100 years since the steam turbine generator was patented by Sir Charles Algernon Parsons. In 1884 he constructed his first machine which ran at 18 000 rpm, the rotor being 75 mm in diameter and the power output 10 hp (7.5 kW) generated from saturated steam at 80 psi/156°C.[1]

Today more than 60% of the world's electrical power requirement is generated by steam turbines, Fig. 1. The source of the steam may be a conventional fossil fired boiler, a nuclear reactor or the waste heat recuperation from a gas turbine in a modern combined cycle power station. In all cases the function of the steam turbine is to convert the maximum amount of the available heat energy into electrical power.

The first steam turbine to enter service in a public power station was a 75 kW Parsons machine at Forth Banks in Newcastle in 1890. Subsequently unit size evolved such that by the 1950s the worldwide average was 50 to 100 MW. There

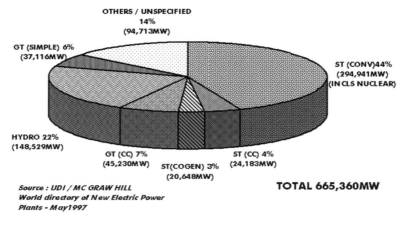

Fig. 1 Worldwide electrical power generation.

was a rapid increase in machine output through to the 1970s, as shown for the UK in Fig. 2. The trend then stabilised until this decade which has seen an increase in the maximum size of nuclear units to 1500 MW, a consolidation of fossil-fired units in the range of 300 to 680 MW with a maximum of 1000 MW and a demand for low output, 100 to 150 MW, machines to operate in combined cycle with gas turbines.

In parallel with this development in unit size has been a gradual increase of unit efficiency from about 10% in 1900 to approaching 50% today, with an expectation of 55% or more by 2020, Fig. 3. Contributing to this improvement

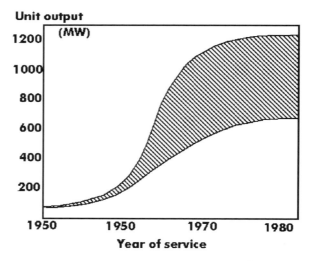

Fig. 2 Increase in turbine unit size.[2]

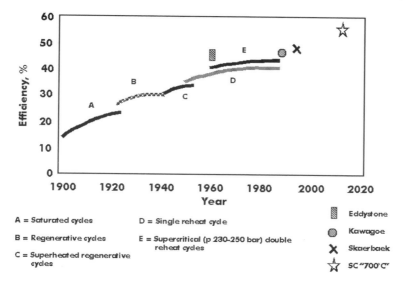

Fig. 3 Steam turbine efficiency trend.[3]

in efficiency of the cycle is the increase in steam parameters, the increase in power of individual machines, an improvement in blading efficiency, a reduction in losses, the introduction of single/double reheat and improved feedwater heating systems. In recent times there has been an increasing awareness of environmental effects and in particular CO_2 emissions. Coal fired power stations are usually identified as being responsible for these emissions although transport is the main CO_2 producer. Nonetheless an increase in the conversion efficiency of coal to electricity from 30 to 50% would reduce CO_2 emissions from coal fired power stations by about 45%. This environmental concern coupled with a reduction in fuel costs (a $+1\%$ improvement in the thermal efficiency of a 660 MW fossil-fired turbine results in a fuel saving amounting to £1.4 million per year) has provided a significant incentive to further improve the efficiency of fossil fired power plant. The inlet steam conditions for fossil-fired plant has been at a level of 540°C or 565°C and sub-critical pressure, typically 170 bar, since the 1960s. It is only recently, with the increased emphasis on the economics and environmental impact of power generation, that there has been serious interest in the use of higher inlet temperatures and pressures in a supercritical cycle to maximise the efficiency of the fuel conversion.

It is the intention of this paper to review the metallurgical developments which have occurred in the fields of steelmaking, forging and foundry technologies, alloy development and fabrication procedures over the past forty years and show that many of the advances which have been made in steam turbine technology in this period are the outcome of these developments.

2   ROTOR FORGINGS

Acid open hearth (AOH) and basic open hearth (BOH) steelmaking was still employed for the manufacture of rotor and disc forgings in the 1950s. Introduced at the end of the 19th century, the refining was by slag/metal reaction and the deoxidation products, MnO and SiO_2, floated into the slag. In the AOH process there was no removal of sulphur or phosphorus, however the oxidising slag in the BOH process did reduce these elements to some degree. The effect of high S and P content on the fracture toughness properties of disc forgings was catastrophically demonstrated in 1969 when a steam turbine disc in a 60 MW turbine at Hinkley Point A suffered a fast fracture, Figs 4 and 5.[4] Subsequent investigation showed that in the high S and P steel, which was of large grain size and contained considerable chemical segregation, the fracture toughness was only about 40 MPa√m. At the service stress level this degree of toughness was only adequate to tolerate a 1.5 mm defect, Fig. 6.

Although electric arc steelmaking was first introduced in 1900, it had not found widespread usage and by the 1950s this type represented only about 10% of all steel produced with furnace capacity at some 50 tonnes. The process uses a double slag procedure and had the ability to remove phosphorus in the first iron-ore containing oxidising lime-rich slag. The furnace was then tilted to remove

Fig. 4 Catastrophic fracture of disc.[4]

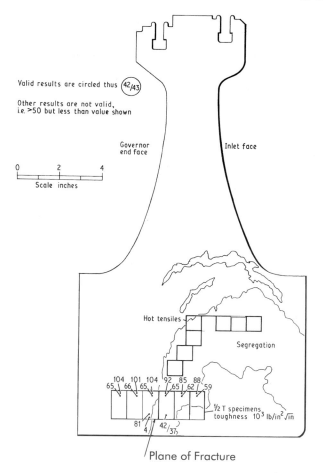

Valid results are circled thus (42/43)

Other results are not valid,
i.e. >50 but less than value shown

Governor
end face

Inlet face

0   2   4
Scale inches

Hot tensiles

Segregation

104 101 104 92 85 88
65   66   65   65 62 59

½ T specimens
toughness 10^3 lb/in^2√in

81   42/37

Plane of Fracture

Fig. 5 Effect of segregation on toughness.[4]

this slag and further refining under a fluorspar containing lime-rich slag reduced the sulphur level, producing a relatively fine grained and cleaner steel with typical S and P levels of 0.010%. Basic electric furnaces of 100 tonnes capacity were commissioned in the 1960s and, with the introduction of ladle holding furnaces, 250 tonne ingots became available for the production of large monobloc forgings in 1970. Later introduction of the ladle-refining furnace has enabled up to 600 tonnes of hot metal to be assembled and poured to form large ingots of a similar weight. The logistics for assembling the metal and pouring an ingot of 570 tonnes weight and 4.2 m diameter is shown in Fig. 7.[5]

Electroslag remelting is used in special cases where the added costs can be justified, particularly to minimise segregation and to improve ingot soundness. The advantages result from the relatively shallow molten pool but this process is

inlet side

0   0.5

1.5"

FRACTURE ORIGIN

Fig. 6 Fast fracture initiating at a stress corrosion crack.[4]

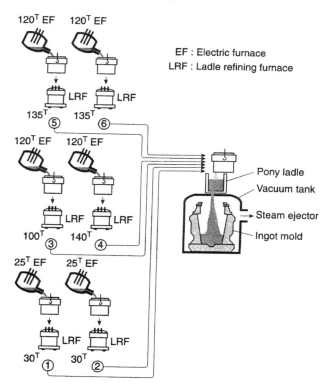

EF : Electric furnace
LRF : Ladle refining furnace

Pony ladle
Vacuum tank
Steam ejector
Ingot mold

Fig. 7 Manufacturing sequence of a 570T ingot.[5]

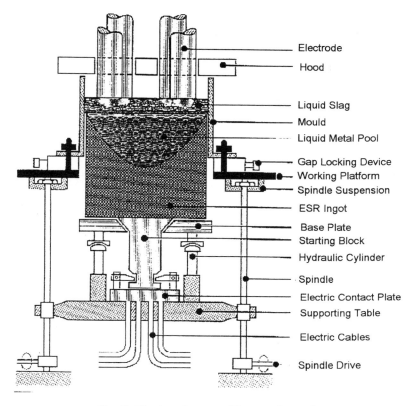

Electrode

Hood

Liquid Slag

Mould

Liquid Metal Pool

Gap Locking Device

Working Platform

Spindle Suspension

ESR Ingot

Base Plate

Starting Block

Hydraulic Cylinder

Spindle

Electric Contact Plate

Supporting Table

Electric Cables

Spindle Drive

Fig. 8 Electro slag remelting equipment.

limited to ingots of typically 2300 mm diameter and ingot weights of about 160 tonnes. However, it is the only method of steelmaking capable of producing certain of the high alloy creep resisting materials in large ingots of the quality required for rotor forgings, Fig. 8.

To maximise toughness it has been realised since the 1960s that phosphorus together with the tramp elements such as arsenic, antimony, tin, etc., have to be reduced to very low levels to minimise grain boundary temper embrittlement during manufacture and also in service, as evidenced by a loss of toughness i.e. an increase in FATT with high temperature ageing. Since the tramp elements cannot be removed by refining there has been the necessity to exert a high level of control on the raw materials and especially in the scrap selection to improve the toughness of rotor quality alloy steel forgings. This philosophy has culminated in the development of Superclean 3.5%NiCrMoV steel which has been shown to be essentially immune to temper embrittlement. The difference in chemistry between Superclean and conventional melts may be seen in Fig. 9. In service if the rotor temperature exceeds about 320°C there may be a decrease in the available toughness of rotors with conventional levels of cleanliness, the effect being a

	Analysis (wt%)	
	Conventional Steel (Average of 100 heats)	**Superclean Steel** (Average of 42 heats)
Mn	0.2250	0.0240
Si	0.0550	0.0190
Al	0.0065	0.0060
P	0.0045	0.0022
S	0.0025	0.0010
Cu	0.0660	0.0240
As	0.0090	0.0046
Sb	0.0010	0.00045
Sn	0.0050	0.0026

	Number of heats	J-factor	J′-factor	X-factor	K-factor
Superclean Steel	42	2.2×10^{-4}	4.3×10^{-4}	39×10^{-3}	1.7×10^{-3}
Conventional Steel	100	27×10^{-4}	53×10^{-4}	79×10^{-3}	22×10^{-3}

$$J = (\%Si + \%Mn)(\%P + \%Sn)$$
$$J' = (\%Si + \%Mn)(\%P + \%Sn + \%As + \%Sb)$$
$$X = (10 \bullet \%P + 5 \bullet \%Sb + 4 \bullet \%Sn + \%As)$$
$$K = (\%Si + \%Mn) \times X$$

Fig. 9 Analysis and embrittlement factors of superclean and conventional 3.5NiCrMoV steels.[6]

maximum at 450°C where the fracture appearance transition temperature may increase by as much as 150°C after about 30 000 hours service, Fig. 10. For rotors operating above 320°C it may be considered worthwhile to pay the 20 to 30% premium for superclean steel which results from the cost of low residual scrap and alloying additions, the electricity costs due to the extra steelmaking residence time and the use of dolomite lined ladles to prevent contamination.

This development was a consequence of the physical metallurgical under-standing of the phenomenon of temper embrittlement derived from the application of newly available techniques such as Auger spectroscopy to identify the elements responsible and the skills of the material producers to manufacture such a steel on an industrial scale.

A most important aspect in ensuring that a large rotor forging will be of the required high standard of integrity in terms of the soundness, cleanliness and chemical uniformity is the ingot technology. There are many classical cut-ups of ingots which show the typical unsoundness, due to primary piping and secondary shrinkage, and the V- and A-segregates as in Fig. 11. The factors controlling these effects became generally known and controlled by experience. As a result larger ingots were gradually introduced. However these developments sometimes

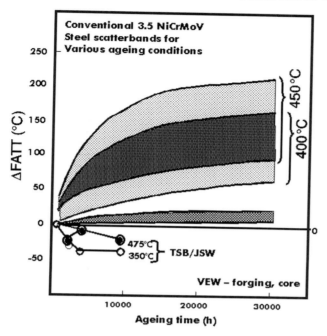

Fig. 10 Temper embrittlement.[7]

took place too rapidly. In the late 1960s developments in Germany were put in place to make an advance to a larger ingot of 250 tonnes. This was achieved by lengthening a standard ingot. The height to diameter (H/D) ratio was increased to 1.7. The ingot was then forged on a 6000 tonnes press to make a rotor 1760 mm in diameter and 7.5 m long in which the forging work was about 3:1. Some years later, in 1987 after 16 years and almost 58 000 hours service, this rotor burst during a routine restart of the machine.[8] The investigation showed that brittle fracture had initiated at a large original defect which comprised planar MnS inclusions and incompletely forged shrinkage, Fig. 12.

Large forging technology has developed since this time and to obtain a sound forging the H/D ratio is now maintained at a maximum of about 1.3, and forging is conducted typically on a 10 000 tonnes press with a minimum direct forging ratio of 3:1. As regards cleanliness the ingot head design and the hot topping practice are most important to ensure adequate feeding to minimise shrinkage and to allow non-metallic inclusions time to float to the top of the ingot. The use of exothermic and insulating powders and the method of application to the top of the ingot is critical in achieving a clean product.

The introduction of vacuum pouring in the late 1960s enabled the gas content of steels to be significantly reduced. After ladle refining, where the alloying additions are made and a vacuum treatment is applied, the molten steel is poured through a pony ladle into the mould which is contained within an evacuated

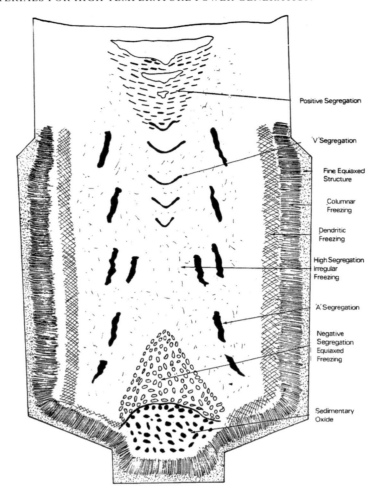

Positive Segregation

'V' Segregation

Fine Equiaxed
Structure

Columnar
Freezing

Dendritic
Freezing

High Segregation
Irregular
Freezing

'A' Segregation

Negative
Segregation
Equiaxed
Freezing

Sedimentary
Oxide

Fig. 11 Classical zone of solidification and segregation in a large killed steel ingot.

chamber. When the stream enters the vacuum it breaks up into liquid droplets and the hydrogen content is reduced to about 1 ppm. Formerly in order to reduce the gas content to less than 2 ppm a dehydrogenation treatment of several hundred hours at 250°C was applied in order to avoid the occurrence of hydrogen flakes.

As stated previously, the press sizes in use for producing the forging have not changed in the last forty years, being typically of 10 000 tonnes capacity. There is now an improved understanding of the forging process and the techniques for ensuring that the work penetrates to the centre of the workpiece. This has led to the development of special types of forging methods, such as broad flat tooling and the JTS procedure where the outside of the forging is allowed to cool prior to

Fig. 12 Catastrophic fracture of a large turbine rotor.[8]

forging so that the work is absorbed by the hotter central region. Through such procedures and with better control of the soaking times and temperatures and ensuring a minimum deformation ratio is applied a consistent level of quality can be achieved over a large range of forging size and material grades. It has been found that the final forging heat when the forging is 'sized' is of great importance in determining the properties which can subsequently be achieved. If the amount of deformation is too low there will be grain growth which does not respond to the grain refining heat treatment. Multiple austenitisations may be used in the refine anneal cycle, and for some grades also a pearlitic transformation, the precise procedure depending on the size of forging and the internal rules of the forgemaster. The objectives of these procedures are to condition the structure so that it responds to the quality heat treatment and produces a forging with fine grain size and thus low ultrasonic attenuation. High sensitivity ultrasonic inspection, to a 1.5 mm diameter minimum detectable defect size, is applied even to the largest of rotor forgings to ensure the integrity of these highly stressed components for their entire service life of 250 000 hours and some 7500 load cycles. The critical condition to be assured is the ability of the rotor to sustain a 120% cold overspeed at the end of service life.

Up to the late 1950s the relatively small forgings were normalised and tempered. As higher strength and hardenability through increasingly larger sections were demanded, initially fan and fog cooling were developed. Some 10 years later full vertical immersion quenching, first in oil and subsequently in water, were introduced in some forges whilst others installed water spray quenching facilities. Today methods are in place to quench and temper even the largest rotor forgings to develop a fully transformed microstucture at the centreline of forgings of almost 3000 mm diameter weighing some 260 tonnes.

These developments in steelmaking, ingot technology, casting, forging and heat treatment practices have enabled significantly larger, stronger, tougher and cleaner rotor forgings to be made available to the turbine design engineer. These improvements have been applied either to increase the rotor size for the design of units of higher output, or to permit higher steam conditions to be used to increase the cycle efficiency. In both cases this has led to an increase in the power density which can be achieved.

3   CASTINGS

Developments in the manufacture of turbine castings over the last forty years have resulted in a significant improvement in the integrity and increase in the size and complexity of components which can now be produced. Figure 13 indicates

Fig. 13 Large cast cylinders.

the large castings utilised on high temperature modules. The largest casting is the outer cylinder, weighing some 25 to 30 tonnes, and made in 2%CrMo. The other components (inner cylinder, diaphragm carriers, thermal shields and nozzle boxes) experience somewhat higher temperatures and are made from improved creep resistance, oil quenched 1%CrMoV steel or 9%CrMoVNbN. High pressure valve chests which receive the steam directly from the boiler and hence are subject to the highest temperatures and pressures and also the maximum thermal cycle are currently manufactured as castings, Fig. 14, whereas formerly closed die forgings were used to obtain the desired level of integrity.

The steelmaking methods are similar to those applied in the manufacture of forgings. Basic electric arc furnaces are used together with ladle refining. A double slag technique provides separate phosphorus and sulphur removal whilst argon bubbling lowers the gas content of the molten steel. The major benefit is that the reduced oxygen level allows lower sulphur contents to be achieved which renders the material less susceptible to hot tearing during solidification. Over the years a better understanding of solidification processes and how these affect defect formation have been developed. Contraction allowances and the degree of taper on wall thickness to promote directional solidification have been established and a significant progressive reduction in shrinkage and tearing type defects has been observed.

Design features which led to foundry defects have been identified and have to a large degree been eliminated. An example of this was in early design of valve chests the bowl was of a cube shape, being separated from the adjacent chamber by ribs. These features provided relatively sharp corners and a high degree of

Fig. 14 HP steam chest assembly.[9]

constraint and formed sites for hot tears. Furthermore this geometry was not conducive to the flotation of non-metallic inclusions. Hence there were significant weld repairs at these positions. In service these same sites concentrated the thermal stresses leading to the development of significantly sized cracks. A palliative was to change the design from a cube shape to an onion shape thereby eliminating many of the geometrical features which promoted casting defects, as shown schematically in Fig. 15. Today, computer modelling of the thermal behaviour of the hot metal entering the mould and its subsequent cooling behaviour is used to modify the methoding features, or even the component design, in order to promote directional solidification and to reduce shrinkage defects and to avoid hot spots which increase the likelihood of tear type defects.

The introduction of Quality procedures has been of major benefit in standardising the various processes which are used in the production of a casting. A marked improvement in the standardisation of the procedures for the preparation of the resin bonded sand and the use of more refractory sands (chromite) protected by a mould wash of zirconia paint particularly in areas subject to erosion by the molten metal stream have led to a marked reduction in sand inclusions and a superior quality of surface finish. Similarly the control of heat treatment has been improved. The only significant change to heat treatment requirement has been the imposition of a faster cool, either oil immersion or forced air cool, to ensure through hardening particularly in the thick section

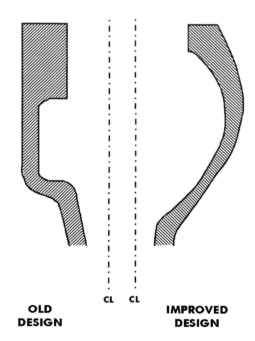

OLD DESIGN   CL   CL   **IMPROVED DESIGN**

Fig. 15 Change of shape to improve castability.

areas of 1%CrMoV cylinders and valve chests. This has to some degree restricted the number of suppliers for certain components to those who have the facilities to immersion quench a 20 tonnes cylinder.

While surface crack detection continues to be by magnetic particle inspection of all surfaces, ultrasonic testing has now almost completely replaced radiographic inspection. The use of 100% volumetric ultrasonic examination with compression wave probes combined with shear-wave probe examination at changes in section and at the sites of methoding features enables all types of defect to be identified and sentenced, either as being acceptable to the specification or requiring to be removed and the casting repaired. In particular, the most serious type of defect with regard to potential for extension in service, the crack-like hot tears, can be readily detected by the shear-wave probes whereas with radiographic inspection the probability of detection is low. For the complex high integrity castings, such as valve chests, the average repair rate is about 1% and may require more than one repair cycle. Some surface defects may be removed by grinding as long as the remaining wall thickness can be maintained above the minimum design requirement but normally the upgrading is by welding. The normal procedure is to remove the defect by grinding, ensuring the whole of the defect has been removed by magnetic crack detection of the excavation, and then the profile of the area to be welded is further modified to eliminate any undercut to enable good access for repair. Because the areas for repair are irregular in shape and of random size and orientation they are not generally amenable to the newer high deposition rate processes and therefore manual metal arc welding continues to be the normal repair process. However, there has been a significant improvement in the quality of weld repair both as a result of improved consumables leading to cleaner welds and the routine application of modern quality requirements. Of particular note are the qualification of welders and welding procedures, attention to control of preheat and improved manipulation of the casting to provide better access for welding.

The consequence of this improved technical ability of the supplier is that from a commercial point of view the maximum amount of fabrication is now conducted at the foundry. It is now rare for an order to be placed which does not require a cylinder casting to have a number of wrought stubs welded on, or the founder to weld together a number of castings and wrought stubs to make a fabricated valve chest assembly. A very high integrity level is demanded for these welds, ultrasonic compression and shear wave inspection to a 1.5 mm side drilled hole standard, in thicknesses up to about 100 mm. Occasionally dissimilar metal joints are involved, particularly with the adoption of the new 9%Cr steels for moderate temperature service where the improved creep strength of these alloys at 540–565°C relative to low alloy steels enables thinner section piping to be used. The inspection of some of these welds when made with Ni-base filler material is then by radiography.

The improvements in foundry technology and the experienced gained in the manufacture of major turbine components over the last forty years together with the incorporation of comprehensive quality procedures enables the most complex castings and fabrications to be made. The recent introduction of a new family of casting alloys, the modified 9–11%CrMo steels, has necessitated new development studies to identify changes to parameters such as the contraction allowance as used for low alloy steels. These new alloys have now been used for a number of contracts and it was observed that there was an increase in repair rate initially at most foundries although these alloys can now be produced as standard components.[9]

A significant change which has been introduced over recent years is the interaction of the design engineer and the foundry specialist at the earliest stage in the design process. This enables the component to be 'designed for manufacture' as a casting thereby eliminating features which would inevitably lead to defects and unnecessary complications to the foundry processes.

4 ALLOY DEVELOPMENT

4.1 Background

Since the late 1950s the highest temperature components of steam turbines have been made from either the low alloy steel CrMoV family or the martensitic 12%CrMoVNb family. The latter materials were used for blading and the low alloy materials for rotor forgings and castings. These alloys had been developed in a largely empirical way over the period since the problem of creep in steam turbines had been identified by Dickenson in 1922.[10]

During this period there has been substantial investigation of the high temperature behaviour of these alloy systems. Long term creep testing programmes were undertaken by the power generation industry to determine the 100 000 hr creep strain and creep rupture values required by the steam turbine design engineers on a number of typical production components of each alloy variant. A collaborative UK programme of work has just concluded after over forty years, whilst the German joint creep testing programme has been in existence since 1950 and new European programmes have recently commenced concentrating on the evaluation of the long term properties of newly developed alloys. In addition to the investigation of the mechanical properties of large turbine components the availability of electron microscopes and the formulation of metallographic procedures such as the carbon extraction replica enabled detailed study of the influence of microstucture on these high temperature properties.

Examination of a number of 1%CrMoV rotors of various sizes which had a range of microstructures revealed the means of obtaining optimum properties even in the larger size of forging which could now be manufactured. Using these newly available electron microscopy techniques, Norton and Strang showed in 1969 that the maximum creep strength is obtained when the microstructure is upper bainitic with a dense dispersion of fine vanadium carbide particles.[11] In order to achieve such a microstructure in the larger sizes it is necessary to control hardenability through appropriate additions of Cr, Mo, Mn and Ni. If the levels of these elements are too high a lower bainitic structure results and if they are too low proeutectoid ferrite is likely to be present in the centre of the forging, in both cases leading to reduced creep strength and in the latter case reduced toughness. An optimum dispersion of carbide particles is obtained by control of the V:C and Mo:V ratios. Large grain size may result in poor ductility and enhanced tendency to embrittlement so grain size is controlled by selection of the appropriate solution treatment temperature. The improvement in unifying the microstructure, and thereby the creep properties, in rotors of significantly different diameters is demonstrated in Fig. 16.

Through regular liaison with the suppliers the increased metallurgical understanding resulting from the studies mentioned above has been used to develop improved production procedures and manufacturing processes and to apply new technology as it became available. This has led to a consolidation and optimisation of procedures for high integrity low alloy steel rotors at sizes up to some 1500 mm diameter. There is a limit to the useful temperature of application, which is 565°C for 1%CrMoV, above which the long-term high temperature creep properties are inadequate to sustain the design loading.

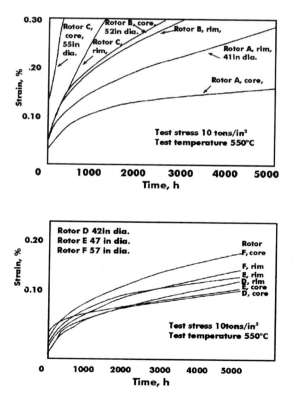

Fig. 16 Control of creep properties in large diameter rotors.[11]

The 12%CrMoVNb family of materials used for high temperature blading and valve components were originally derived from materials used for jet engine applications. The high alloy content of these steels gives them high hardenability so that on cooling after solution treatment martensite is formed even in thick section and at relatively low cooling rate. This martensitic matrix contains a dense network of dislocations which is stabilised by a dispersion of carbides formed during tempering. Dislocation movement is inhibited by interaction with the particle dispersion and with the high density of other dislocations, resulting in high creep strength. These steels contain a high level of carbon which combines with the large amounts of carbide forming elements such as Mo and V but especially Nb. Although these steels are stronger than the low alloy 1%CrMoV used for large rotors and castings they were not considered appropriate for such components as they were susceptible to segregation, had poor forgeability, poor weldability and poor fracture toughness. Even their advantage in creep strength is limited; whilst the high carbon 12%CrMoVNb steel has very high creep strength at durations of 10 000 hours, appropriate to their jet engine application, at temperatures above about 560°C their creep strength falls rapidly at durations above 10 000 hours. Due to the sigmoidal creep rupture behaviour of these steels extrapolation of short term creep test data would lead to a considerable overestimate of their long term creep strength, see Fig. 17. This reduction in creep strength occurs as a result of microstructural instability whereby the carbide dispersion becomes much coarser after long durations and is then less effective in inhibiting the movement of dislocations, Nonetheless the creep strength has been satisfactory for both the low alloy 1%CrMoV and the 12%CrMoVNb blading materials to have provided a reliable basis for the increased size of the major steam turbine components introduced over the last 30–40 years.

In the early 1960s an attempt was made to introduce supercritical steam turbines with higher steam temperatures, up to 649°C through the exploitation of

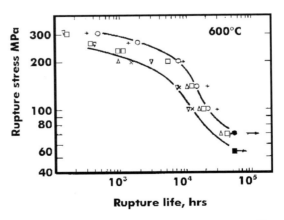

Fig. 17 Sigmoidal creep rupture behaviour.[12]

the greater creep strength of austenitic steels. However, these steels present significant problems for application as thick section components in steam turbines and boilers. Their high coefficients of thermal expansion, low thermal conductivities and low yield strength gave rise to high thermal stresses during starting and cyclic load operation and creep damage due to relaxation of these stresses during operation. This behaviour has resulted in thermal fatigue cracking, dimensional instability and structural collapse and although certain of these machines have continued to operate, often with down-rated steam conditions and operational limitations, interest in the application of advanced steam conditions was lost for sometime. As a result of this experience there has been no interest in the further development of austenitic alloy steels as their physical characteristics do not allow the turbine design engineer to meet the increasing demand for operational flexibility of units demanded by the modern pattern of power generation. However, austenitic steels are regularly used in the form of thin walled superheater components in the hottest regions of modern boilers.[13]

4.2 New Ferritic Steel Developments

In the mid 1970s ORNL were charged with the development of a ferritic steel with high creep strength and good weldability for application in fast breeder reactors for the US Department of Energy. They took as their basis a composition similar to the UK tubing alloy 9%Cr1%Mo and modified it through the addition of V, Nb and N. The resultant alloy, known as modified 9%CrMo, bore some resemblance to the high temperature blading alloy used for many years, 12%CrMoVNb. The Nb level was reduced from 0.3 to <0.1% which eliminated the occurrence of large particles of primary NbC and improved weldability by reducing hardness in the as welded condition and also by raising the martensite transformation temperature. The other objective of high creep strength was also readily met, the alloy having almost twice that of the previous generation of martensitic stainless materials at 600°C.

As interest in higher steam inlet conditions re-emerged in the 1980s, as a means of improving the efficiency of fuel conversion to gain economic advantage and also as a means of meeting the increasing pressures to reduce the environmental impact associated with power generation, significant interest was focused on this new ORNL steel. The original application was as a pipework material and therefore additional effort had to be applied to demonstrate the material in the form required for use in steam turbines. Work was undertaken with suppliers to manufacture trial castings and thick section cast pipes. All necessary mechanical properties, including creep rupture and high strain fatigue values, were evaluated and appropriate welding procedures established.[14] For modern turbine requirements cast modified 9%CrMo, with its significantly higher creep strength and improved thermal fatigue characteristics, has now been established for inlet temperatures up to 600°C.

Some time after this work at ORNL was underway the Japanese began to develop alloys suitable for the manufacture of high temperature rotor forgings.

These studies led to the formulation of a steel designated TMK1 which is similar to modified 9%CrMo but contains slightly more Cr (about 10%) and C (about 0.14%) and Mo increased from 1.0 to 1.5%. The increased C content together with a lower tempering temperature enables a proof strength of > 700 MPa to be achieved whilst at the same time fracture toughness is significantly better than the low alloy 1%CrMoV rotor forgings currently used. The creep strength at 600°C remains at the same level as modified 9%CrMo pipe.

Under the auspices of COST 501 the European turbine makers and their suppliers have worked together in parallel with the Japanese studies to develop turbine forging and casting alloys based upon modified 9%CrMo for use in supercritical steam turbines.[15] Two materials which have been fully evaluated and already put into service in advanced power plant are an optimised 10%CrMoVNbN alloy (steel F) very similar to TMK1 and an alloy with an additional 1% tungsten (steel E) which have properties suitable for operation with 600°C steam, Fig. 18. A comparison of these new 10%Cr rotor steels with the previously used 1CrMoV alloy is shown in Fig. 19. In addition to offering the turbine design engineer improved creep strength, these materials have higher yield strength, superior toughness and improved thermal fatigue resistance.

Further studies are continuing both in Japan and within the European COST 501 activity to develop alloys mainly on an empirical basis centred on two principles, (i) the addition of tungsten in substitution of molybdenum and (ii) the addition of boron, Fig. 20.

The use of tungsten to elevate creep strength results from the work of Fujita and he concluded that the optimum composition was when Mo + 0.5W equalled

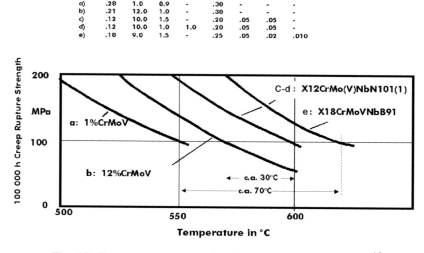

Fig. 18 Creep rupture strength of new turbine rotor steels.[15]

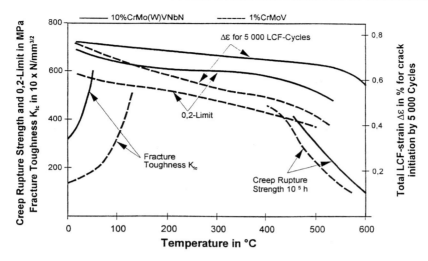

Fig. 19 Comparison of design criteria available from1%CrMoV and 10%CrMo(W)VNbN.[15]

1.5%.[16] Steel F has 1.5%Mo and no W whereas steel E has 1%W and 1%Mo and in long term testing they have almost identical creep strength as measured on full-scale rotor forgings. Similar trials on castings containing 1%W have shown no advantage over modified 9%CrMo cast material. Another COST steel with 0.5%Mo and 1.8%W was found to be very unstable and had significantly poorer creep properties.[15] The European experience is mirrored by that with a similar

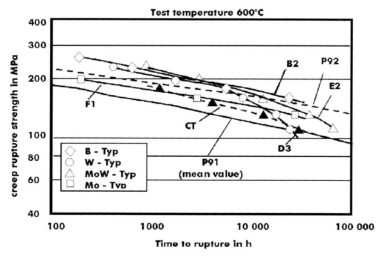

Fig. 20 Creep rupture strength of new steels at 600°C.[15]

Japanese rotor steel (TMK2) which offers at best only a marginal advantage in the long term over the tungsten free alloys TMK1 and steel F. This behaviour is in contrast to the Japanese steel NF616, a lower tensile strength pipe material tempered at a higher temperature in thinner sections which contains 1.8%W and 0.5%Mo and which has an estimated 10^5 hour rupture strength some 30% higher than the tungsten free alloys.

The mechanism by which tungsten enhances creep strength has not been established and as stated the effect of tungsten is not consistent in that the beneficial effect observed in low strength pipe steel is not displayed in higher strength rotor steels. In steels containing above about 1%Mo significant quantities of Laves phase, $Fe_2(Mo,W)$, are formed on long term exposure at the potential service temperatures around 600°C and the extent of this precipitation is intensified when tungsten is increased. It has been suggested that the precipitation of Laves phase is responsible for enhanced creep strength but the kinetics of this process have shown that the precipitation process is complete within 30 000 hours at 600°C. The rapid coarsening rate of Laves phase and the associated removal of Mo and W from solid solution suggest that W may only give rise to improved properties in the relatively short term.

Large amounts of tungsten also promote the formation of delta ferrite unless the chromium level is reduced to less than 9%. Application of such alloys at temperatures above 600°C gives rise to concern about oxidation resistance and chromium levels nearer to 12% are desirable. Combination of high Cr with high W would result in large amounts of delta ferrite which has led to the addition of elements such as cobalt or copper to increase the austenite stability so as to depress formation of delta ferrite. Alloys of this type are currently being evaluated for application as large forgings both in Japan and in Europe.

The development of alloys containing boron for steam turbine applications began in the COST 501 project in the early 1980s. Large additions of this element led to severe segregation problems and also to difficulties during forging and a sensible maximum level was established at about 100 ppm. A steel containing this level was subsequently produced as a 800 mm diameter forging (steel B) and was shown to have a creep strength at $600°C/10^5$ hours some 10 to 15% higher than the boron free materials. A full size forging has recently been manufactured by ESR steel making, using a special slag technique which has restricted boron segregation, and will be fully evaluated in a VGB long term testing programme.

The use of boron has also been adopted in Japan. A rotor alloy (HR 1200) containing 2.5%W balanced with 2.5%Co as well as high boron, initially at a level of 180 ppm but subsequently reduced to 100 ppm as a result of segregation, shows promise in short term tests but may be shown to weaken in longer time due to structural instability. Long term data on NF616, a piping alloy containing high tungsten and 30 ppm boron, shows it to be some 30% stronger than modified 9%CrMo. However, it is uncertain whether it is the tungsten or the boron providing the enhanced creep strength. The mechanism by which tungsten results in improved creep strength, particularly when the steel is tempered to a low strength condition, is not established nor is that of the boron addition. Detailed metallographic studies have shown that B is present in the $M_{23}C_6$ carbides and it is possible that it modifies the coarsening behaviour of these particles.

Within the COST activity there has been a considerable effort to obtain quantitative metallographic information on the evolution of the microstructure of these new alloys[17] as a means of providing a fundamental understanding of the contribution of the various phases to creep strength. As heat-treated samples and the long term creep specimens are continuing to be examined to educate the selection of the composition of further alloy steels for evaluation. It is anticipated that ferritic materials can be developed capable of operating at inlet steam temperatures of 650°C.

4.3 Alloys for Advanced Steam Conditions

A collaborative European Thermie project to develop a steam turbine to operate at 720°C has recently started.[18] Under these conditions the conventional combination of pulverised coal fired boilers and steam turbines in the Rankine cycle has potential for efficiencies as high as 55%. For the highest temperature components of this machine it is envisaged that Ni-based materials can be developed. The major challenge is to develop alloys which will be stable for long term, 250 000 hours, service at these temperatures and to develop forging, casting and joining techniques to provide the large pieces required for the manufacture of the steam turbine components.

5 FABRICATION

5.1 Welding of Low Alloy Steels

The welding of high temperature materials for construction of steam turbine components and assemblies, apart from the weld repair of castings mentioned above, was initially conducted by the turbine manufacturers. Materials used since the late 1950s for the cylinder and valve castings, and the associated stubs and pipework, were low alloy ferritic steels such as 1%CrMo, 2.25%CrMo and 0.5 to 1%CrMoV and these are still used for machines operating in the inlet temperature range of 540 to 565°C.

A number of generic problems were experienced particularly in thick butt welds between CrMoV parent materials welded with 2.25%CrMo by the manual metal arc process. A major problem which initiated during the post weld heat treatment cycle was reheat cracking. This revealed itself as both heat affected zone and transverse weld metal cracking either during the production ultrasonic inspection or subsequently in service, Fig. 21. Numerous studies were conducted and a large number of contributing causes were identified arising from deficiencies both in the process and in the materials. Reheat cracking in the coarse grained heat affected zone was attributed to the levels of elements such as P, Sn, As and Sb which segregate to the grain boundaries and significantly reduce their

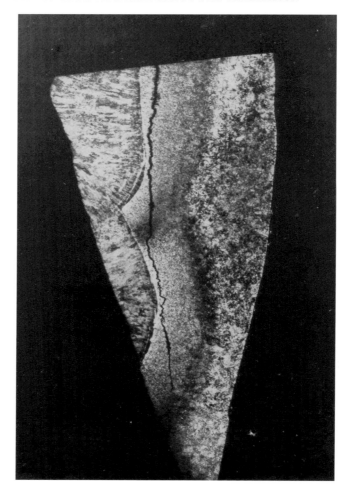

Fig. 21 Stress relief cracking initiating in coarse grained HAZ.

creep ductility and to the high level of vanadium which produced a high strength inside the grains by carbide precipitation, and the deoxidation practice.[19] Changes to the welding procedure to produce grain refinement of the coarse grained region of the heat affected zone such as the use of smaller gauge electrodes and reduced angle of attack were beneficial. Control of the heating rate and an increase in the temperature of the stress relief heat treatment from 630°C to 690–720°C were introduced. The weld metal was also subject to a reduction in the residual and tramp elements, an increase in manganese and a reduction in silicon content. Although 'new' welding techniques have been evaluated over the years, e.g. narrow gap TIG, to increase the efficiency of the process or to introduce some degree of automation, they have not been adopted perhaps due to the essentially 'one off' nature of the welds which are non-uniformly distributed and in different orientations even on one assembly. However, even though manual metal arc welding is still used, there has been a change in the geometry of the weld preparation which significantly reduces the weld volume for thick butt welds and has led to increased productivity. The old 40°/10° included angle has effectively been reduced to 10°/10° which also means that the welder is encouraged to use a higher angle of attack. Today production problems with these materials have been virtually eliminated and for commercial reasons almost all of this fabrication welding is now routinely carried out in the foundry.

5.2   Fabrication of Diaphragms

Diaphragms, are manufactured in a number of ways and often involve welding. In some cases this is only fillet welding of the blade sections between an inner and outer ring for the low temperature duty stages whereas high temperature stages in creep resistant ferritic steels require full section welds up to 200 mm thick in geometries which are highly constrained and contain many notches. There has been little change to the construction methods over the years but the diaphragms have increased in thickness and more creep resistant, less weld friendly, materials have been demanded to meet the higher duty of the modern machines. The aerofoil blade sections are fillet welded to spacer bands before being attached to the rim and centre by submerged arc welding, Fig. 22. The secret to obtain a successful outcome is to rigorously apply strict control of the process. A balanced weld sequence between the four welds joining the two sides of the rim and centre to the spacer band assembly together with control of both pre- and post-heat are essential parts of the overall good housekeeping necessary. The application of other welding techniques to improve the efficiency of the process by introducing more automation are being assessed.

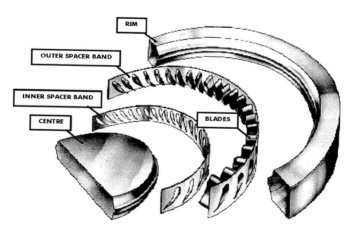

Fig. 22 Assembly of spacer band diaphragm.

5.3   *Stelliting*

Stellite weld surfacing has been used on valve components for many years, on sliding surfaces and also valve seats. The process originally used was oxy-acetylene gas welding. The high preheat of 500°C together with the weld heat input caused some degradation of the ferritic base material properties and the process was difficult to control. Considerable rework was necessary to produce a deposit of the required standard and it was not at all unusual for the deposits on valve spindles and valve seats to experience cracking in service, Fig. 23. In the late 1960s a TIG deposition procedure was established which had the benefit of lower heat input and the process was more controllable and the service behaviour was improved. Minor defects present in the deposit led to the initiation of thermal fatigue cracks and it was established that control of the carbon level to the lower end of the permitted range of 0.8 to 1.2% led to a significant improvement in thermal fatigue resistance. The deposition method currently employed is plasma transferred arc using a powder consumable which enables a thinner, 2 mm, layer to be deposited without the need for any preheat although slow cooling of the of the component after deposition is still applied.

5.4   *Welded Rotors*

Alternative methods for the manufacture of rotor forgings are the shrunk-on disc and shaft method, the use of monobloc forgings or the welding together of a number of smaller forgings, Fig. 24. This technique is by no means new and production of welded rotors commenced at our Le Bourget factory in 1946, and has competed with the progress in the development of the available maximum

Fig. 23 Cracking of stellited valve seat.

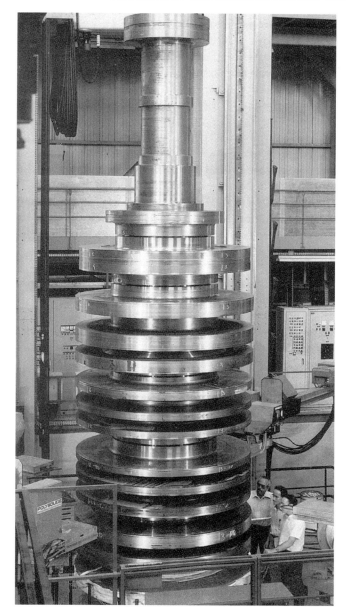

Fig. 24 TIG welding of LP rotor.

size of monobloc forging. Since 1978 over 130 large welded rotors have been made including HP rotors for 1300 MW, HIP rotors for 1500 MW turbines, Fig. 25, in addition to LP rotors. Many of these rotors which have been in service since 1981 have now accumulated over 100 000 hours of service.

Fig. 25 1500 MW HP/IP rotor.

As an example we can consider the LP rotors of some slow speed, 1500 rpm, French nuclear machines. The original design was made using the shrunk-on disc and shaft technology. After some time in service the rotors were observed to suffer stress corrosion cracking. For the retrofit of these rotors two alternative methods of construction were considered: (a) monobloc rotors in 3.5%NiCr-MoV steel, the same material as the shrunk-on disc but with a lower proof strength, and (b) welded rotors in 3%NiCrMoV with a proof stress similar to the monobloc forging. The machined dimensions of the rotor are a total length of 9636 mm, body length of 4356 mm, and a maximum body diameter of 2920 mm. As a monobloc forging this would require an ingot of 550 to 600 tonnes in weight and there are few forgemasters throughout the world with the ability to melt and collect this amount of high quality steel or produce such an ingot. Furthermore the necessary forging sequence and the number of austenitising heat treatments to refine the grain size to achieve the required ultrasonic permeability all lead to a high cost of the forging and a high risk to the delivery schedule should the forging be rejected. These considerations led to the adoption of the welded rotor production route.

The welded rotor was designed in four parts; two shaft ends and two discs. The two shaft ends are forged together from a 167.5 tonnes ingot and the two discs come from a single 136.5 tonne ingot but are forged separately. After rough machining the four pieces are vertically stacked and after preheating the weld roots are made using the TIG process with argon shielding of the central bore and the weld pool. Several layers are deposited to provide sufficient weld strength to enable the rotor to be lifted after the quality of the TIG root has been checked by radiography. With the rotor in the horizontal position the grooves are filled

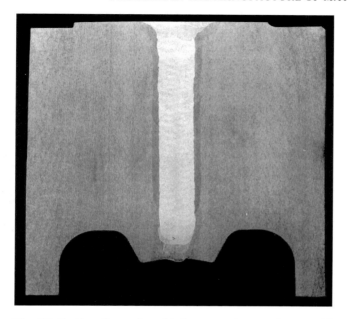

Fig. 26 Section through welded rotor submerged arc weld.

by submerged arc welding. The weld is typically about 100 mm in depth shown in section in Fig. 26. The holes used for the argon inlet and to insert the gamma source are plugged and welded. The welds are inspected by ultrasonic test in the axial, radial and tangential directions using a range of compression and shear wave probes to provide complete coverage of all welds. Today more than 10 of these retrofit rotors are in service.

6   CONCLUSIONS

Over the last forty years considerable progress has been made in the fields of steelmaking, forging, foundry and welding technologies to facilitate the manufacture of larger critical steam turbine components. Alloy development has markedly improved the available mechanical properties of materials for both low and high temperature service. These enhanced material characteristics have been used by the steam turbine design engineer to introduce machines of increased unit size and of increased thermal efficiency which have the required operational flexibility and long term reliability. Materials which have been developed over the last decade are now in service in supercritical units operating at 600°C and offer the potential for service at even higher temperatures.

During this period the new developments introduced by the metallurgist have enabled significant advances to be achieved in steam turbine technology without any radical changes to the basic turbine design. For the future it will be necessary to increase the thermal efficiency of fossil fired steam turbines still further to meet more severe environmental pressures and to continue to compete with other power generation technologies. A new collaborative European project to design and manufacture an ultra supercritical turbine with steam inlet temperatures as high as 720°C has just started. This will provide an even bigger challenge to the metallurgist than that of the last forty years and the turbine engineer should not rely on materials of similar properties and size equivalent to those now available for use at 600°C being capable of development. It may well be necessary for radical design changes to be introduced in order to meet such a demanding target.

REFERENCES

1. F.R. Harris: 'The Parsons Centenary – a hundred years of steam turbines', *Proc. IMechE*, 1984, **198**(53), 1–42.
2. M.G. Gemmill: 'Materials for the power industry', *Metals and Materials*, December 1985, 759–763.
3. D.H. Allen *et al*: *Requirements for Materials Research and Development for Coal-Fired Power Plant: Into the 21st Century*, DTI, September 1997.
4. D. Kalderon: 'Steam turbine failure at Hinkley Point' A', *Proc. IMechE*, 1972, **186**(32), 341–377.
5. Y. Ikeda, H. Yoshida, Y. Tanaka and T. Fukuda: 'Production and properties of superclean monoblock LP turbine rotor forgings', *Clean Steel: Superclean Steel*, 1995, 71–87.
6. E. Potthast, K. Langer and F. Tince: 'Manufacture of superclean 3–3.5% NiCrMoV steels for gas turbine components', *ibid.*, 59–69.
7. R. Viswanathan: 'Application of clean steel/superclean steel technology in the Electric Power Industry-Overview of EPRI Research and Products', *ibid.*, 1–31.
8. J. Ewald *et al*: 'Untersuchung an einer geborstenen Niederdruckwelle', VGB-Werkstoffagung, 1989, Vortrag 12.
9. M. Taylor and D.V. Thornton: 'Experience in the manufacture of steam turbine components in advanced 9–12%Cr steels', *Proc. IMechE Conf on Advanced Steam Plant*, London, 1997.
10. J.H.S. Dickenson: 'Some experiments on the flow of steels at a low red heat with a note on the scaling of heated steels', *J. Iron & Steel Inst.*, 1922, CVI.
11. J.F. Norton and A. Strang: 'Improvements of creep and rupture properties of large 1%CrMoV steam turbine forgings', *ibid.*, February 1969.
12. P. Greenfield: 'A review of the properties of 9–12%Cr steels for use as HP/ IP rotors in advanced steam turbines', *EUR 11887 EN*, COST 1988.
13. R. Blum: 'Materials development for power plants with advanced steam parameters-utility point of view', *Proc. of COST 501 Conf on Materials for Advanced Power Engineering*, Liège, October 1994.

14. D.V. Thornton and R. Hill: 'The fabrication and properties of high temperature, high strength steel castings', *Third EPRI International Conference on Improved Coal Fired Power Plants (ICCP)*, San Francisco, April 1991.

15. D.V. Thornton and K.-H. Mayer: 'European high temperature materials development for advanced steam turbines', *Advanced Heat Resisting Steels for Power Generation*, IOM Communications, London, 1999, 349–364.

16. T. Fujita: 'Effect of molybdenum, vanadium, niobium and nitrogen on creep rupture strength of TAF steel (12% Chromium heat resistant steel)', *Trans. JIM*, 1986, 9 supplement, 176–169.

17. R.W. Vanstone: 'Microstructure and creep mechanisms in advanced 9–12%Cr creep resisting steels. A collaborative investigation in COST 501/3 WP11', *Proc. of COST 501 Conf on Materials for Advanced Power Engineering*, Liège, October 1994.

18. Elsamprojekt Homepage: 'http//www.elsamprojekt.dk/700'.

19. P. Harris and K.E. Jones: 'The effect of composition and deoxidation practice on the reheat cracking tendencies of CrMoV steel', *CEGB Int. Conf on Welding Research related to Power Plant*, Southampton, 1972.

Material Developments for Supercritical Boilers and Pipework

A. FLEMING, R. V. MASKELL, L. W. BUCHANAN
and T. WILSON

*Materials Dept. of the Mitsui Babcock Technology Centre, High St,
Renfrew PA4 8UW, UK*

ABSTRACT

This paper describes the link between increased power plant efficiency and the need for higher operating temperatures and pressures. The implications for improved designs of supercritical plant and the need for the development of new creep resistant alloys are discussed. Special attention is given to the development and metallurgy of the martensitic creep resistant 9%Cr and 12%Cr steels which are currently the preferred materials for the high temperature areas of supercritical plant. Likely advances in plant design and operating conditions over the next 20 years are considered together with the current development initiatives. The oxidation and corrosion problems associated with high efficiency supercritical plant are discussed. Fireside and waterside corrosion problems are considered separately and the impact of the move from subcritical to supercritical plant operation explored. The new austenitic alloys currently under development and the potential use of high nickel alloys are also discussed.

1   HISTORY AND TRENDS

1.1   General

The drive for ever-increasing efficiency of power plant has always been present but in recent years it has been given greater impetus by the concerns over polluting emissions of SOx, NOx and CO_2. Apart from simple economics demanding lower electrical energy unit costs, there is the need to reduce the amount of emissions per megawatt generated.

Combined cycle gasification processes offer inherently lower emission levels over pulverised coal steam cycle power plant, but these technologies are comparatively new, whilst the now traditional pulverised fuel (PF) plant still makes it attractive to power producers throughout the world.[1,2] Designers of steam raising PF equipment are well aware of the alternatives and are presently striving to develop coal fired steam cycles with efficiencies matching or even surpassing other systems.[3]

The historical trend in PF steam cycle efficiency and the planned future is shown in Fig. 1. The increase in efficiency on moving from Castle Peak Power Station to Meri Pori PS, Hemweg PS etc was accompanied by changing the steam cycle from subcritical to supercritical. In these examples of supercritical

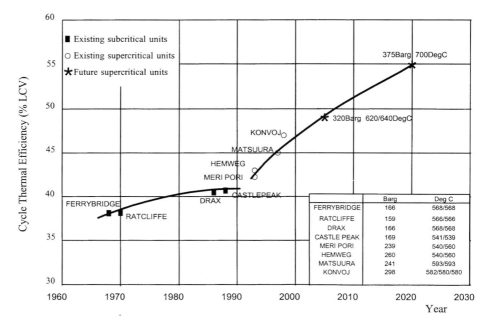

Fig. 1 Developments in thermal efficiency in coal power plant.

plant the main difference in operating conditions was an increase in pressure. This was due to the limits placed on operating temperatures by the materials available for their construction. In the future, improved materials currently in development will allow significant increases in operating temperatures resulting in large increases in plant efficiency. These higher efficiency stations are known as ultrasupercritical (USC) plant.

The present subcritical plant design evolved from the water tube boiler in which water was boiled within tubes running through a firebox. In 1856 Stephen Wilcox proposed a design incorporating inclined water tubes, heated by the flue gases, connecting water spaces at the front and rear of the boiler. Later (1867) a steam space was provided in a pressure vessel situated above the water tubes (Fig. 2).[4] This design allowed better water circulation, increased the heating surface and reduced the explosion hazard. In 1880 Allan Stirling developed a design connecting the steam generating tubes directly to a steam separating drum.[4]

Steam from these boilers was principally used to provide heat and power for local industry. With the advent of practical electricity generation and distribution, the utility companies adopted steam as a major source of electricity generation. Initially, as electricity generating stations increased in size, the number of individual boilers was simply increased. In the 1920s the economic benefits of building larger boilers were realised. This led in the 1930s to the introduction of fusion welding in the UK boiler industry and allowed the evolution of the subcritical cycle which requires the use of very large pressure vessels for steam drums.[5]

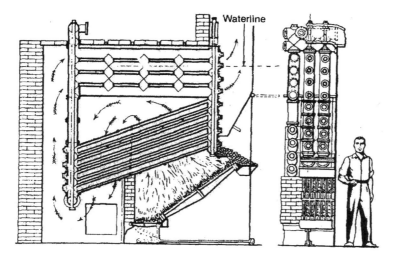

Fig. 2 First Babcock & Wilcox Boiler (patented 1867).

However, in other countries, the once-through cycle was favoured: this is based on the Benson concept and can be operated either subcritically or supercritically.[6]

The size of individual boilers rapidly became so large that existing designs of boiler and methods of coal firing (such as stokers) were unable to cope which led to a new design of boiler which remains the basis for coal fired power generation boilers to this day.[4] In these subcritical boilers (Fig. 3) pulverised coal in the form of a fine powder is fluidised by a flow of air and fired through a series of burners to produce a flame within the main section of the boiler known as the combustion chamber. This section consists of a large vertical box with the walls constructed from water tubes and known as 'water-walls'. The heat from the burners is absorbed by the water within the tubes which boils under pressure to give steam at around 360°C. The steam/water mixture is fed by natural circulation into a steam drum at the top of the boiler where the steam and water are separated. The water is mixed with more feed water and is directed back to the bottom of the water-wall to recirculate through the system. The hot combustion gases rise to the top of the boiler where they pass over a series of tubes attached to manifolds suspended from the boiler roof. Steam from the steam drum is further heated in these final superheater and reheater tubes and passed on to the steam turbine. The gases then descend through a another section of the boiler, known as the second pass, where more heat is extracted from them and is used to preheat the steam destined for the final superheater and also preheat the feed water in the economiser. This system both gives a high heat transfer surface and provides water cooling for the boiler walls. Currently in the UK the largest such plant, is at Drax, where the 6 × 660 MW units operate at a steam temperature of 568°C and a pressure of 166 bar.

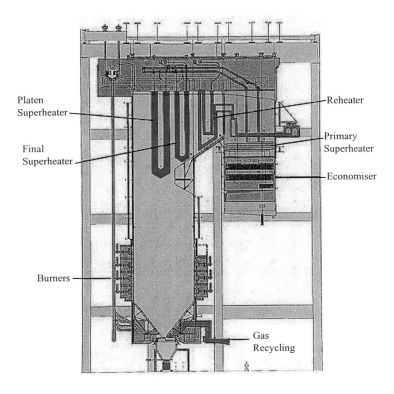

Fig. 3 Typical modern subcritical PF boiler.

The efficiency of boiler plant is controlled by the laws of thermodynamics; effectively, the higher the pressure and temperature achieved, the higher the thermal efficiency. This is illustrated for a typical subcritical steam cycle in Fig. 4(a), where the thermal efficiency is proportional to the ratio of the two

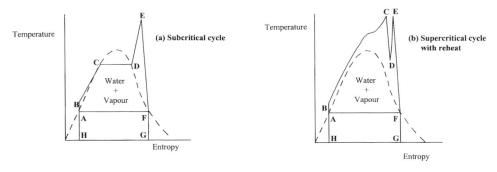

Fig. 4 Temperature–Entropy Diagrams for subcritical and supercritical cycles.

areas ABCDEF and HABCDEFG. Increasing operating temperature increases the height of the points CDE with a resultant increase in efficiency. Much larger increases in efficiency can be achieved if, again by increasing the operating temperatures and pressures, the boiler can be made to operate in the supercritical region (i.e. where liquid and vapour phases no longer exist as separate entities and there is no latent heat of evaporation). This is shown in Fig. 4(b) where the efficiency of the system (ratio of areas ABCDEF and HABCDEFG) can be seen to have increased significantly. This increase in efficiency required further development of the boiler from the natural convection recirculating boilers still presently used, to forced flow once-through boilers for supercritical operation. 'Once-through' means that there is no recirculation within the boiler.

The realisation of such high efficiency supercritical plant will result in overall increases in thermal efficiency estimated at about 6%, corresponding to a change in conditions from 305 bar at 585/602°C to 320 bar at 620/640°C (in each case the first temperature quoted refers to the *main steam* temperature and the second to the temperature of the *reheated steam*, which is at a lower pressure). In terms of coal consumed and emissions produced, the savings are even larger and are estimated to be above 14%. This has resulted in a world-wide incentive to increase steam conditions in plant.

Recently constructed plant such as Hemweg and Meri Pori, designed by Mitsui Babcock Energy Limited, operate in the supercritical range at steam temperatures of 560°C and at much higher pressures of 239 and 260 bar. Even though the choice of materials limited the operating temperature to 560°C these plants are considerably more efficient. The highest temperatures and pressures achieved in the UK were at Drakelow 'C' which was built in 1966 and had a terminal steam pressure of 252 bar and a steam temperature of 599°C. This plant was in many ways ahead of its time and the limitations of the ferritic steels then available were such that much of the plant was fabricated in austenitic steels. As will be explained later this can lead to restrictions in boiler flexibility and potential thermal fatigue problems.

Current supercritical plant has the advantage of using established PF technology and its construction is not radically different from sub-critical units, Fig. 5. As such, the technological risks are less than for the development of other high efficiency generation systems (such as Integrated Gasification Combined Cycle) and the operational problems are expected to be fewer. However, the supercritical conditions and once-through design mean that the boiler design has features different from subcritical units.

In a subcritical unit the main process taking place within the water walls of the combustion chamber is evaporation. Thus, while the enthalpy of the fluid within the water wall increases, its temperature does not. This restricts the temperatures which the water walls must withstand and results in uniform thermal expansion. In contrast there is no phase change within the water wall of a supercritical unit resulting in a progressive tube temperature increase across the height of the water

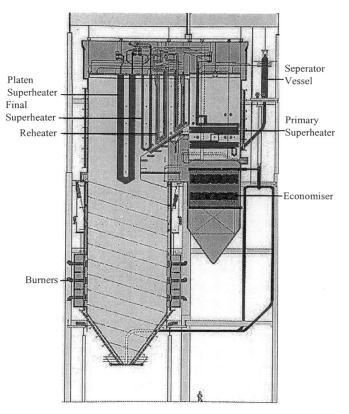

Fig. 5 Typical modern supercritical PF boiler (Note the helical water walls).

wall with a corresponding increase in thermal expansion. In order to limit this escalation in tube wall temperature it is necessary to achieve a suitable flow of fluid in all of the water wall tubes. When the flow rate is too low, conditions are such that vapour blanketing occurs on the inside of the tube, and the rate of heat transfer on the unwetted tube wall decreases, with a resulting increase in the tube wall temperature. Until recently the most favoured solution to prevent the onset of vapour blanketing was to build the water wall using inclined tubes wound round the combustion chamber in a shallow helix. The number and inclination of these tubes is selected to achieve the required flow rate. This arrangement of spiral water walls allows compensation for different heat fluxes around the boiler and thus ensures approximately equal heat input to each tube. In addition, each tube is also much longer than if it had taken a direct vertical route as well as being of smaller diameter. The required flow rate for tube cooling and the length of the tubes result in the pressure losses being much higher than in a natural circulation boiler.[3] This is important as it means spiral water walls exhibit a negative flow response, whereby an increase in heat absorption by the fluid results in a decrease in the flow and this feed back results in an exaggerated increase in the fluid temperature.

A further difference is that no steam drum is required in supercritical operation. However, in order to deal with steam generation during start-up and shut-down when subcritical conditions exist, it is necessary to have some form of steam separation and water recirculation. Once-through boilers require a minimum flow through the water wall tubes at all times. This flow is known as the Benson Load which is the load at which the minimum flow equals the boiler evaporation. At loads below this, and in order to ensure this minimum waterwall flow, it is necessary to separate the steam from water during subcritical operation and a separator vessel is employed to achieve this. The steam continues through the remainder of the circuit whilst the water is returned via a recirculation pump to the economiser inlet.[7] The supercritical fluid from the top of the water-wall is collected in headers and passed through further superheater and reheater surfaces hanging from the boiler roof as in conventional subcritical plant, although the temperatures and pressures are of course higher.

This spiral wall concept has been used in Japan in the Matsuura Thermal Power Station No 2 which commenced operation in July 1997 and has been designed to run at steam temperatures of 593°C and 241 bar, and this represents the current state of the art of supercritical plant in operation.

One important development is the return to vertical tubes in the water walls. This has been possible by the use of rifle bored tubing. Siemens Research Centre in Erlangen has developed, over many years, a new design philosophy for Benson boilers based on the use of rifle bore tube which exhibits very good heat transfer.[8] The rifling imparts an angular momentum to the flow of fluid which generates a centrifugal force separating any water from the steam and forcing the fluid against the inner surface of the tube. This gives good heat transfer even at low flow rates. In addition, if the boiler operates at an appropriately low flow rate the vertical tubes show a positive flow characteristic, at predetermined temperature and pressure conditions, with an increase in heat absorption developing a pressure difference which increases flow through the tube. This behaviour mimics the flow characteristics found in the water walls of subcritical plant and results in a self compensating effect in each tube, which means that the spiral construction can be replaced with the much less expensive vertical construction.[9]

The essential lesson is that boiler efficiencies and thus operating temperatures and pressures have historically increased as shown in Fig. 1. Thus despite the changed economics of the energy market place the future development of boilers will continue the trend of ever higher temperatures and pressures seen in the past. Just as the introduction of wrought iron plate in the 1800s led to a new generation of boilers it is the development and introduction of new materials which will enable the construction of a new generation of high efficiency supercritical plant.[4]

1.2   Future Trends

Mitsui Babcock Energy Limited have identified possible efficiency goals for the short, medium and long term with regard to final steam temperature and pressure as shown below,

	Short term	Medium term (5 years approx.)	Long term (20 years approx.)
Efficiency	42–46%	48–50%	50–55%
Steam temp.	585–602°C	620–640°C	700°C
Steam pressure	305 bar	320 bar	375–400 bar

As with all design and subsequent material selection problems the choice of material for use in high temperature power plant is a compromise between a number of important factors. The thermal properties (conduction and expansion), the creep properties, resistance to high temperature corrosion, formability, weldability and not least cost must all be considered.

2   MATERIAL PROPERTIES

The capability of a boiler is limited by the properties of the materials used in the construction of those parts subjected to the most onerous conditions of stress and temperature. In effect it is the ability of the final superheater tubing and associated headers and pipework to cope with these severe conditions that governs the boiler operating parameters of temperature and pressure. The development of materials is therefore focused primarily on the improvement of creep strength and on the resistance to high temperature corrosion and, as will be shown later, this has led to the evolution from low alloy steels, through martensitic steels to 'super' austenitic steels and will eventually arrive at nickel-base alloys for these high temperature components.

One other area of the boiler has also benefited from material developments and this is the water walls, where, for supercritical boilers, there is the need to operate at temperatures much higher than with subcritical units (e.g. up to 550°C compared with 360°C). This is also discussed later.

2.1   Creep Strength

For plant operating at temperatures around 600°C creep is a critical factor and plant is designed to operate for a finite life (for example: up to 250 000 hrs, i.e. nearly 30 years). With the move to higher temperatures and pressures required for more advanced supercritical plant, the materials generally used for operation in subcritical plant were unable to provide the performance required for the high temperature sections of the boiler. The ferritic steels then available were principally 0.5Cr0.5Mo0.25V, 1Cr0.5Mo and 2.25Cr1Mo. These steels lacked sufficient creep strength and corrosion resistance for use at the higher temperatures required for supercritical operation. Although the austenitic steels offer good creep and corrosion properties their thermal properties are relatively poor with a low thermal conductivity coupled with a high coefficient of expansion. During thermal cycling, high stresses can develop in thick walled components, leading to thermal fatigue. It has been calculated that to avoid such unacceptable thermal stresses thick walled components such as headers and pipework would require to be heated to the operating temperature over an extended period which would be unacceptable to plant operators meeting real-time electricity demands. Thus severe design and operating restrictions are imposed on plant using thick section austenitic steels. Hence the main use of these steels is in thin section components such as superheater/reheater tubing and stub headers where full benefit can be taken of their improved creep and corrosion properties.

The solution adopted was the introduction of martensitic creep resistant steels containing either 9% or 12%Cr, as these show high thermal conductivity with relatively low thermal expansion. These steels are fully air hardening and develop a fully martensitic structure even after slow air cooling from the solution treatment temperature and, in part, their improved creep strength is derived from this martensitic matrix. To control the high hardness and brittle behaviour normally associated with martensitic materials low carbon levels were employed. Specified nitrogen levels were also introduced as this element increases both the hardenability and the creep strength. Currently these creep resistant martensitic steels are the favoured materials for the construction of components in supercritical boilers. In many cases the parent material properties of the alloy may be ideal but it must be possible to produce the required product forms, normally tube and pipe, and to join and form them under both factory and site conditions. Thus fabricability plays as important a role in the development of a material as obtaining improved mechanical properties.

2.1.1   Creep Resistant Martensitic Steels

The first of the creep resistant martensitic steels to be commonly used in power generation plant was the European 12Cr steel known as 'X20' (from the DIN designation X20CrMoV 12 1). The chemical composition for X20 is given in Table 1 where it can be seen that it differs from the steels now preferred in that it has a higher carbon content and the nitrogen level is not specified.[10] This steel exhibits very good creep strength and has been used in continental Europe in high temperature areas of subcritical boilers. However concerns exist over the fabrication and welding of this material. Welding has to be performed using a high preheat temperature (up to 400°C) and, on thick section weldments, cracking can occur when the weld is cooled to room temperature. In addition, the hardness of weldments is very high in the as-welded condition and stress corrosion cracking may develop in such welds if they experience damp conditions prior to post weld heat treatment. As a result special precautions are necessary, such as placing the completed weldment in a furnace before it cools down from preheat temperature. It is allowed to cool under controlled conditions to about 100°C and then post weld heat treated. This is to allow any austenite, present

Table 1 Chemical compositions of 9%Cr and 12%Cr martensitic steels.

	X20	P91	E911	NF616	HCM12A	TB12M	NF12
C %	0.17–0.23	0.08–0.12	0.10–0.13	0.07–0.13	0.07–0.14	0.10–0.15	0.08
Si %	0.5 max.	0.20–0.50	0.10–0.30	0.5 max.	0.5 max.	0.5 max.	0.05
Mn %	1.00 max.	0.30–0.60	0.30–0.60	0.30–0.60	–	0.40–0.60	0.5
P %	0.030 max.	0.020 max.	0.020 max.	0.020 max.	0.020 max.	0.020 max.	–
S %	0.030 max.	0.010 max.	0.010 max.	0.010 max.	0.010 max.	0.010 max.	–
Cr %	10.00–12.50	8.00–9.50	8.50–9.50	8.50–9.50	10.00–12.50	11.0–11.3	11.0
Mo %	0.80–1.20	0.85–1.05	0.90–1.10	0.30–0.60	0.25–0.60	0.40–0.60	0.15
Ni %	0.30–0.80	0.40 max.	0.40 max.	0.40 max.	0.50 max.	0.70–1.00	0.5
Nb %	–	0.06–0.10	0.06–0.10	0.04–0.10	0.04–0.10	0.04–0.09	0.07
V %	0.25–0.35	0.18–0.25	0.15–0.25	0.15–0.25	0.15–0.30	0.15–0.25	0.2
W %	–	–	0.90–1.10	1.50–2.00	1.50–2.50	1.60–1.90	2.6
Al %	–	0.04 max.	–	0.04 max.	0.04 max.	0.010 max.	–
N %	–	0.030–0.070	0.05–0.08	0.030–0.070	0.040–0.100	0.04–0.09	0.05
B %	–	–	–	0.001–0.006	0.005 max.	–	0.004
Co %	–	–	–	–	–	–	2.5

after welding, to transform to martensite at a relatively high temperature to reduce the risk of cracking.[11] A procedure such as this poses problems in both the factory based and site fabrication of X20 weldments. Nonetheless, many perfectly acceptable X20 weldments have been completed and have performed satisfactorily in service.[12] Because of these fabrication difficulties there was great interest in the production of a material with equal, or improved creep properties which is easier to fabricate and this led to the development of the 9Cr martensitic steels which have now supplanted X20 as the preferred material. This highlights the importance of the fabricability of materials for use in such complex fabrications as boilers.

A basic 9Cr1Mo steel had originally been developed as early as 1936 for use in the oil industry to provide improved corrosion resistance over 2.25Cr1Mo. In the 1970s this steel was used, mainly in the form of tubes, in the UK nuclear power programme. Although the creep strength was adequate for these applications, it was inferior to that of X20 and therefore efforts were made to improve it. An early attempt resulted in a 9Cr2Mo steel known as EM2 which had improved creep strength but suffered from in service embrittlement and low creep ductility.[13] In the 1970s the desire for an improved 9Cr1Mo steel without the problems associated with EM2 led to the start of an alloy development programme aimed at supplying material for the US fast reactor programme. This resulted in the production of a steel known as Grade 91 or T91/P91 where the initial letter refers to the product form (<u>T</u>ube or <u>P</u>ipe). The steel had a lower carbon content than the original 9Cr1Mo steel coupled with small additions of vanadium and niobium to develop strengthening precipitates.[14] It was also the first steel in this class to specify a nitrogen level to further improve both the hardenability and the creep strength, (Table 1).

With the relatively slow cooling rate experienced in industrial normalising, Grade 91 steel has a 'martensite-start' temperature, Ms, of about 385°C and a 'martensite finish' temperature, Mf, of about 120°C. However, both the Ms and Mf temperatures are a function of the cooling rate due to precipitation altering the composition of the matrix. This in turn controls the hardness of Grade 91 steel which varies from 410 HV10 to 385 HV10 in the range of cooling rates of commercial interest. The relatively high Ms temperature and low martensite hardness of Grade 91 steel (compared to X20) are a direct result of the reduced carbon level and it is these properties that give Grade 91 steel its improved weldability and facilitates the processing of the steel into tube and pipe.[15]

When ferrite appears in the matrix there is a marked decrease in hardness and the production of ferrite is a sign of an inadequate cooling rate. Pipes up to 80 mm thick can be air cooled without the formation of ferrite, and hence this size is the likely commercial limit on component thickness in P91.

Grade 91 steel is generally used in the normalised and tempered condition.[16] On cooling from the normalising temperature of around 1050°C, the steel transforms to martensite containing niobium, titanium and vanadium rich carbonitrides of the MX type. Fine acicular precipitates, identified as ε-carbides, have also been found in the interior of the martensite laths. Tempering is normally carried out in the range 730–780°C and results in further precipitation of MX carbonitrides together with the precipitation of $M_{23}C_6$ type chromium-rich carbides; these precipitates increase the creep strength of the material.[15] Impact properties are available for P91 in a range of thicknesses and show the steel to have good toughness down to −20°C.

Grade 91 is welded with a much lower preheat than used for X20 (~200°C compared with ~400°C) and may be cooled to room temperature after welding without risk of cracking.[17,18] The improved mechanical properties and weldability helped to make Grade 91 the technically preferred steel for high temperature applications in the UK and, in the late 1980s, a collaborative project began between UK power plant fabricators and the now privatised CEGB. It had been recognised that the adoption of Grade 91 as a power plant material required the demonstration of its properties in a power station operating at high temperature and pressure. It was therefore agreed to carry out a development plan aimed at using this steel in the replacement of existing 0.5CrMoV superheater headers at boiler No 9 at Drakelow C Power Station. This required the development of welding techniques to join Grade 91 steel not only to itself but also to 2.25Cr 1Mo and 316 austenitic steels in transition joints. Although provisional design data for the steel were then available, a thorough assessment of the fabricability of the material was required, to prove its suitability for long term service. This encompassed the steel-making route, production of pipe and forged bar, weld consumable selection and weld procedure qualification. Following the completion of the initial development programme, three separate UK manufacturers each produced two of the six headers required. These were fabricated under commercial manufacturing conditions employing only normal safeguards. As a result of this rigorous approach no unexpected difficulties were experienced and the headers were installed at Drakelow C without incident.[19] Subsequent monitoring of the headers has confirmed the good performance of the P91 components in service.

Following the success of its adoption at Drakelow C, Grade 91 has been used for replacement headers and also for the production of superheater tubing. It is also in use as a replacement for lower alloy materials, where thermal fatigue problems had caused ligament cracking in service. Since the higher creep strength of Grade 91 allows the construction of much thinner headers, this reduces the impact of thermal cycling during service.

While Grade 91 steel has been a great success and is now an established material for operation at temperatures in the 550°C–610°C range for tube, but with a greater range for headers and pipework, its development revealed some features that currently appear to be inherent in the use of such creep resistant martensitic steels. During the cross-weld creep testing of Grade 91 steel it became apparent that, for long duration tests, Type IV failure occurred ie failure in a softened zone between the HAZ and parent metal, Fig. 6.[15,20] The expression 'Type IV failure' was coined from the categorisation of the various forms of cracking found in failed weldments in power plant. Failures of this kind had previously been found

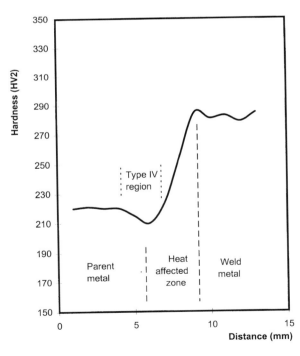

Fig. 6 Hardness profile over the Type IV region of a weld.

in ferritic 0.5Cr0.5Mo0.25V and 2.25Cr1Mo welds in pipework systems as well as in martensitic X20 weldments, and can occur in many other materials

The importance of Type IV cracking lies in its effect on cross-weld creep strength. Cross-weld test specimens fail far more rapidly than either parent or all-weld metal tests. Although the exact effect varies with weld procedure, it is now generally accepted that the cross-weld creep properties are reduced to some 80–50% of those of the parent metal. Currently it is believed that this behaviour is a general feature of all martensitic creep resistant steels and will not improve as new steels or improved welding consumables are developed. This problem has been successfully circumvented by careful design of components. As these are typically cylindrical such as headers, pipes or tubes, it has been possible to avoid the use of longitudinal weldments, leaving only circumferential joints which are subject only to axial loads in the cross-weld direction. These pressure induced axial loads are only 50% of the hoop stress which is the design stress for the remainder of the component. Consequently, the parent material strength is normally used as a reference point for design purposes, although design consideration must still be given to features where weldments may be subjected to other than pressure induced stresses.

Although Type IV failure was at first thought to occur in the sub-critical heat affected zone (HAZ) due to over-ageing of the parent plate it is now generally accepted that it occurs in the intercritical HAZ.[15,21] During welding, the peak temperatures in this area allow partial transformation to austenite and may result in depletion of the martensitic matrix in important alloying elements. The overall result is to produce a small but 'creep weak' zone associated within the HAZ which can be detected by hardness surveys (Fig. 6). Monitoring of strain along the length of cross weld tests has shown that strain accumulates in these creep weak zones leading to local necking and eventual failure (Fig. 7). The detailed distribution of strain in this area is also dependent on the creep strength and ductility of the surrounding weld and parent metals which tend to physically support the weaker Type IV zone.

With the acceptance of Grade 91 steel, attention focused on possible further improvements in the martensitic steels. Parallel to the acceptance of Grade 91 in commercial plant, steels containing tungsten additions were also being investigated in both Europe and Japan.[23,24] In Europe, alloy development was continued by the steelmakers under European Coal and Steel Community contracts and several casts of tungsten-bearing martensitic steels were produced, with potential application to high temperature plant. These new martensitic steels are under varying degrees of development and codification and can be divided into two main groups: the 9Cr and 12Cr steels.

There are principally two tungsten-bearing 9Cr steels which both claim to have creep strengths approximately 30% higher than Grade 91:

(a) the Japanese developed NF616 (Table 1) which has been codified in ASME Code Case 2179 and designated as P92 and T92, and

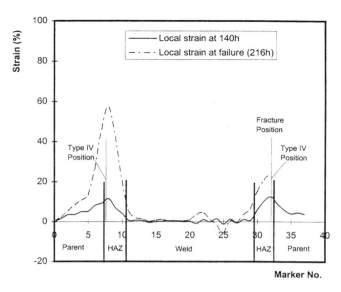

Fig. 7 Local strain distributions on a P91 stress rupture specimen (600°C at 150N/mm²).

(b) the European equivalent E911 (Table 1) which was being validated in COST 501 by long term creep testing.

While both molybdenum and tungsten contribute in some part to the formation of precipitation strengthening their main effect is one of solution strengthening.[24] This has the advantage that, whilst precipitates tend to coarsen and lose their strengthening effect at long times, solution hardening of the matrix does not decrease with ageing, except when depletion of the strengthening elements occurs as a result of in-service precipitation. Thus, at least in theory, solution hardening offers a near permanent increase in creep strength. Tungsten is only about half as effective a solution hardening element as molybdenum and so twice as much tungsten is required for the same effect.[25] The low diffusion rate of tungsten means that tungsten carbides are, however, more stable and contribute more effectively to long term strength.[26] In Japan it was found that the optimum creep strength could be obtained when the molybdenum equivalent is equal to 1.5 (the molybdenum equivalent is defined as: Mo% + 0.5W%).[27] At higher molybdenum equivalents the long term creep strength was seen to decrease irrespective of the individual molybdenum and tungsten levels. This was attributed to the formation of δ-ferrite in these alloys during the normalising heat treatment.

It has been noted that the creep strength of these alloys increases as the normalising temperature is increased as this results in the dissolution of increasing amounts of vanadium and niobium carbonitrides which can then precipitate as very fine particles during subsequent tempering. However the dissolution of these carbonitrides also results in increased austenite grain growth which has a deleterious effect on toughness. The minimum austenitising temperature for NF616 has therefore been specified as 1040°C.

NF616 containing almost 2% tungsten exhibits improved creep strength over Grade 91 steel and retains its strength better to higher temperatures, while maintaining the ease of formability and weldability.[20] Similar tests to those conducted on Grade 91 steel have been completed showing NF616 to have good weldment properties though Type IV failure can again cause large decreases in cross-weld creep strength.[28,29] The material has now been installed in one German, two Danish and two Japanese power stations and appears to be performing well.[30,31] Thus NF616 is currently the state of the art material in supercritical plant design.

One potential problem which has been the subject of extensive investigation is the in-service precipitation of Laves phase. This precipitate contains tungsten and could result in both embrittlement of the alloy at longer times and in the depletion of the tungsten content in the matrix resulting in a decrease in the long term creep strength.[32] This subject is still rooted in controversy with different opinions being held by different authorities. One problem is that precipitation of Laves phase only occurs below about 720°C and above this temperature Laves phase will redissolve. Thus it is not possible to determine the effects of Laves phase by testing material aged at higher temperatures and the effects of precipitation of this phase will only become apparent at long test times. While it is generally agreed that the precipitation of Laves phase results in a decrease in toughness, it is believed by some authorities that this precipitation phase is a strengthening mechanism whilst others regard it as weakening the steel by the depletion of tungsten, thus reducing the solution strengthening.[33] Recent work has suggested that after relatively short term exposure fine Laves phase precipitates form which result in an increase in creep strength but that these Laves precipitates coarsen faster than other precipitates and this may result in longer term decreases in creep strength.[34]

Partly due to concerns over Laves phase, the COST 501 project in Europe concentrated on the validation of E911 steel containing only 1% tungsten (compared with the 2% tungsten in NF616) whilst optimising the other alloying additions to give increased creep properties. This steel is still under test but appears to be as strong as NF616. Testing of welding consumables and development of welding and bending procedures is well underway and complete with respect to the time-independent properties. The creep rupture testing is continuing and as expected, Type IV failures have also been observed in this steel.

Whilst NF616 has ASME Code Case status, the stress rupture testing durations that were required for this are much shorter than would be the case in Europe. Due to this, and the ever present difficulties and assumptions associated with creep extrapolations, the long term design stress values for NF616 are questioned by many. In the case of E911 validation, the COST 501 partners have committed themselves to real data collection via long term testing.

Although the 9Cr steels have been the subject of most effort through E911 and NF616, their maximum temperature of operation may be restricted by steam oxidation. This aspect is discussed further in Section 2.2.2, where it is indicated that this steam side phenomenon is associated with the chromium level; greater resistance being conferred with higher free chromium content. For higher temperatures of operation the new steels under development generally contain in the region of 12%Cr and, like P91, have low carbon levels and controlled nitrogen levels.

Current examples of 12Cr steel are HCM12A and TB12M (Table 1). These have been under development since 1989 in the EPRI 1403–50 programme which brings together steel manufacturers and utilities.[35] Both these steels contain a minimum of 11%Cr with around 0.5%Mo and 2%W. TB12M contains one of the highest levels of nitrogen yet specified (0.09%) but, despite this, may also contain small amounts (1–3%) of δ-ferrite. The ferrite is not believed to seriously affect the properties of the alloy as it exists as small isolated islands in the martensitic matrix.[36] HCM12A contains a lower level of nitrogen but a small addition of boron and is fully martensitic. Bending and welding trials have been completed and there currently exists an ASME code case for HCM12A, which is now referred to as P122 (Code Case 2180).[37]

Further development of high strength martensitic boiler steels is underway in both Japan and Europe with target creep properties which will allow the use of such steels at metal temperatures up to 650°C. One such steel (NF12 – Table 1) makes use of cobalt as an alloying addition and has no delta ferrite even in a commercially sized ingot.

Table 2 gives the extrapolated 10^5 hour rupture strength for this material and Table 3 shows the ratio of the 10^5 hour data for P91, NF616 and NF12 to that of X20CrMoV 12 1 steel.[38,39] Based on these predictions NF12 is 2.5 and 3.6 times stronger than X20 at 550 and 650°C respectively. Fujita[39] expresses the opinion that NF12 would be capable of being applied to boiler plant under service conditions of 650°C, 34.3 MPa based on stress-rupture testing, though this would appear to extend only to approximately 5 000 hours.

A comparison of the minimum pipe thicknesses required for a steam pipe operating at nominal conditions are given in Fig. 8. This demonstrates the reduction in dimensions potentially possible with these materials.

Table 2 Extrapotential 100 000 hour Rupture Strength of NF12 (MPa).

Temperature	Mean Value
550°C	316
600°C	216
650°C	122
700°C	(60)

Table 3 Ratios of 100 000 hour Rupture Strengths of: P91, NF616 and NF12 to that of X20.

Material	Temperature		
	550°C	600°C	650°C
X20	1.0	1.0	1.0
P91	1.3	1.6	1.5
NF616	1.6	2.2	2.2
NF12	2.5	3.5	3.6

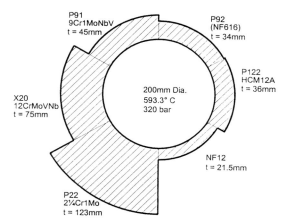

Fig. 8 Comparison of nominal pipe thickness of different steels for the same design conditions.

In Europe, under the guidance of COST 522, various European steel-makers are preparing different ideas for alloy development for 12Cr-based steels and these naturally centre on chemical composition and heat treatment options. There is no doubt that the task of alloy design is being made easier by new tools in the designer's armoury such as computerised thermodynamic equilibrium calculations and a deeper understanding of strengthening effects brought about by microstructural investigations in, for example, COST 501 III WP11 Metallography and Alloy Design Group.

2.1.2 Austenitic Steels

Austenitics offer higher creep strength and corrosion resistance and are therefore suitable for tubes and small diameter headers for superheaters/reheaters. Their use for large headers and pipework is restricted because of their thermal fatigue properties.

Extensive experience has been gained with austenitic materials in plant such as Drax, which has employed the very strong Esshete1250 material for 25 years in

Table 4 Chemical compositions of the 'standard' austenitic steels.

	TP321H	TP347H	Super 304H	Esshete 1250
C %	0.04–0.10	0.04–0.10	0.07–0.13	0.06–0.10
Si %	0.75 max.	0.75 max.	0.30 max.	0.20–1.00
Mn %	2.00 max.	2.00 max.	1.00 max.	5.50–7.00
P %	0.040 max.	0.040 max.	0.040 max.	0.040 max.
S %	0.030 max.	0.030 max.	0.010 max.	0.030 max.
Cr %	17.0–20.0	17.0–20.0	17.0–19.0	14.0–16.0
Mo %	–	–	–	0.80–1.20
Ni %	9.0–13.0	9.0–13.0	7.5–10.5	9.0–11.0
V %	–	–	–	0.15–0.40
Cu %	–	–	2.50–3.50	–
Nb %	–	8 × C	0.30–0.60	0.75–1.25
Ti %	4 × C	–	–	–
N %	–	–	0.05–0.12	–
B %	–	–	–	0.003–0.009

superheater tube and stub headers. The 18–8 type stainless steels such as TP321H and TP347H (Table 4) have also been in use for many years and, whilst in their conventional form, they are not as creep strong as the 15Cr–15Ni types, such as Esshete 1250, they do have superior corrosion resistance.[40]

High strength austenitic steels have recently been developed, especially in Japan, based on 18–8 chemistry with additions to provide precipitation strengthening in the creep range (Table 5). Super 304H, containing copper, niobium and nitrogen is an example of an improved 18–8 type steel.[40] The

Table 5 Chemical compositions of the 'super' austenitic steels.

	NF709	HR3C	X7NiCrCeNb 32 27 (AC66)	HR6W
C %	0.04–0.10	0.04–0.10	0.04–0.08	0.10 max.
Si %	0.75 max.	0.75 max.	0.30 max.	1.00 max.
Mn %	1.50 max.	2.00 max.	1.00 max.	2.00 max.
P %	0.030 max.	0.030 max.	0.015 max.	0.030 max.
S %	0.010 max.	0.030 max.	0.010 max.	0.030 max.
Cr %	19.0–22.0	24.0–26.0	26.0–28.0	21.0–25.0
Mo %	1.00–2.00	–	–	–
Ni %	23.0–27.0	17.0–23.0	31.0–33.0	35.0–45.0
Nb %	0.10–0.40	0.20–0.60	0.60–1.00	0.40 max.
Ti %	0.02–0.20	–	–	0.20 max.
N %	0.10–0.20	0.15–0.35	–	–
B %	0.002–0.008	–	–	–
Ce %	–	–	0.05–0.10	–
Al %	–	–	0.025 max.	–
W %	–	–	–	4.0–8.0

addition of the 3% Cu produces fine coherent precipitates of a copper-rich phase in the austenitic matrix and confers improved creep resistance. The further additions of niobium and nitrogen also promote high elevated temperature strength through the precipitation of $M_{23}C_6$, $Nb(C,N)$ and $NbCrN$, this increased strength being obtained without loss of ductility.

Sawaragi *et al.* have plotted the variation of corrosion resistance, high temperature rupture strength and ductility and toughness as a function of Nb and Cu content, as shown in Fig. 9.[41] The steam oxidation resistance of various austenitic materials is given in Fig. 10. It will be noted that the data is based on 1 000 hour tests and the validity of extrapolation must be questioned, if such data are to be used to predict the 250 000 design life of such materials.

Resistance to fireside corrosion is obviously important as the quest for efficiency leads to higher temperatures and, as is pointed out in Section 2.2.3, fireside corrosion is notoriously difficult to investigate experimentally in the laboratory.

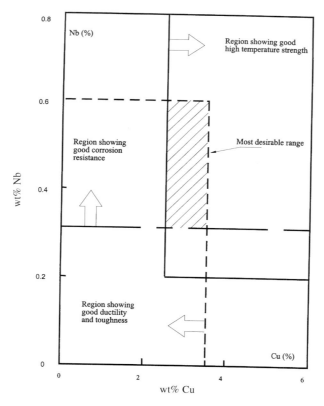

Fig. 9 Relationship between various properties and Cu, Nb content in 304 H type steels.

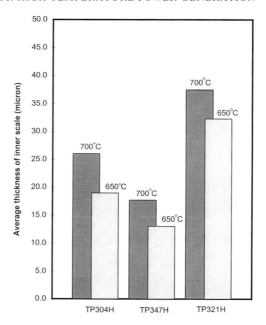

Fig. 10 Steam oxidation resistance of various steels (1 000 hours).

Beyond the 18–8 variants, a further series of higher chromium/nickel alloys has been developed capable of coping with steam temperatures up to 620°C, in terms both of corrosion resistance and creep strength: the creep strengths of several of these alloys are plotted in Figs 11 and 12.

These alloys are particularly high in chromium, with NF709 at ∼20% being at the lower end and AC66 at the higher end with ∼27%. HR6W and AC66 are not steels in the strictest sense, having a high nickel content and an iron level below 50%. The oxide formed by high pressure steam in such materials is composed of two layers, the inner being $(Cr,Fe)_2O_3$, a compound of the spinel type which is considered to act as an oxidation-resistant protective film. The higher the chromium content of the alloy, the stronger the spinel layer and the higher the steam oxidation resistance.[42] Furthermore, it has been reported that steam oxidation resistance is improved with increasing nickel content.[43]

Potentially, harmful phases may precipitate in such materials during long term heating and consequently it is necessary to design the alloy such that σ-phase does not precipitate in the nickel-chromium matrix. Such phases for examples can result in reduced toughness as shown by the Charpy-V notch impact values for NF709 after long term ageing which are given in Table 6.[39] The Charpy values tend to reduce with time over 1 000 hours and then increase, but even at 37 kgf.cm^{-2} are considered sufficient for boiler tube application. The loss of toughness is ascribed to the precipitation of $M_{23}C_6$ and the resultant hardening due to increased internal strain in the matrix, combined with grain boundary

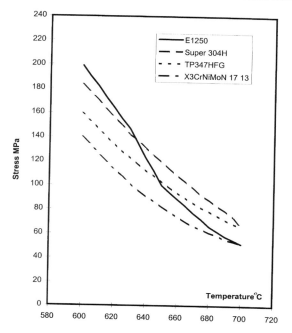

Fig. 11 Creep data (100 000 hour mean values) of the 'standard' austenitic steels.

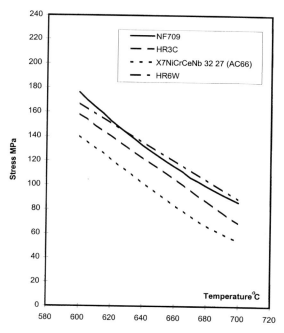

Fig. 12 Creep data (100 000 hour mean values) of the 'super' austenitic steels.

Table 6 Charpy V-notch toughness at 20°C of NF709 tube after long term ageing at 700°C.

	Ageing time at 700°C (hours)				
	0*	10	100	1 000	3 000
Impact value (kgf.cm^{-2})	180	105	44	27	33

* As solution heat treated

embrittlement from the same precipitate. The toughness recovery after 1 000 hours is attributed to thermal reduction of the internal strain in the matrix surrounding the precipitates.

2.1.3   Nickel-Based Alloys

As the need for further increase in boiler operating temperatures and efficiency continues, it is apparent that the improved properties deriving from the super austenitic steels described above will not suffice and their potential for further evolution is limited. This will probably produce another quantum leap in alloy design with a move to nickel-based alloys which demonstrate better high temperature creep strength and corrosion resistance than steels. This subject is treated later in more detail in Section 3.

2.2   Corrosion

2.2.1   Introduction

Corrosion is an aspect of materials degradation which has always existed, from the earliest boilers to the present day. Indeed such concerns extend into the future, where the higher temperatures indicated in Section 1.2 Future Trends, will bring their own set of challenges to be overcome. Whereas in earlier days it was usual to be reactive to corrosion problems, the large amount of experience gained over the years means that many of them can now be foreseen, and avoided at the boiler design stage.

Not all corrosion is detrimental. The economic use of carbon and low alloy steels depends usually on the formation of oxide films which, as they thicken, provide greater protection against further damage, i.e. the reaction rate is slowed down. Higher alloyed steels can form protective films at low temperatures, e.g. stainless steels. However this very attribute can make them particularly vulnerable to certain localised corrosion mechanisms such as pitting and stress corrosion cracking.

Corrosion phenomena in boilers can be conveniently divided into two major areas; those occurring on the material in contact with the water, steam or supercritical fluid, and those occurring on material in contact with the gaseous products of combustion. In this section these divisions are broadly followed, looking first at corrosion in water in sub-critical units (both recirculating and once-through), and considering how this will differ in supercritical units. Similar consideration is then given to corrosion in steam and supercritical fluid, and finally fireside corrosion is addressed.

2.2.2 Waterside Corrosion

Excessive waterside corrosion is prevented by control of the water chemistry, to ensure the integrity of the protective oxide film. Typical problems occur when small quantities of dissolved salt species are able to concentrate. This can occur, for example at the base of porous deposits or oxide films under heat flux conditions, at local areas of film damage, or in regions of steam blanketing. Particular care is required when using a solids dosing regime to control pH, where free NaOH is present or formed as it can be concentrated by this mechanism. Likewise the presence of oxygen and chloride can lead to acid pitting. Hence correct water chemistry is of extreme importance in minimising corrosion[45] in both boilers and subsequently in steam turbines.

Higher temperatures, pressures and heat fluxes require more stringent controls, e.g. the concentration factor in porous deposits increases markedly with film thickness and heat flux. Trends in larger, higher pressure and temperature boilers are to utilise zero solids water treatments for more rigorous control and cleaner boilers. This can be achieved by using hydrazine instead of sodium sulphite as an oxygen scavenger, and employing ammonia or amines rather than phosphates/NaOH for pH control.

A subcritical once-through boiler will contain a water-steam interface where a potential concentration effect can take place (in the same manner as the steam blanketing phenomenon). In the latter case, the boundary is relatively static, but in the boiling zone of a once-through system this boundary tends to oscillate over a short distance in the tubes, which alleviates the situation. No particular problems appear to have arisen in this area. Beyond the boiling zone, some degree of superheat is employed, and the higher temperatures here may require austenitic steels for strength and oxidation resistance. It is essential to design the boiler such that the boiling zone occurs on the ferritic sections of the tube run, and water does not come into contact with the stainless steel sections, i.e. they see steam only. This is to prevent the possibility of stress corrosion cracking occurring, which is one of the localised corrosion phenomena mentioned earlier. Chlorides are particularly active in causing this on stainless steels.

In general, supercritical units employ the same zero solids water chemistry regimes as used in subcritical boilers. The physical situation at the supercritical point is largely unknown, however it is unlikely that this will be any more onerous than that in a sub-critical unit, and indeed in operating units based on a standard steam cycle of 240 bar 540°C no particular problems have been experienced over a 20 year period in either Europe, the USA or Japan.[46]

In once through units there is a basic difference compared with the recirculation units in that any dissolved salts, which are present in the water (and are not carried through with the steam) will remain in the system. This is not the case in the recirculating systems where a proportion of the water can be removed by 'blow down' and treated before returning to the system. There is another water chemistry regime which can offer advantages, known as the neutral, oxygenated, low conductivity (NOLC) regime. Although the pH is neutral rather than alkaline, and significant oxygen is present, the feedwater is rigorously stripped of all contaminating salts to give very low conductivity in the boiler. This negates the damaging effect of oxygen and the lower than normal pH, produces good protective oxide films and reduces deposition resulting in a very clean boiler.[47] Obviously maintaining the ultra-low conductivity is of major importance. However there could be one note of concern when using stainless steels in this regime. It is well known that high oxygen environments can cause cracking of stainless steels when the latter are in a sensitised condition and under very high tensile stresses. Sensitisation is a condition in which areas adjacent to the grain boundaries in stainless steels are depleted in chromium due to precipitation of chromium carbides on the grain boundaries themselves. This condition can be very slowly remedied by diffusion of chromium from the bulk of the metal, but whilst in a sensitised condition the lack of chromium in the grain boundary areas renders the steels susceptible to damage in aggressive environments. The sensitisation effect tends to peak at around 650°C which is a target temperature for future plant (see Section 1.2). Hence it is almost certain that the materials used will become sensitised during operation. This is not a problem in steam as a liquid environment is required for cracking, however the behaviour of supercritical fluid, with respect to oxygen potential, is not known. The very high stresses required for cracking are unlikely on plain tube, but areas of discontinuity such as welds and bends may require special consideration.

In considering the steam, or higher temperature supercritical fluid regions, the main aspect is the thickness of the oxide films which are formed. With increasing temperatures, the films form more quickly and grow to a greater thickness (in an equivalent time), leading to two potential concerns. The first is the increased insulation of the tube material from the cooling fluid by the low thermal conductivity oxide film. With hot flue gas on the outside of the tube, this leads to an increase in metal temperature, which must be allowed for at the design stage in terms of material creep properties.

The second concern is that thicker oxides can spall more easily, this generally occurring when the unit is cooled down putting the oxide films, formed during running, into compression. The spalled material may lodge somewhere in the system, and, depending on the particle sizes, tube dimensions etc., may eventually cause tube blockage. Alternatively, it may be swept out with the working fluid and eventually impinge upon the turbine component, particularly blades, when erosion damage can be caused. In the UK, turbine erosion does not appear to be a particular problem.

Careful boiler design (including material selection) and boiler operation can be used to combat these concerns of internal oxide blanketing and oxide spalling.

As is explained elsewhere in this paper, whenever possible, ferritic steels are preferred over austenitic due to their higher thermal conductivity and lower coefficients of thermal expansion. When, however, the oxide spalls from such steels, usually the complete scale is involved. This exposes fresh metal surface to the working fluid, which is re-oxidised at an initially fast rate. Hence instead of the process of oxidation following a parabolic curve, with the rate gradually slowing down as the film thickens, the onset of regular spallation can result in an almost linear rate of metal loss. As temperatures increase, and scales grow more quickly, the process of spallation/regrowth occurring many times over the boiler lifetime can result in significant metal loss.

On austenitic steels the oxide and metal have a greater mismatch of thermal expansion coefficients which could lead to more cracking and spalling. However generally only the outer layer of scale becomes detached, leaving the inner rate controlling layer intact, hence the effect of a single spalling event is less.

The development of the 9% and 12% Cr steels was primarily aimed at increasing the creep strength, however the higher chromium contents when compared, e.g. to T22, also impart greater resistance to steam oxidation. Nevertheless the increase in oxidation resistance may not be necessarily sufficient to accommodate the higher temperature allowed by the increase in creep strength.

A substantial programme of work to investigate this has been conducted by Zabelt *et al.* on X20, T91, E911 and NF616 steels, the latter two being possible candidates for the next generation of plant.[48] Laboratory testing was carried out in air and steam, and tube sections exposed to actual hard coal-fired plant conditions at Wilhelmshaven Power Station. The latter were incorporated in the final superheater stage at steam and metal temperatures up to 605 and 630°C respectively. Control valves ensured the average mid-wall metal temperatures were in the range 621–636°C. In addition, T91 and NF616 tube sections were also exposed in the Avedøre Power Station.

Air exposure tests were conducted up to 4 000 hours at 630 and 680°C, and the results, extrapolated to 100 000 hours are shown in Figs 13 and 14. These indicate that the X20 steel exhibits greater resistance to air oxidation, and the extrapolated values do not appear to change substantially with temperature. In contrast, the 9%Cr group of steels have higher oxidation losses, and, whilst at 630°C the extrapolated values are relatively low, these increase to a significant figure of 0.3–0.6 mm metal loss after 100 000 hours at 680°C.

Autoclave tests were made in 6 bar steam at 500°C for times up to 3936 hours, and again there was little difference between the 9%Cr variants but scale thicknesses were about twice those found on X20 (Fig. 15).

The test section at Wilhelmshaven Power Station was in service for 28 990 hours, with scale thicknesses measured at intervals. On the steam side, with a mean steam temperature of 600–605°C, after 11 000 hours these thicknesses were 330, 460 and 325 µm for T91, E911 and NF616 respectively, and on the flue gas side 210, 165 and 205 µm. The present authors would consider these differences to be insignificant and within the scatter of an industrial sized field test. Me-

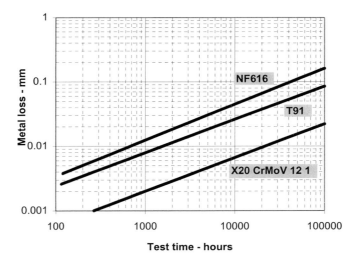

Fig. 13 Extrapolated metal losses of martensitic steels in air at 630°C. (after Zabelt *et al.*[48])

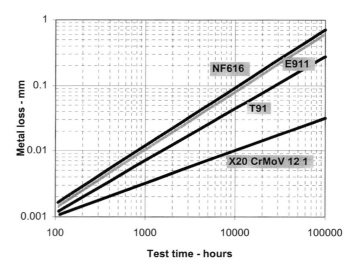

Fig. 14 Extrapolated metal losses of martensitic steels in air at 680°C. (after Zabelt *et al.*[48])

tallography of various samples at different temperatures indicated that, whilst the scale on X20 was homogeneous and protective, that on T91 had many pores, and no outer scale was seen on E911.

The metal loss results were treated on a parametric basis, using the plot derived by Nitikin, employing a time and temperature parameter similar to the

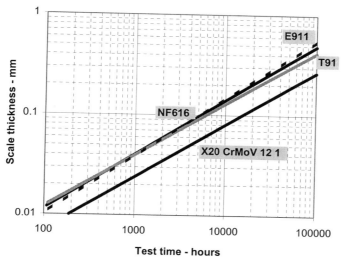

Fig. 15 Extrapolated metal losses of martensitic steels in steam at 500°C. (after Zabelt *et al.*[48])

Larson–Miller parameter.[49] Results, including those for steam and air tests, are shown in Figure 16 which gives the upper bound lines for the three environments. The constant in the equation was that proposed by Nitikin for a 12Cr1MoV steel.

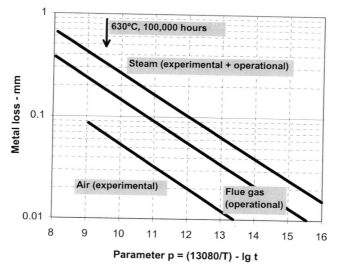

Fig. 16 Upper limit of metal loss for 9%–12% Cr steels using the time/temperature parameter p. (after Zabelt *et al.*[48])

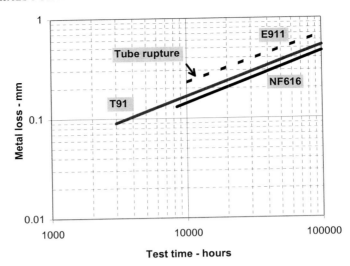

Fig. 17 Extrapolated metal losses on boiler tubes. (after Zabelt *et al.*[48])

Using the steam and flue gas data, a maximum total metal loss of 0.6 mm may be expected after 100 000 hours (Fig. 17). However the Zabelt *et al.*[48] suggest that this value is misleading as the blanketing effect of the bore oxide scale leads to higher and higher metal temperatures as the scale thickness increases. The final metal temperatures after 11 000 hours were estimated, from metallographic considerations, to be 675–700°C rather than 630°C.

It can be seen that the results from the field trials show far more oxidation damage than the laboratory tests in steam may suggest. The authors conclude that none of the 9%Cr steels are suitable for long term use as boiler tubes above ~600°C. The limitations on temperature are not so much due to the metal loss, but to the decrease in heat transfer which the steamside scale produces. The resulting increase in tube wall temperature means that the strength advantages of the developing 9%Cr steels (E911 and NF616) cannot be fully realised. The metal loss however also contributes to this; the predicted 0.6 mm reduction in wall thickness means that after 100 000 hours stresses will have increased by 12%.

These results and conclusions mean that the new generation of 9%Cr steels will be limited in temperature because of oxidation and therefore inadequate for the near and mid-term improvements envisaged in Section 1.2. On the other hand, while it is more oxidation resistant, X20 is limited because of inadequate creep strength. It is thus considered that steels such as HCM12, HCM12A and possibly NF12 (11%Cr), which combine high strength with greater scaling resistance, will be the main ferritic contenders for these conditions of service. Further advances already referred to being developed in COST 522, also recognise that higher chromium contents are necessary from the corrosion point of view.

Despite the above, it should be noted that the restriction of 600°C on 9%Cr steels does not apply when heat transfer is not involved, e.g. in headers and pipework. In this case the full creep potential of the these lower chromium variants can be realised.

Other methods of increasing the resistance of ferritic steels to steam oxidation which may be considered are chromising or surface cold working. The former would be very difficult on the inside of tube bores and may be impractical. With regard to the latter, surface cold work (or the use of fine grained material) has been successful in reducing oxidation rates in austenitic materials, due to the easier access of chromium by diffusion to the surface. Whether this is a method applicable to the higher chromium ferritics is not clear.

Austenitic steels will be and are used for thin section heat transfer tubing at temperatures above the metal temperature limit for ferritic steels. Field trials at the (sub-critical) Mizushima No 2 unit, using test sections welded into a secondary superheater, give an indication of the relative steam oxidation rates of ferritic and austenitic materials, albeit at the relatively low temperatures of 545°C for the former and 565°C for the latter.[50] After 10 149 hours of testing, approximate steam side oxide thicknesses were 105 and 60 µm for the ferritic materials NF616 and HCM12 respectively, but around only 40–50, 30 and 20–25 µm for the austenitics Super 304H, HR3C and NF 709, even though these were at the higher temperature. The effect of shotblasting the austenitic materials was to reduce the oxide to only a few microns, indicating as noted above the beneficial effect of cold working.

The present preferred steel for high temperature service in superheaters in the UK is Esshete 1250, due to its high creep strength. Early trials were held in two UK power stations, North Wilford and Skelton Grange, where sections including this material were made up into replacement superheater loops.[51,52] In the former station, steamside metal losses were measured after 12 719 hours at 660°C and ranged from 50–74 µm (100–148 µm oxide). The oxide thicknesses at Skelton Grange, after 6840 hours at 633–677°C ranged respectively between 200 and 230 µm.

Esshete 1250 is also one of the three austenitic steels being tested in the Wilhelmshaven Power Station as a follow on to the 9%Cr steels tests, together with 1.4910 (which is similar to a TP 316 with nitrogen), and AC66, a much more highly alloyed steel.[53] Results so far after 12 463 hours exposure in 600°C steam show negligible steamside oxide scale on AC66. Thicknesses for Esshete 1250 are 45–55 µm (on the side facing the flue gas) and 75–80 µm on the other side. Similarly, the corresponding values for 1.4910 are 90–120 µm and 100–120 µm respectively.

Although the metal temperature at Wilhelmshaven was not known, the results for Esshete 1250 generally appear to be lower than found in the UK power station trials. Given the high cost of the highly alloyed material AC66, it appears that for the temperature range considered, Esshete 1250 is the most economical material to provide the service required, as has been proven over many years in UK boiler plant.

It can be seen therefore that even though several items of information are available, it is difficult to compare them as the temperatures are different. The summary given in Table 7 appears to indicate generally that as tube metal temperatures rise towards 700°C then enhanced versions of the austenitic steels, such as NF709, HR3C, AC66 and HR6W, will be required to minimise the boreside oxide film thicknesses.

Finally, a comparison of tests in steam at 630°C for 10 000 hours exposure is shown in Fig. 18.[53] Here the improvement due to an increase in chromium content is illustrated.

The above has looked at the effect of temperature; the other factor in super-critical units is pressure. Very little has been reported on this, in fact, certainly in the sub-critical region it was considered that pressure had little effect. However work by Nippon Steel Corporation has indicated that for NF709 there is a relationship between the square root of the pressure and the logarithmic values of oxidation weight loss found. This is not so marked for the steel TP 347H.[54] It should be noted however, that these tests were only for 500 hours. Moreover, even if these results are indicative of a real trend, the effect will not be significant as the changes in plant pressure are not anticipated to be large in the short to medium term (~300 bar to ~320 bar – see Section 1.2).

To sum up waterside/steamside considerations for both ferritic and austenitic materials for supercritical units, existing water chemistry regimes, as long as they are rigorously controlled, would appear to be, and are indeed being used, successfully. The major effect in such units appears to be the effect of temperature on oxidation rates, although some work does indicate a small pressure effect.

Table 7 Bore oxide thicknesses from plant trials in steam.

Material	Station	Steam temperature (Metal temp) °C	Exposure time hours	Bore oxide thickness–μm
T91	Wilhelmshaven	600–605 (621–636)*	11 000	330
E911	Wilhelmshaven	600–605 (621–636)*	11 000	460
NF616	Mizushima 2	(545)	10 149	105
NF616	Wilhelmshaven	600–605 (621–636)*	11 000	325
HCM12	Mizushima 2	(545)	10 149	60
Super 304H	Mizushima 2	(565)	10 149	40–50
HR3C	Mizushima 2	(565)	10 149	30
NF709	Mizushima 2	(565)	10 149	20–25
Esshete 1250	Wilhelmshaven	600	12 463	40–80
Esshete 1250	North Wilford	(660)	12 719	100–148
Esshete 1250	Skelton Grange	(633–677)	6 840	200–230
1.4910	Wilhelmshaven	600	12 463	90–120
AC66	Wilhelmshaven	600	12 463	Negligible

* Mid wall, others unspecified

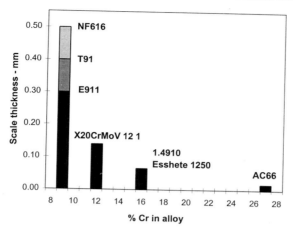

Fig. 18 Scale thicknesses in steam at 630°C and 10 000 hours. (after Zabelt *et al.*[48])

2.2.3   *Fireside Corrosion*

Considering now fireside corrosion, the main factors influencing this (as well as temperature, which affects all corrosion reactions) are the chemical species produced from the indigenous elements in the coal such as sulphur and chlorine. Although gaseous corrosion in flue gas is not normally a problem, due to the formation again of protective films, there are two main areas in a boiler which warrant special attention, viz. relatively low temperature areas near the burner belt, and the superheaters/reheaters.

In the first area, earlier problems arose (and still occasionally do) from burner flame impingement on the furnace walls. The majority of the boiler surfaces see oxidising conditions, where, as noted previously, protective films are formed in an atmosphere containing predominantly N_2, O_2, CO_2, H_2O, and SO_2. However, at the edges of the burner flame where impingement can occur, the gas atmosphere is primarily reducing, with a change in composition to N_2, CO, COS and H_2S. There is competition between oxygen and sulphur species in the reaction with iron. The result is that, as the atmosphere becomes more reducing, instead of compact, resistant magnetite/haematite films being formed, a fast growing magnetite containing iron sulphide occurs, which is far less protective than the normal oxide films. The situation is exacerbated by the presence of incompletely burnt coal particles which can stick to the outside of this film and release volatile compounds, particularly CO and HCl. The latter can diffuse through the oxide and cause intergranular attack at the metal surface. Metal loss rates can be as high as 5 mm per year.[55]

The most efficient remedy is to prevent flame impingement by careful attention to burner alignment, and fuel and air distributions to the burners.[56] A material orientated solution may be the use of co-extruded tubes comprising, e.g., a ferritic material on the inside and stainless steel on the outside.[57] However, there is still a question mark over the prediction of the behaviour of stainless steels, even TP 310 (25%Cr 20%Ni), under the varying conditions containing sulphur and chlorine species.

A more recent aspect of this type of corrosion has occurred under certain circumstances when low NOx staged combustion systems have been used. In these systems, the gas atmosphere in the burner belt is inherently sub-stoichiometric. Although such circumstances may not include the influence of smouldering coal particles, nevertheless examples of significantly high corrosion rates have been found. Again high CO and H_2S levels were detected at the corrosion sites.[58]

Potential solutions to this corrosion include changing the stoichiometry of the gas near the furnace walls. This can be done by biasing the wing burners to a higher stoichiometry, or, more drastically, by introducing an air curtain next to the walls.

Mention was made earlier of the provision of co-extruded tubes, however, it was noted that prediction of the behaviour of stainless steels was difficult. A further option is to use protective coatings, either in the form of diffusion or sprayed coatings. The former includes chromising, which has the advantage that there are no adhesion problems as the coating occurs by diffusion. However it is a high temperature process, performed usually on a batch basis, and is expensive. Sprayed coatings may offer an advantage in this respect, subject to obtaining good adhesion with the substrate (which is possible with modern coating methods such as plasma or high velocity oxy-fuel). In this respect, iron aluminides look to be a promising material as they have good resistance to attack by sulphur species. Methods have also been developed to apply these by weld cladding rather than spraying, which would ensure no problems with adhesion.[59,60]

The second area of concern, corrosion in superheaters and reheaters, has great relevance to proposed supercritical units due to the higher steam (and hence metal) temperatures encountered.

The alumino-silicate materials in the coal result in the formation of ash. However there are other species in coal which, although far less abundant, are extremely important from a corrosion aspect. Mention has already been made of sulphur and chlorine, and to these must be added sodium and potassium. The sulphur in the coal is, during combustion, converted to sulphur dioxide (SO_2) and in addition a further very small proportion of this is further oxidised to sulphur trioxide (SO_3). The presence of the latter together with water and the sodium and potassium constituents can lead to the formation of various sulphate and perhaps pyrosulphate species.

Deposition of ash occurs on the superheater and reheater surfaces, particularly on the leading tube in a tube bank. Where this ash deposit is thinner (and has a lower insulating effect) metal temperatures are the highest, generally just around the sides of the tube from the leading crown. At high metal temperatures, the sulphate constituents can accumulate at the base of the ash layer, and, as they have very significantly lower melting points than the ash, remain molten on the tube surface. The layer contains free SO_3 which attacks and removes the protective oxide scale, hence leading to high corrosion rates. The temperature gradient across the deposit causes the migration of dissolved iron and chromium outwards and consequently ensures a continuous dissolution process.[56,61]

The role of chlorine in this process has been found, by measurements of corrosion rates at different stations, to play an important role, but the exact mechanism of this is unclear. The characteristic intergranular attack occurring in the flame impingement mechanism is absent. It is thought that chlorine acts mainly (but not entirely) to aid the release of species, particularly potassium, from the coal which in turn contributes to the low melting point of the deposits. The other important factor in potassium release is the ash content of the coal.[62]

The metal temperature at which this type of corrosion starts to become significant is around 600°C, and it follows a 'bell shaped' curve peaking at around 670°C, and then diminishing due to stability effects on the dissolved sulphates. Corrosion rates however can remain substantial to well above 700°C.

The main variables affecting the rate of corrosion are flue gas and metal temperatures, chlorine content of the coal, tube position (leading or non-leading) and tube material, hence their interaction is complex.

In plant to date, this type of corrosion was avoided by restricting steam temperatures to around 565°C to ensure low metal temperatures in the superheaters and reheaters. However, as the aim of modern supercritical units is to increase the steam temperature, then the problem of this molten ash corrosion needs to be addressed. With metal temperatures increasing, the remaining important variables are coal chlorine content and material composition.

Coal chlorine has a very powerful influence, with coals containing less than 0.1% chlorine generally showing relatively low corrosion rates.[55] However, it is not always possible, physically or economically, to ensure that coals of these levels are used. Under these circumstances, material composition becomes very important.

In the past, the changeover surface metal temperature in superheaters from ferritic to austenitic materials has been around 580°C, governed by the creep properties. This meant that the ferritic material was never in the temperature range in which molten ash corrosion was active. With the new generation of materials (T91 etc.) the tube outside surface temperatures will reach those where enhanced corrosion can take place. However because ferritic materials were not previously used under these circumstances, information is sparse.

Early work in synthetic ash/synthetic gas indicated that T22 ferritic material had twice the corrosion rate of TP321 austenitic steel, and at temperatures between 600°C and 700°C the rate increased from around 20 to 50 mm/year. The tests lasted for only 5 days, and were using synthetic constituents, hence the corrosion rate values must be questionable for application to real plant, however they do indicate the trend of rapid escalation of corrosion with temperature.[63] The trial at North Wilford Power Station referred to earlier indicated that,

although the temperature ranges tested did not overlap, the low chromium ferritics fared rather worse than the austenitics, whereas a single result on 11%Cr
material was low, (but the latter was only tested at 575°C).[51] The similar trial at
Skelton Grange suggested that the low chromium ferritics should be limited to a
surface temperature of 580°C, but that the figure for the 11%Cr steel could be
around 620°C.[52]

In contrast, it has been reported that, for oil firing, where a similar molten
deposit mechanism takes place associated with sodium and vanadium, ferritic
steels perform better than austenitics as the temperature increases.[64] Clearly at
higher outside tube metal temperatures there could be a problem with ferritic
materials, but precise data are lacking and need to be addressed.

The situation with austenitic steels is well known as these have been used at
temperatures where the phenomenon is active. There does not appear to be a
great deal of difference in the behaviour of Esshete 1250, TP347 and TP316,
(TP347 may be marginally better).[61] TP310, with higher chromium and nickel
appeared to show more resistance, however this, after use in plant, is now being
questioned as to its actual degree of advantage. Testing in Japan in synthetic ash
mixtures compared the steels TP347H, NF709 and TP310, and showed the benefit
of increasing chromium content. However it is interesting to note that there was
only a small improvement between TP347 and TP310 (with NF709 lying between
these), perhaps bearing out the earlier statements regarding TP310.[65]

The situation is of potentially great concern when using higher chlorine coals
at the higher steam temperatures of supercritical units, and it is possible that co-
extruded tubes may have to be used. The chromium nickel alloy variants of
approximate composition 50%Cr 50%Ni appear to show much improved resistance (and are frequently used as uncooled components in boilers). They have
low strength, but would provide the corrosion resistance required when used as
an outer layer over a creep strong alloy in a co-extruded tube.

This corrosion phenomenon decreases in severity at temperatures above 670–
700°C but is still significant. The 'tail' of this bell-shaped curve may well overlap
the region in which the phenomenon of hot corrosion experienced in gas turbines
occurs, which is also associated with sodium sulphate rich deposits. Hence it is by
no means certain that increasing the temperature significantly above the peak of
the bell shaped curve will result in acceptable corrosion rates.

With regard to the possible use of nickel alloys, there are lot of data in the
temperature range where gas turbine hot corrosion occurs, but little in the lower
temperature molten ash region. Again if severe problems occur, the solution is
almost certain to be through the use of co-extruded tubing.

2.3   Water-Walls

Even where operating conditions are not severe, improved materials are required
to allow supercritical plant to be fabricated economically. One problem exists in
the water-wall design. These walls whether spiral or vertical are fabricated at site
by the joining of panels comprised of many tubes welded together. The dimen-

sional accuracy of these panels is critical to the site erection of the boiler and great care must be taken to minimise the effects of welding distortion during the fabrication of these panels. Post weld heat treatment (PWHT) of such panels can increase welding distortion considerably and increase the complexity of on-site erection. PWHT of the water-wall panels is therefore to be avoided for economic reasons if possible.

In natural circulation subcritical plant the water-wall operates at a fixed (~360°C) temperature and the wall panels are made from CMn tubing which does not require PWHT. However in supercritical plant the water-wall temperature increases with the height of the boiler and different materials are required at different levels. Currently the highest grade steel which does not require PWHT is 1Cr0.5Mo material.

Three main alloys have been developed for water-walls. The first two of these alloys are separate developments of the 1940s steel T22 containing 2.25Cr, but both have reduced carbon contents in order to aid weldability and require neither preheat nor post weld heat treatment. Further, by appropriate alloy additions, the creep strength of these steels has been enhanced by a factor of approximately 1.8 over the traditional T22 and approaches that of T91. The chemical compositions of these two steels:

HCM2S, developed by Sumitomo and Mitsubishi Heavy Industries Limited;[66]
7CrMoVTiB 9 10, developed by Mannesmann,[67]

are given in Table 8 along with that of the traditional T22.

In the Mannesmann alloy the loss of carbon has been compensated for by addition of vanadium, titanium and boron, whilst HCM2S also virtually dispenses with molybdenum but utilises 1.6% tungsten for solid solution strengthening, along with vanadium and niobium for precipitation hardening.

Whilst these steels show potential for thick section components, their attraction for water-wall application lies in their good weldability with a low hardness, high toughness HAZ being developed without the need for PWHT. A typical hardness traverse across a weld in 7CrMoVTiB 9 10 is given in Fig. 19, which shows that the maximum hardness is in the region of 350HV.[67]

The third steel with potential for use in water-walls is the Sumitomo and MHI developed steel HCM12. This 11%Cr steel, unlike HCM12A, has had its chemical composition engineered such that the material contains between 20% and 30% delta ferrite. The delta ferrite is usually avoided in martensitic steels for reasons of high temperature workability, however, the presence of this phase appears to suppress the hardness increases usually developed in the HAZ, which allows the material to be used without post weld heat treatment. It is claimed that HCM12 offers a further advantage over other 9–12% chromium steels in that the creep strength reduction due to Type IV cracking is less severe. Iseda et al.[68] have compared T91 and HCM12 and suggested that for T91 the creep strain in cross-weld testing is concentrated in the softened zone of the HAZ and

Table 8 Chemical compositions of water-wall steels for supercritical boilers.

	A213 T22	HCM2S	7CrMoVTi B 9 10	HCM12
C %	0.05–0.15	0.04–0.10	0.05–0.095	0.14 max.
Si %	0.50 max.	< 0.5	0.15–0.45	0.50 max.
Mn %	0.30–0.60	0.10–0.60	0.30–0.70	0.30–0.70
P %	0.025 max.	0.030 max.	0.020 max.	0.030 max.
S %	0.025 max.	0.010 max.	0.010 max.	0.030 max.
Cr %	1.90–2.60	1.90–2.60	2.20–2.60	11.0–13.0
Mo %	0.87–1.13	0.05–0.30	0.90–1.10	0.80–1.20
W %	–	1.45–1.75	–	0.80–1.20
V %	–	0.20–0.30	0.20–0.30	0.20–0.30
Nb %	–	–	–	0.20 max.
Ti %	–	–	0.05–0.10	–
B %	–	–	0.0015–0.0070	–
Al %	–	–	0.020 max.	–
N %	–	0.030 max.	0.01 max.	–

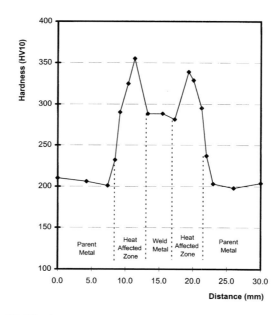

Fig. 19 Hardness traverse across a 7 CrMoVTiB 9 10 weld.

the strength of this zone is inferior to the surrounding areas. Due to the presence of delta ferrite, the softened zone in HCM12 is not so prominent as with T91 and hence creep strain does not concentrate in any particular zone. Cross-section hardness distributions for creep-ruptured cross-weld specimens from HCM12 and T91 weldments are shown in Fig. 20. The modes of failure were completely different with the HCM12 specimens showing very good ductility in contrast to the very brittle nature of the T91 failure.

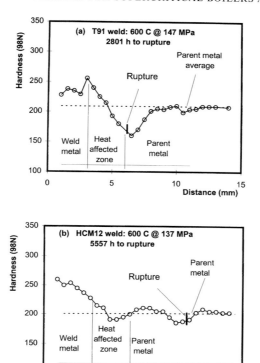

Fig. 20 Hardness profiles across failure region of stress rupture specimens.

2.4   Candidate Materials for Future Plant

The candidate materials for short and medium term goals for supercritical plant are shown in Table 9.

Table 9 Candidate materials for supercritical plant.

Component	Short Term 585–602°C Steam temperature	Medium Term 620–640°C Steam temperature
Water-walls	HCM2S or 7CrMoVTiB 9 10	HCM12
Superheater	Esshete 1250 or TP347HFG	NF709 or 'Super' austenitic or Sandvik*
Reheater	HCM12A	NF12 or BS *
S/H Stub headers	Esshete 1250	NF709 or Sandvik *
Main Headers & Pipework	NF616 or E911	NF12 or BS *

* Undesignated steels currently being developed by British Steel and Sandvik

According to Franke *et al.*[7] the timescale for the availability of steels for large plant construction is as follows:
1997: NF616, HCM12, HCM12A and HCM2S
1998: 7CrMoVTiB 9 10,
2000: E911
2004: Further advanced ferritics

3 THE THERMIE ADVANCED (700°C) PF PROGRAMME

A group of European steel-makers, boiler-makers, turbine makers and a utilities have considered, under the auspices of a Thermie B project, the feasibility of taking the considerable step of designing and building a complete demonstration power plant operating at 700–720°C and up to 375–400 bar.

Many of the materials already discussed would be employed in such an undertaking as well as further developments of both ferritic and austenitic steels. However, it is obvious that the material for the final superheaters of such plant would have to operate at metal temperatures of around 750°C and the final steam pipework would experience temperatures in the range 700–720°C. The difference in metal temperature of these two components is due to the flow of heat from the furnace in the superheater tube, that is not present in the pipework which has the sole function of working fluid transportation.

At these high temperatures it was quickly appreciated that even the super-austenitic steels are insufficient in terms of both mechanical and corrosion properties and it was judged that nickel-based alloys would be required.

The project group realised that, apart from the necessary creep strength requirements, many other aspects would require very detailed consideration including formability, weldability, corrosion resistance (both flue gas side and steam side), and particularly the cost. Even before the identification of a specific material having the desired properties, it was considered that there was a danger that the cost of nickel materials could make the capital build costs of the plant such that they would outweigh the increased efficiency benefits. This is a valid material issue and the boiler design sub-group within the project group recognised that radical plant layout changes would be necessary to minimise the use of such materials.

In power plant operating at 700°C the material temperatures are relatively low in comparison to other applications for nickel-based alloys, such as gas turbines for example. However, the high pressure of around 400 bar does make creep a dominating factor. Furthermore, the size of the ultimate product forms required by the boiler industry and the fabricated shapes (bent tube and pipe) would present forming challenges quite different from other high temperature nickel applications. Welding of large sections is also considered a problem and threatens to eliminate those alloys that depend on specific precipitation heat treatments to develop their properties, since it is not considered practicable at present to heat treat whole pipework systems.

Table 10 Compositions of two candidate 'Ni-base Alloys.

	Alloy 625	NiCr23Co12MoAlloy 617
C %	0.10 max.	0.05–0.10
Si %	0.50 max.	–
Mn %	0.50 max.	–
P %	0.015 max.	–
S %	0.015 max.	–
Fe %	5.00 max.	2.00 max.
Cr %	20.00–23.00	20.00–23.00
Mo %	8.00–10.00	8.00–10.00
Ni %	Rem.	Rem.
Co %	1.0 max.	10.00–13.00
Nb %	3.15–4.15	–
Ti %	0.4 max.	0.20–0.60
Al %	0.40 max.	0.60–1.50

Several candidate materials were considered by the group.[69,70,71,72,73] However, the main conclusions were that no alloy was available which could be proven to meet the material property demands created by the design.

The compositions and creep strengths of Alloys 625 and 617, two of the possible candidate materials considered capable of being rolled and forged, are listed in Table 10 and shown in Fig. 21. Alloy 625 has excellent creep strength but

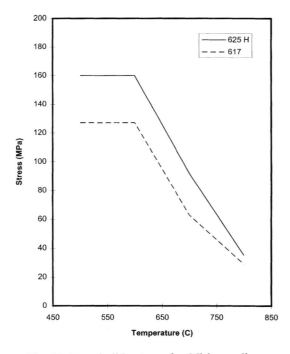

Fig. 21 Permissible stress for Ni-base alloys.

more information is needed on its formability while Alloy 617 is relatively weak but the very good experience in Germany in producing forgings and seamless pipe does make this material attractive.

It was considered that the development of an alloy which can meet all the requirements could be accomplished in a period of between 4 to 8 years, but that before experimental work is started theoretical consideration should be given to:

(i) formability
(ii) effect of alloy addition
(iii) influence of heat treatment
(iv) creep properties of weldments.

The original Thermie B project has now been followed up by a new THER-MIE contract for the first phases of a project with the ultimate goal of the commissioning of a 700°C plant by the year 2015.

4   CONCLUSION

The need for higher efficiencies has resulted in a period of rapid development of new materials for use in supercritical boilers. The challenge for the boiler manufacturer is to select the alloys that give the correct balance of properties at a suitable cost for each component. The well known difficulties in extrapolating the short term creep properties of new alloys to the 250 000 hour design life of a boiler and the variation in corrosion conditions between different types of boiler make this a complex, expensive and time consuming task. The complexities of plant construction also require materials with good fabricability though, in some cases, elements of the boiler design may need to be altered to make the best use of the materials available. The high costs and the need for extensive technical and metallurgical investigations of these problems has led to the formation of collaborations and joint development initiatives throughout the world. This has in turn led to the development of a range of new exciting materials. The past twenty years have seen an explosive growth in the number of new alloys being developed and evaluated and the power generation industry is now demonstrating the high technology approach to designer alloys previously associated with the aerospace industries.

REFERENCES

1. D.H. Scott: 'Emerging coal-fired power generation technologies', *IMechE Conference Trans. International Conference on Advanced Steam Plant*, Mechanical Eng. Publications Ltd, Bury St Edmunds and London, 1997.
2. *Coal gasification for IGCC power generation,* Toshi'ichi Takematsu & Chris Maude eds, IEA Coal Research, London, 1991.

3. I. Torkington, M. Upton and F. George: 'Two pass boiler for advanced steam conditions', *IMechE Conference Trans. International Conference on Advanced Steam Plant*, Mechanical Eng. Publications Ltd, Bury St Edmunds and London, 1997.

4. *STEAM: Its generation and use*, Babcock & Wilcox Co, J. Ray McDermott Inc, 1010 Common St, PO Box 60035, New Orleans, Louisianna 70160.

5. A.K. Smith, G.M. Kellet, J.L. Harper and D.E. Upton: 'The development of design and technology of boilers', *Fusion Welded Steel Pressure Vessels*, Lloyds Register of Shipping 50th Anniversary International Conference 1984.

6. J. Franke, and E. Wittchow: 'The Benson Boiler: A Key Component for Modern Coal-Fired Power Plants', presented at the *9th CEPSI Conference*, Hong Kong, 23–27 Nov, 1992.

7. J. Franke, R. Cossmann and H. Huschauer: 'Benson Steam Generator with Vertically-Tubed Furnace', *VGB Kraftwerkstechnik*, 1995, **75**(4).

8. J. Franke, R. Kral and E. Wittchow: 'Benson Boiler with Vertical Tube Water Walls-Principles and Advantages', *IMechE Conference Trans. International Conference on Advanced Steam Plant*, Mechanical Eng. Publications Ltd, Bury St Edmunds and London, 1997.

9. K. Sakai and S. Morita: 'The design of a 1000 MW coal fired boiler with advanced steam conditions of 593°C/593°C', *IMechE Conference Trans. International Conference on Advanced Steam Plant*, Mechanical Eng. Publications Ltd, Bury St Edmunds and London, 1997.

10. R. Sandstrom, S. Karlsson and S. Modin: 'The Residual Lifetime of Creep Deformed Material - Microstructural changes occurring during creep of a 12% Cr Mo V W steel', *High Temperature Technology*, 1985, **3**(2), 71–78.

11. C.H. Kreischer, J. Cothren and A.E. Near: 'Welding 12% Chromium Martensitic steels', *Welding Research Supplement*, Nov 1961, 489s–496s.

12. H. Naoi, H. Mimura, M. Ohgami, H. Morimoto, T. Tanaka, Y. Yazaki and T. Fujita: 'NF616 Pipe Production and Properties and Welding Consumable Development', *New Steels for Advanced Plant up to 620°C*, The EPRI/National Power Conference, 11th May 1995 London, The Society of Chemical Industry.

13. M. Caubo and J. Mathonel: 'Properties and Industrial Applications of a Superheater Tube Steel Containing 9%Cr 2%MoV Nb', *Revue de Metallurgy*, **69**(66), 345.

14. C. Rasche, W. Bendicke and J. Orr: 'Physical Properties, Transformation Behaviour and Microstructures of Grade 91', *ECSC Information Day – The Manufacture and Properties of Steel 91 for the Power Plant and Process Industries*, VDEh Dusseldorf, 5th Nov 1992.

15. (a) F. Bruhl, K. Haarmann, G. Kalwa, H. Weber and M. Zschau: 'Behaviour of the 9% Chromium Steel P91 in Short and Long Term Tests - Part I: Base Material', *VGB Kraftwerkstechnik*, 1989, (12).

(b) F. Bruhl, H. Cerjak, H. Musch, K. Niederhoft and M. Zschau: 'Behaviour of the 9% Chromium Steel P91 in Short and Long Term Tests - Part II: Weldments', *VGB Kraftwerkstechnik*, 1989, (12).

16. J. Orr, D. Burton and C. Rasche: 'The Sensitivity of Microstructure and Mechanical Properties of Steel 91 to Initial Heat Treatments', *ECSC Information Day – The Manufacture and Properties of Steel 91 for the Power Plant and Process Industries,* VDEh Dusseldorf, 5th Nov 1992.

17. P.J. McFadden and L.G. Taylor: 'Welding of 12Cr Mo V Pipework', *Mitsui Babcock Energy Ltd. Report No: 800108*, Jan 1980.

18. G. Guntz, M. Julien, G. Kottmann, F. Pellicani, A. Pouilly, and J.C. Vaillant: *The T91 Book, - Ferritic tubes and pipe for high temperature use in boilers*, Vallourec Industries, France, 1990.

19. W.M. Ham and B.S. Greenwell: 'The Application of P91 Steel to Boiler Components in the UK', *ECSC Information Day – The Manufacture and Properties of Steel 91 for the Power Plant and Process Industries*, VDEh Dusseldorf, 5th Nov 1992.

20. *Data Package for NF616 Ferritic Steel (9Cr-0.5Mo-1.8W-Nb-V), Second Edition*, March 94, Technical Development Bureau, Nippon Steel Corporation, 20-1 Shintomi, Chiba-ken, 299-12 Japan (English).

22. T. Fujita: 'Advanced High- Chromium Steels for High Temperatures', *Metals Progress*, August 1998.

23. T. Fujita: 'Current progress in Advanced High Cr ferritic Steels for High Temperature Applications', *ISIJ International*, 1992, **32**(2), 175–181.

24. T. Fujita, T. Sato and N. Takahashi: 'Effect of Mo and W on Long Term creep Rupture Strength of 12%Cr Heat-resisting Steel Containing V, Nb and B', *Transactions of the Iron and Steel Institute of Japan*, 1978, **18**, 115–124.

25. K.J. Irvine, D.J. and F.B. Pickering: 'The Physical metallurgy of 12%Cr Steels', *Journal of the Iron and Steel Institute*, 1960, **195**(8), 386–405.

26. R. Blum, J. Hald and E. Land: 'Untersuchung von Grundwerkstoff und Schweissverbindung der neu entwickelten Stahle NF616 und P91 (9% Cr-Stahle)', *VGB_Komferenz Werkstoffe und Schweisstechnik im Kraftwerk 1991*, (jan 1991), Essen.

27. H. Masumoto, H. Naoi, T. Takahashi, S. Araki, T. Ogawa and T. Fujita: 'Development of a 9Cr-0.5Mo 1.8W Steel for Boiler Tubes and Piping', *2nd Int. Conf. Improved Coal Fired Power-Plants*, Palo Alto, 1988.

28. F. Masuyama and T. Yokoyama: 'NF616 Fabrication Trials- In Comparison with HCM12A', *New Steels for Advanced Plant up to 620°C,* The EPRI/ National Power Conference, 11th May 1995 London, The Society of Chemical Industry.

29. B. Nath: 'Materials Comparisons Between NF616, HCM12A and TB12A-I: Dissimilar metal welds', *New Steels for Advanced Plant up to 620°C*, The EPRI/ National Power Conference, 11th May 1995 London, The Society of Chemical Industry.

30. E. Metcalfe, Bakker, R. Blum, R.P. Bygate, T.B. Gibbons, J. Hald, F. Masuyama, H. Naoi, S. Price and Y. Sawaragi: 'New steels for advanced coal fired plant up to 620°C', *New Steels for Advanced Plant up to 620°C*, The EPRI/ National Power Conference, 11th May 1995 London, The Society of Chemical Industry.

31. H. Naoi, H. Mimura, M. Ohgami, H. Morimoto, T. Tanaka, Y. Yazaki and T. Fujita: 'NF616 Pipe Production and Properties and Welding Consumable Development', *New Steels for Advanced Plant up to 620°C*, The EPRI/ National Power Conference, 11th May 1995 London, The Society of Chemical Industry.

32. J. Orr and L. Woolard: 'The Development of 9Cr 1Mo steels from Steel91 to E911', *Microstructural Development and Stability in High Chromium Ferritic Power Plant Steels*, The Institute of Materials, 1996, 53–72.

33. J. Hald: 'Materials Comparison Between NF616, HCM12A and TB12M-III: Microstructural Stability and Ageing', *New Steels for Advanced Plant up to 620°C*, The EPRI/ National Power Conference, 11th May 1995 London, The Society of Chemical Industry.

34. V. Foldyna, Z. Kubon, A. Jakabova and V. Vodarek: '*Development of Advanced High Chromium Ferritic Steels, Microstructural Development and Stability in High Chromium Ferritic Power Plant Steels*', The Institute of Materials, 1997, 73–92.

35. *EPRI Research Project 1403-50.*

36. G.A. Honeyman: 'TB12M Pipe Production and Properties', *New Steels for Advanced Plant up to 620°C*, The EPRI/ National Power Conference, 11th May 1995 London, The Society of Chemical Industry.

37. F. Masuyama: 'ASME Code Approval for NF616 and HCM12A', *New Steels for Advanced Plant up to 620°C*, The EPRI/ National Power Conference, 11th May 1995 London, The Society of Chemical Industry.

38. T. Fujita: 'Creep rupture strength of advanced high Cr ferritic heat resistant steels', *7th Meeting of EPRI Research Project (1403-5O)*, Nagasaki, March 1994.

39. T. Fujita: 'Future Ferritic Steels for High-Temperature Service' *New Steels for Advanced Plant up to 620°C*, The EPRI/ National Power Conference, 11th May 1995 London, The Society of Chemical Industry.

40. Y. Sawaragi, K. Ogawa, S. Kato, A. Natori and S. Hirano: 'Development of the economical 18-8 stainless steel (Super 304H) having high elevated temperature strength for fossil fire boilers', The Sumitomo Search 1992 (48), pp 50–58.

41. Y. Sawaragi, N. Otsuka, H. Senba and S. Yamamoto: 'Properties of a new 18-8 austenitic steel tube (Super 304H) for fossil fired boilers after service exposure with high elevated temperature strength', *The Sumitomo Search*, 1992 (56), 34–43.

42. K. Suzuki: *Thermal and Nuclear Power*, 1976, **27**(1), 27.

43. H. Coriou: 4th International Conference on the Peaceful Use of Atomic Energy, *A/Cobnf.*, 1979, 49, 576.
44. *Data package for NF709 austenitic stainless steel*, Nippon Steel Corporation 1994.
45. *Principles of Industrial Water Treatment*, Drew Chemical Corporation, New Jersey, 1981.
46. R.D. Townsend: 'Final Report of DTI Project B', *CEGB Report: TPRD/L/ MT0451/M87*, 1987.
47. 'Guidelines for Oxygenated Treatment for Fossil Plants', *TR-102285*, Electric Power Research Institute, USA.
48. K. Zabelt, O. Wachter and B. Melzer: 'The Corrosion of 9% Chromium Steels under Typical Boiler Conditions', *VGB Kraftwerkstechnik*, 1996, **76**(12), 936–940.
49. W.I. Nikitin: *The Calculation of the Heat Resistance of Metals*, Metallurgical Press, 1976, Moscow.
50. K. Tamura, T. Sato, Y. Fukuda and T. Izumi: 'Field Test Results of High Strength Heat Resistant Boiler Tube Materials', *International Conference on Power Engineering*, 1997, Tokyo.
51. A.M. Edwards, P.J. Jackson and L.S. Howes: 'Operational Trial of Superheater Steels in a C.E.G.B. Pulverized-fuel-fired Boiler Burning East Midlands Coal', *J. Inst. Fuel*, 1962, **35**(252), 16–28.
52. A.M. Edwards, G.J. Evans and L.S. Howes: 'Operational Trial of Superheater Steels in a C.E.G.B. Pulverized-fuel-fired Boiler Burning Yorkshire Coal', *J. Inst. Fuel*, 1962, **35**(254), 121–131.
53. K. Zabelt: 'Corrosion Behaviour of Martensitic and Austenitic Steels in Water Vapour and Flue Gases', *Corrosion Damage in Power Stations*, 9th Annual VDI Damage Analysis Meeting, October 1997, Würzburg.
54. T. Takahashi, M. Kikuchi, H. Sakurai, K. Nagao, M. Sakakibara, T. Ogawa, S. Araki and H. Yasuda: 'Development of High Strength 20Cr-25Ni (NF 709) Steel for USC Boiler Tubes', *Nippon Steel Technical Report No. 38*, 1988, 26.
55. P.J. James and L.W. Pinder: 'The Impact of Coal Chlorine on the Fireside Corrosion Behaviour of Boiler Tubing: A United Kingdom Perspective', *Corrosion '97*, NACE Conference, New Orleans March 1997, Paper 97133.
56. F. Clarke and C.W. Morris: 'Combustion Aspects of Furnace Wall Corrosion', *Corrosion Resistant Materials for Coal Conversion Systems*, D.B. Meadowcroft and M.I. Manning, eds, Applied Science Publishers Ltd, London, 1983.
57. S. Brooks, D.B. Meadowcroft, C.W. Morris and T. Flatley: 'A Laboratory Evaluation of Alloys Potentially Resistant to Furnace Wall Corrosion', *Corrosion Resistant Materials for Coal Conversion Systems*, D.B. Meadowcroft and M.I. Manning, eds, Applied Science Publishers Ltd, London, 1983.

58. 'Solutions for Low NOx-related Waterwall Corrosion', *Materials and Components in Fossil Energy Applications*, DoE/EPRI Newsletter No. 129, August 1997, 6.

59. G.M. Goodwin: 'Weld Overlay Cladding with Iron Aluminides', paper presented at the *11th Annual Conference on Fossil Energy Materials*, Knoxville, USA, May 1997.

60. S.W. Banovic, J.N. DuPont and A.R. Marder: 'Iron Aluminide Weld Overlay Coatings for Boiler Tube Protection in Coal-fired Low-NOx Boilers', paper presented at the *11th Annual Conference on Fossil Energy Materials*, Knoxville, USA, May 1997.

61. E.P. Latham, T. Flatley and C.W. Morris: 'Comparative Performance of Superheated Steam Tube Materials in Pulverised Fuel Fired Plant Environments', *Corrosion Resistant Materials for Coal Conversion Systems*, D.B. Meadowcroft and M.I. Manning, eds, Applied Science Publishers Ltd., London, 1983.

62. W.H. Gibb and J.G. Angus: 'The Release of Potassium from Coal during Bomb Combustion', *J. Inst. Energy*, 1983, **56**(428), 149–157.

63. W. Nelson and C. Cain Jr.: 'Corrosion of Superheaters and Reheaters of Pulverized-coal-fired Boilers', *J. Eng. Power, Trans. ASME*, 1960, **82**, 194–204.

64. J.C. Parker, D.F. Rosborough and M.J. Virr: 'High Temperature Corrosion Trials at Marchwood Power Station', *J. Inst. Fuel*, 1972, **45**(372), 95–104.

65. Nippon Steel Corporation: *Data Package for NF709 Austenitic Steel* (20Cr-25Ni-1.5Mo-Nb, Ti, B, N), First Edition, November 1994.

66. Y. Sawaragi, A. Iseda, S. Yamamoto and F. Masuyama: 'Development of high strength 2%Cr Steel Tubes (HCM2S) for Boilers', *The Sumitomo Search*, 1997, n59, 113–123.

67. *Private Communication*, Mannesmann, 1996

68. A. Iseda, Y. Sawaragi, K. Ogawa, M, Kubota and Y. Hayase: 'High Temperature Properties of 12%Cr Steel Tube (HCM12) for Boilers', *The Sumitomo Search*, 1992, n48, 21–32.

69. Schilke, Foster, Pepe and Beltran: 'Advanced Materials Propel Progress in Land-Based Gas Turbines', *Advanced Materials & Processes*, 1992, No:4, 22–30.

70. *Data Sheet INCONEL Alloy 617.*

71. *Data Sheet INCONEL Alloy 625.*

72. Hardt: *Materials for Steam Generators with Ultra-Super Advanced Steam Conditions*, COST-WP11 Meeting 1995, Vienna.

73. Brill, Heubner, Drefal and Henrich: 'Zetstandarte von Hochtemperaturwerkstoffen', *Ingenieur-Werkstoffe*, 1991, **3**(4), 59–62.

Advanced Materials for Advanced Heat Exchangers

F. STARR

ETD Ltd, Ashstead, Surrey, UK

ABSTRACT

A variety of new austenitic and nickel based alloys have been developed in the 1980s and 90s, many of which are finding application in advanced energy conversion plant. The paper briefly examines the historical background to these new materials. They can be traced back to the simple austenitic stainless steels of the 1930s, in both their wrought and spun cast forms, and in the development, in the late 1940s, of super stainless alloys such as Incoloy 800.

Most of these alloys have been developed outside of the boiler plant materials programmes aimed at conventional power generation systems. These programmes have largely focused on low alloy ferritic, bainitic, and martensitic superheater tube materials, intended for use at fairly modest temperatures and in moderate corrosive conditions. The advanced alloys, covered in the paper, are used in much more aggressive conditions, in terms of temperatures and in the severity of corrosion. Both are characteristic of novel energy conversion processes requiring advanced forms of heat exchanger. Examples of alloys from Europe, America, and Japan are used to describe some of the thinking behind the development of these new materials.

The background to commercial and near commercial energy conversion systems, which require advanced heat exchangers, is also described. These include indirect fired gas turbines; recuperative gas turbines of various types, fluidised bed combustion, coal gasification, and waste incineration.

To optimise strength over a given temperature range, each of the new alloys needs to utilise a specific type of precipitation or dispersion strengthening mechanism. In the older austenitic alloys the principal precipitation processes involved gamma prime and chromium carbides. The newer alloys utilise yttria for dispersion strengthening, and nitrides, carbo-nitrides, complex carbides, and copper rich phases for precipitation hardening.

It is emphasised that the general fall off in stress rupture properties, with temperature, is governed by the ease of dislocation climb over strengthening precipitates. It is therefore possible to draw an 'average' stress rupture strength line for austenitic and nickel based alloys, the gradient of which is probably governed by the diffusion rate and elastic modulus of such materials.

In order to maximise resistance to attack in advanced heat exchanger environments, the newer materials need higher levels of alloying elements. In general, chromium contents are very much greater, particularly where the attack involves a molten salt phase. However, at very high temperatures, to resist simple oxidation, the newer alloys need to contain significant amounts of aluminium, so that in that in some cases they form scales of alumina rather than chromia. This feature has been made possible by the addition of active elements, so reducing the propensity to scale spallation. Recent progress with the understanding of coal ash corrosion is described.

Two major problems remain to be overcome in a satisfactory manner. These are corrosion in coal gasification and waste incineration systems. In both processes there is now a reasonably good understanding of the complex mechanisms that lead to such high rates of corrosion at such comparatively low temperatures. The presence of high levels of chlorine creates new difficulties. This points to the need for development of purpose designed alloy systems and coatings. Spun cast production of high alloy tubing may give new opportunities in this respect.

1 INTRODUCTION

The past decade has shown a refreshing change in the development and use of materials for heat exchangers. In some cases materials have been specifically developed to resist new and harsher conditions. In others, experience with new alloys has been shared between diverse industries and technologies. A principal aim of this paper is to identify these common features and to indicate how progress may be further accelerated.

Rather than to produce a compendium of commercial, near commercial, and experimental alloys and coatings, a more selective approach will be adopted in this paper, focusing on the alloy design aspects of high temperature heat exchanger materials. The chief needs for such a material, if it is to be used to construct a heat exchanger, is that it must have adequate creep strength at temperature, must have reasonable resistance to high temperature corrosion, and must, normally, be capable of being fabricated in a tube form. These requirements are often in conflict. It is one reason why, with the need for exchangers to work in ever more aggressive conditions, corrosion resistant coatings are often perceived to be the most viable solution.

Authors of other papers at this first ever Institute of Materials (IOM) Materials Congress will be examining the present status of materials in state-of-the-art power generation. Here we are concerned with a rather different set of materials, intended for higher temperatures or more aggressive conditions.

The difference between these advanced materials, and the current set of low alloy and stainless steels for power plant superheater and reheater applications, is best shown in the variety of strengthening mechanisms utilised. The full range of these will be explored in the text, but they include strengthening by oxide dispersions, gamma prime, and complex carbides and nitrides. In terms of corrosion resistance, the advanced alloys are characterised by much higher levels of chromium, aluminium, silicon, and nickel than the common or garden stainless steels. Again, in this paper, we will be examining the principal environments that these advanced materials are intended to withstand, and indicating what factors govern good corrosion resistance.

In most types of energy conversion and process plant, the development of materials for steam boiler pipework and superheaters has been, and still is, evolutionary rather than revolutionary. This is particularly true of electricity generation, where the increase in steam temperatures has been relatively slow, particularly over the last thirty years. To a certain extent, this has been caused by the need to ensure that other sections of the generating plant, for example steam turbine casings, rotors, and stop valves, etc, can take the higher steam conditions. The introduction of new materials will be faster if an increase of temperatures and pressures in a high temperature heat exchanger has no serious implications for the rest of the equipment. This is often the case in the oil refining and petrochemical sectors, where the upstream and downstream plant consists of pipework and low technology heat transfer equipment, rather than complex, highly stressed machinery.

This would suggest that there will be greater incentives to introduce a new material if *the heat exchanger itself is of a new type*, or if it is of a conventional form that is working under extreme conditions. In considering the prospects for advanced materials in heat exchangers, one needs to consider the prospects for advanced heat exchangers and processes *themselves*, and this will form the first section of this paper.

There is no set definition for what can be described as an advanced heat exchanger, but it is one that will be used in:

A simple development of a well known process, operating in conditions which are significantly more severe than in standard equipment.
A process which is extremely conventional in many respects, but requiring a fairly radical change in heat exchanger materials or design.
A completely new process, which is only viable as the direct result of improvements in heat exchanger design, fabrication techniques, and/or new materials.

Obviously, these classifications overlap. For example, an advanced pulverised fuel steam plant, operating at 750°C and 450 bar pressure, would undoubtedly require a quantum leap in the performance of superheater materials. Some, however, would regard a pressurised fluidised bed boiler, as being a simple variation of conventional steam plant technology, despite the fact that it is subject to new forms of metals wastage. Even exchangers for indirect coal fired gas turbine cycles, which are only possible now that we have materials which can operate above 1100°C, in corrosive environments, would be seen by many of those who work in the petrochemical field, as being a straightforward development of steam reformer or ethylene plant technology.

Nevertheless, in the past decade, the desire for compact, high temperature heat exchangers has led to concepts which fall almost completely into the third classification. That is, heat exchanger designs and materials, without which a new system will fail. Of the concepts that are about to reach the market, the best example is that of the recuperative gas turbine. Whether in the form of the micro turbine giving out less than a 100 kW, or one on a rather larger scale for shipborne propulsion or industrial use, if this, and related developments are successful, we may be at the start of a new era in power generation and transport.

2 LESSONS OF HISTORY

Although most papers in this section of the Materials Congress focus on low alloy and bainitic steels for generating plant, most of the initiatives, for high temperature heat exchanger design, and for their materials of construction, have come from outside of the power generation sector. This history goes back a long way, almost to the start of the Industrial Revolution. The problems that engineers faced at that time were basically those we have today. The very first high temperature heat exchanger, the Neilson hot blast stove, was introduced just over 150 years ago, Fig. 1. This, by heating the charge air to a blast furnace to about 350°C, reduced coke consumption by a factor of 2, revolutionising iron making practice.[1]

Neilson's stove used cast iron exchanger tubing, and two of the main problems which bedevilled the very first gas to gas heat exchanger, high temperature corrosion, and what we would regard as thermal stress, are still, in some respects, the major challenges in the advanced heat exchanger field.

Real progress in the high temperature heat exchanger field only recommenced, in the oil industry, during the 1920s, with the need for thermal cracking of high molecular weight hydrocarbons. Despite an absolute paucity of stress rupture data, low alloy carbon and austenitic stainless steels were used in pipe stills and other forms of oil heater, at metal temperatures of up to 800°C! Not surprisingly, sulphide corrosion and coke formation were critical issues. There are also indications that elevated temperature embrittlement, due to carburisation or sigma phase formation, was a problem.[2] These are still major issues in the oil refining and petrochemical sectors.

The most dramatic step forward, in terms of really high temperature operation, came about with the development of centrifugal casting of high carbon austenitic stainless steels in the late forties.[3] It is difficult to conceive of steam reforming or ethylene production without tubes of this type. Pressures in these

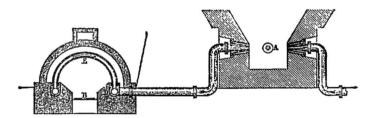

Fig. 1 Prototype Neilson Blast Stove, circa 1838. Blast air, at atmosphere pressure, is heated to about 350°C in an arch of cast iron tubes in furnace on right, and passed to blast furnace on left.[1]

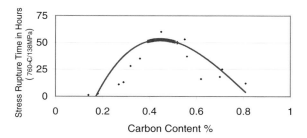

Fig. 2 Effect of carbon content on short term stress rupture life of 18Cr/37Ni type spun cast stainless alloy. (Redrawn from Ref. 2)

processes are still low by power plant standards, even today not exceeding 40 bar for reforming and about 5 bar for ethylene plant, but metal temperatures are in the 900°–1100°C region. It was quickly determined that carbon levels of around 0.4–0.45% were optimum in terms of stress rupture properties, Fig. 2.

The main shortcoming of the early cast stainless steels was embrittlement due to the formation of fine carbides during service. The high chromium, low nickel, sigma forming alloys have had little use for pressurised tubing. Whilst precipitation of carbides does not cause serious problems at working temperature, below 600°C ductility would fall to under 4%, giving the risk of embrittlement and catastrophic failure. Initially, outlet headers on reforming plant were also made from cast stainless steels of 18Cr–37Ni type, but in the mid 1960s there were some very nasty accidents in the UK gas industry with this type of design. At least two serious explosions were caused by severe thermal stressing of the outlet headers, during boiler priming incidents.[4] Accordingly, cast headers quickly went out of fashion, being replaced by those of the wrought variety. Having been involved in such a boiler priming 'incident', the writer can write with feeling about the efficacy of wrought headers.

The principal high temperature wrought alloy on offer, at that time, was Incoloy 800, itself a late 1940s material. This had been devised as an iron based substitute for the Inconel series. Chromium and nickel contents were set to give a good combination of high temperature strength and *static* high temperature oxidation resistance (that is in the absence of frequent temperature cycling), whilst avoiding the sigma phase propensity of conventional stainless steels such as Type 310. Incoloy was a key alloy in the history of high temperature materials, and was clearly formulated in the light of experience. Electron/atom ratio concepts, such as PHACOMP, to control the alloy composition, so as to prevent the formation of sigma, were not then in use. And obviously, at that time, computer prediction of phase diagrams was not even a pipe dream.

A major conference was held on the Incoloy alloys in Europe in 1978.[5] Incoloy 800 itself has long been superseded for high temperature use by Incoloy 800 HT, also from the Inco stable, but this alloy now has many competitors, and some of these will be given separate consideration in this paper.

Incoloy 800 and most of the alloys of that time were, by modern standards, quite low in chromium. These materials were intended for use in high temperature furnaces burning natural gas or naphtha, giving combustion products that were oxidising, low in sulphur dioxide and free from ash. Conversely, in a 1950s style conventional steam plant, where ash and SO_2 were present, there was little call for an alloy which could withstand fireside corrosion, since temperatures at that time were low. However, it is understood that a more resistant variety of Incoloy 800 was developed for an indirect fired closed cycle gas turbine.[6]

It was in the UK a little later, in the 1960s, where the problem of fireside corrosion received most attention. UK coals are more aggressive than those on the Continent, and average steam temperatures, in this country, were at that time higher.[7,8] Some aspects of fireside corrosion will be covered in Section 5.3, since it will be a critical issue in the development of very advanced pulverised fuel plant.

Most forms of fossil fuel plant for power production can be expected to encounter thermal fatigue. This seems to have been a major reason why the generating industry has avoided cast stainless materials. At some time in the life of generating plant it will move to 'two shifting', so that it is essential that superheater tubing should retain high ductility and fracture toughness, even after long periods of service. Latterly there has been a real effort to develop cast stainless alloys with lower carbon contents, having enhanced resistance to embrittlement. Despite this, there still seems to be a reluctance to consider the merits of the spun cast stainless tubing. None of these materials have been incorporated into very long term generating plant test programmes to date.

There is, it would seem, in the refinery and petrochemical fields, a greater willingness to introduce higher temperature materials, without long-term stress rupture and low cycle fatigue data to underpin their use. This is not to say that new high temperature alloys are used in a cavalier fashion, or that there are no institutional obstacles to overcome in their introduction. Manufacturers have to provide reasonably extensive stress rupture data for design purposes, but often this is extrapolated from medium term tests using log time-log stress, or simple parametric methods. A suitable safety factor is then applied to give a design stress. This makes for a conservative approach in the stressing of high temperature heaters and furnace tubing, and a number organisations, including ERA Technology, have become quite skilful at indicating how much extra life can be wrung out of apparently time expired equipment.[9]

The power plant sector insists on ever more stringent criteria for new materials, particularly when advanced processes are under consideration. When European organisations began to consider the prospects for nuclear heated gas turbine cycles and steam reforming processes, it was considered vital to put in hand major programmes on HK40, Incoloy 800, and Inconel 617, among other materials, even though some of these had been in commercial use for many years.

Today, even in the oil refining and processing sector, it seems more difficult to get new materials introduced than it was once. Health and Safety legislation is a major concern, and there is less tolerance of unscheduled shutdowns due to a new material not coming up to expectations. There is also, perhaps, a perception that the newer alloys do not have much to offer compared to standard materials.

Despite these obstacles, mainstream alloy development is still continuing, and it is the arrival of processes, which require new types of heat exchanger, which give advanced materials their best opportunity. In exchanger designs which are close to the conventional, the possibility of introducing truly innovative *tube alloys* will be low, particularly if there are misgivings of any kind about the newer materials. To help allay reasonable concerns, the paper will attempt to show how these new materials are superior to those of old. It is also intended, where possible, to indicate the thinking behind the new alloys, so that engineers and designers can have greater confidence in the claims of alloy developers.

3   ADVANCED HEAT EXCHANGER CONCEPTS FOR THE NINETIES AND BEYOND

As indicated in the Introduction, to give focus to the paper, a number of nearer term technologies that need heat exchange will be briefly described. These include indirect fired and recuperative gas turbines, very advanced pulverised fuel plant, fluidised bed combustion, coal gasification, and waste incineration. Table 1 summarises the technologies that require advanced heat exchange systems. The scale varies tremendously, from 100 kW up to 1000 MW.

3.1   Indirect Coal Fuelled Fired Gas Turbine

In an indirect fired gas turbine, the combustion chamber of the gas turbine is replaced by a large, externally fired, heat exchanger, into which flows pressurised air from a compressor. The air for the gas turbine is therefore heated 'indirectly' by the combustion of coal, biomass, or any other fuel, which is burnt on the furnace side of the exchanger. The pressurised air, now heated to a suitably high temperature, is expanded through the turbine section where it produces sufficient energy to drive a generator and the compressor,[10] Fig. 3.

In most types of indirect fired gas turbine cycle, the temperature to which the air must be heated has to be in excess of 1100°C, if these cycles are to be competitive with advanced pulverised combustion or fluidised bed coal plant. Ideally, turbine inlet temperatures should be equal to that of the best natural gas fired combined cycle plant. This today is approaching 1450°C, but such a temperature would be impracticable with metallic materials of construction, and perhaps even with ceramics. Hence high efficiencies can only be obtained by using a relatively sophisticated thermodynamic cycle, and ensuring that component design is better than in an open cycle plant. Fortunately, these are quite

Table 1 Characteristics of energy conversion system materials requiring advanced heat exchange equipment.

System	Output	Fuel	Critical heat exchanger metal temperatures	Pressure	Challenge	Comparable equipment	Fuel for comparable equipment
Indirect fired gas turbine cycle	500–1000 MW	Coal	600–1150°C	10–60 bar	Fireside corrosion HT oxidation strength	Closed cycle GT Reforming plant Ethylene plant	Lignite Gas Oil
Advanced PF steam plant	500 MW	Coal	600–750°C	400 bar	Fireside corrosion Steamside corrosion Strength	Conventional PF plant	Coal
Innovative PF steam plant	500 MW	Coal	600–1000°C	60 bar	Fireside corrosion Steamside volatilisation Strength	Conventional PF plant Reforming plant	Coal Oil
Fluidised bed combustion	100–500 MW	Coal	300–600°C	250 bar	Erosion Corrosion Steamside oxidation	Conventional PF plant	Coal
Recuperative gas turbines	10–50 MW	Natural gas Diesel fuel	300–600°C	10–25 bar	Thermal fatigue Condensate Corrosion	Compact heat exchanger practice	N/A
Recuperative gas turbines	<10 MW	Natural gas	200–750°C		Strength Thermal fatigue Condensate corrosion	Compact heat exchanger practice	N/A
Entrained bed gasification	500 MW	Coal	200–750°C	30–50 bar	Sulphidation/ Chloridation Corrosion/creep interactions	Refinery heat exchangers	High Sulphur Oils
Waste incineration	50 MW	Waste or biomass	200–600°C	100 bar max	Chloridation/ Sulphidation	Conventional chain grate stokers	Coal

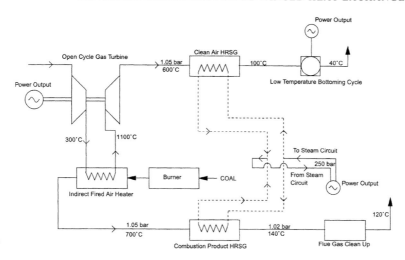

Fig. 3 Coal fuelled indirect fired open cycle gas turbine. Air flows from compressor to high temperature heater, then to turbine. It exits the system through the clean air HRSG and low temperature bottoming cycle.

practical aims, since the working fluid in the cycle is clean air, free from combustion products and moisture, and whose temperature never exceeds that of turbine inlet.

Operating pressures, and stresses within the exchanger, or air heater as it is called, depend on the cycle parameters, but are significantly lower than in conventional steam plant. For a 'once through' open cycle type plant, pressures would be between 10–30 bar. The lower pressure system would correspond to the simplest non-reheat form of gas turbine cycle. The higher pressure system would be for an open cycle, reheat type of plant, using two sets of air heaters, one working at about 30 bar, the other at about 6 bar. These numbers follow from an elementary study of gas turbine cycles.[11,12]

In any form of indirect fired gas turbine there will be a considerable amount of waste heat in the exhaust from the gas turbine, but there is also a large amount of heat in the 'spent' combustion products from the furnace side of the air heater. This waste heat must be utilised to maximise thermal efficiency and, in an open cycle indirect fired plant this will normally be used for steam raising via a HRSG (heat recovery steam generator). These combustion products will contain dust and acid gases, so that this section of the equipment will be similar to that of a conventional pulverised fuel plant. Although this form of indirect fired cycle may not be the most efficient, it has the merit of confining the most serious materials issues to the indirect fired heater itself.

Since the exhaust, from the gas turbine itself, is simply air at around 600°C, this too can be put through a very efficient HRSG and feed water heater, with the outlet stack temperature being very low by normal coal fired power station standards. Alternatively, since there is no dewpoint problem, it is possible to envisage the use of an organic Rankine cycle utilising the last fraction of the heat. This arrangement would necessitate the use of low pressure drop heat exchangers of the compact type.

So far little has been published on either the design or materials challenges of the indirect fired heater, although Wright and Stringer have produced an initial review of the overall stress and corrosion requirements.[10] However an air heater, using coal as a fuel, is likely to resemble those in closed cycle gas turbine plants built in Germany, France, and Japan during the nineteen fifties and sixties. These units used blast furnace gas or low grade liquid fuels for firing.[6] Tubing was placed vertically along the walls of a kind pulverised fuel boiler, the tubes containing air at 40 bar pressure, rather than water or steam at much higher pressures. Tube diameter was much smaller than in pulverised fuel boilers, at around 30–40 mm, a typical size when high value alloys are specified, since this minimises materials costs. In the Continental plants higher grade stainless steels and Incoloy 800 type materials were used.

In these units and, indeed, in any form of indirect fired plant, the tubing, throughout its length, is made to operate very close to its temperature limits. To ensure a constant tube temperature, a parallel flow design is specified, so that relatively cool air, or other 'working fluid', enters the heater in the region of the burner. In a multi megawatt plant, sidewall firing burners, at various heights, should make it even more practical to aim for a uniform but high tube temperature. Obviously the width of the furnace should be large enough to avoid direct impingement by ash bearing flames.

Figure 4 shows a schematic of another form of air heater, as conceived by the author, with a tentative materials schedule. The 'S' bend form of the combustion chamber is intended to promote separation of molten ash and particulates before

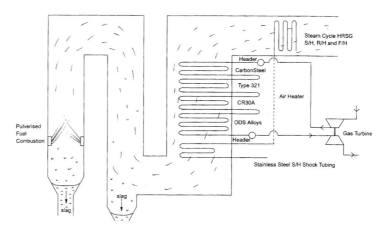

Fig. 4 Schematic of indirect fired heat exchanger, showing possible materials schedule and tube arrangement.

the flue gas enters the air heater section proper. The sides of the combustion chamber would be water cooled refractory, covered with solidified slag, which would build up to an 'equilibrium' thickness.

The low pressure of the working fluid result results in hoop stresses in the range 10–60 MPa, depending on the operating pressure, for a tube wall to diameter ratio of 1/10. There will be higher stresses, but of a short term variety only, arising from tube wall temperature gradients, and from bending due to lengthways expansion of the tubing. Hence, tube materials and welds need to have good short term ductility at temperature, especially at tube-to-manifold joints. Without this, cracks will form during periods of start up and shut down, these defects growing by creep during steady periods of plant operation.

In moderate temperature, refinery process heaters, minor variations in tube to tube temperature can be accommodated by coiling the tubes in a helical fashion around the walls of the furnace chamber. This can lead to heavy bending stresses due to sagging and radial expansion. In a practical indirect fired heater, tubes would have to be suspended vertically, individual differences in thermal expansion being accommodated by forming a complex pigtail of pipe at each end of the straight length. Essentially, many of the practices used in the design of reforming or ethylene pyrolysis plant will be used.

Clearly pressure and thermal stresses will be the first concern in the design of an indirect coal fired system, but high temperature corrosion will present significant problems. If we are to believe the evidence, the most serious form of attack, caused by deposition of ash, peaks between 600° and 800°C,[7,13] Fig. 5. The attack begins to decline at higher temperatures for two reasons. The phases

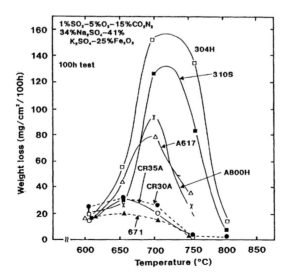

Fig. 5 Fireside corrosion resistance versus temperature. Note high rate of attack on alloys containing 25% chromium or less.[64]

in the deposit which cause molten salt attack, as it is sometimes called, are no longer stable, and the deposits themselves are less likely to form, since they are unable to condense out on the tube surface.

At around 700°C metal temperature, there is a critical need to balance resistance to 'fireside corrosion' against mechanical properties. Fortunately the low operating pressures in indirect fired systems imply that the stress levels can be about a fifth of those in conventional steam plant. Hence a number of the newer austenitic stainless and super stainless steels, such as the 20–32 spun cast alloys, and 253MA should be suitable, these having acceptable strength up to 850°C, and having been designed to minimise sigma phase embrittlement. The chromium and nickel levels in many of these somewhat older materials are too low for good resistance against coal ash corrosion, and they would need a 50Cr–50Ni overlay. Alternatively there are some new alloys, such as Tempaloy Cr30A, which do offer good resistance against coal ash attack, whilst retaining adequate high temperature strength.[14]

At still higher temperatures, 'hot corrosion' another form of molten salt attack, also involving sulphur compounds, could be a possibility. Even with a resistant overlay the rate of corrosion might be too severe. If this were the case, the only solution would be to shield the tubing with silicon carbide based castable refractory or brickwork.

3.2 Advanced Pulverised Fuel Steam Plant

The next major step in conventional power generation will have pulverised fuel plant running at steam temperatures of 630°C and pressures of 300 bar. This will give efficiencies of around 47%, but will require tube alloys to operate in the most critical temperature regime in terms of fireside corrosion. Higher alloy materials, overlays, or duplex tubing seem essential, and here the UK experience should be invaluable. What, however, are the prospects for a further improvement to 700°C and beyond?

Extrapolating from the papers by Blum, Birks and Smith, and Nakabayashi et al., to reach an efficiency of over 50%, would imply operating with a steam temperatures a little in excess of 700°C at pressures of about 400 bar.[15,16,17] Such a unit would be operating at ultra supercritical steam conditions. Given the rates of heat transfer possible when heating steam at these pressures, and the inferior conductivity of highly alloyed materials, tube temperatures are likely to be at least 50°C hotter than the steam. Using the criteria suggested by Metcalf et al., of a hoop stress in the superheaters of 100 MPa, and combining this with the data from the paper by Wright and Stringer, the implication is that there are no off-the-shelf materials which could be used for superheater tubing in ultra supercritical systems.[10,18] The stress rupture curves in Figs 6 and 7 indicate the nature of the problem. If tube walls were to be thickened up so as to reduce hoop stresses to the 40–50 MPa level, as suggested by Scott, alloys such as NF709, Inco 617 or Haynes 230 may be acceptable.[19] These materials would have reasonable tolerance to fireside corrosion, but tube walls may be unacceptably thick,

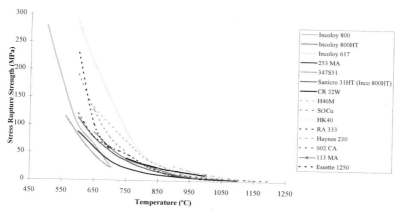

Fig. 6 100 000 hour stress rupture values of commercial and near commercial wrought and cast austenitic and nickel based alloys.

particularly for the lower conductivity, iron based, austenitic materials. Ought we then to be starting to think about other methods of using coal and steam?

There are innovative forms of steam plant, which offer another way of achieving plus 50% efficiencies. The 'Figure of Eight' cycle is an Austrian concept that consists of two separate steam cycles, the low temperature section (up to 500°C) operates at normal temperatures and pressures. Linked to this is a high temperature cycle, operating with steam temperatures of 850°–950°C, but at pressures of under 60 bar.[20] The reduction of stress levels gives a much wider

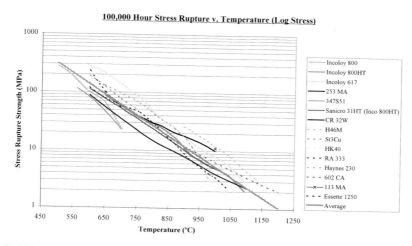

Fig. 7 100 000 hour log stress/linear temperature values of wrought and cast austenitic stainless steels and nickel based alloys. Note straight-line average curve.

choice of materials. This concept has been subject to a rigorous analysis by the Austrian power plant producers.

3.3 Fluidised Bed Combustion

Power generation systems, built around one of the various types of fluidised bed combustor (FBC) systems, are proving very attractive to potential operators, since apart from the furnace and boiler, the rest of the plant is standard technology. The FBC, whether atmospheric bubbling bed, pressurised bubbling bed, or circulating bed, is basically a straightforward steam plant tied to a new design of boiler and superheater.[21,22] As such it can offer little in terms of energy efficiency over pulverised fuel systems in its present form.

The main attraction of the FBC is that it is inherently cleaner than pulverised fuel firing. Sulphur dioxide is captured in the fluidised bed itself, through the addition of limestone, eliminating, in most circumstances, the need for flue gas desulphurisation. Furthermore, coal preparation costs are lower, particularly with the latest development, the circulating fluidised bed (CFBC), since this is tolerant of a large range of particle sizes, Figs 8 and 9.

The principle behind the fluidised bed is that coal and ash particles are supported by a stream of up-flowing combustion air, with the heat being removed from the bed sufficiently quickly to keep the temperature to between 800° to 900°C. The temperature is high enough to give a reasonable rate of combustion, but low enough for the limestone to capture the sulphur and to stop local sintering in the bed. In the bubbling bed, the density of particles is quite high, so

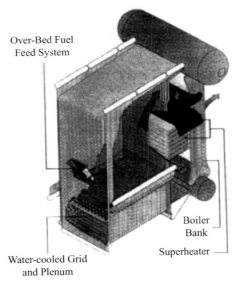

Over-Bed Fuel
Feed System

Boiler
Bank

Superheater

Water-cooled Grid
and Plenum

Fig. 8 Schematic of atmospheric pressure fluidised bed combustor. Note in-bed boiler tubes. (Courtesy of EPRI)

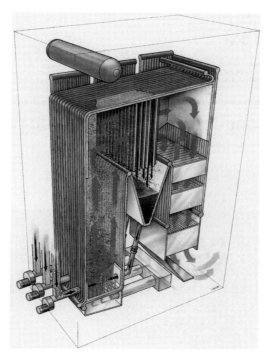

Fig. 9 Schematic of Sandvik circulating fluidised bed combustor. Note sidewall boiler tubes in furnace region. (Courtesy of Studsvik Energiteknik AB)

that a relatively low fluidising velocity, between 2–4 m.s^{-1} for atmospheric pressure bubbling combustors (AFBC) and 1–2 m.s^{-1} for pressurised bubbling bed combustors (PFBC), is adequate. Particle size in such beds is around 0.1 mm. These beds, therefore, have the appearance of a bath of bubbling red-hot quicksand. The CFBC operates at a much lower density, but works with larger particles, up to 6 mm in diameter, and at higher velocities, 7–10 m.s^{-1}. This would look more like spray fountain or shower.

In bubbling bed designs, heat is removed via small to medium diameter in-bed exchanger tubing. In the CFBC, water wall tubing is used, so that the design is not too different to a pulverised fuel boiler.

Much of the early work on fluidised bed systems is now considered to be obsolete, but has given us valuable insights into corrosion mechanisms. Initial studies focused on atmospheric fluidised bed combustors (AFBC), using in-bed exchanger tubing, sometimes heating high pressure air rather than water or steam, (for use in an 'early model' indirect fired gas turbine). Unfortunately, despite the high heat transfer *coefficients* of fluidised beds, the heat transfer *rates* of the AFBC were little higher than with pulverised fuel furnaces. Hence there was no real cost saving, and attention has now shifted to PFBC and CFBC designs.

Although for larger scale plant the AFBC system is being superseded, the materials work done in these programmes is highly relevant to modern designs, and should not be neglected. The early studies concentrated on the sulphidation of in-bed superheater and air heater materials. It is an elegant illustration how thermodynamics can be used to explain unexpected corrosion phenomena. The studies showed that in reducing conditions, deposits of calcium sulphate, resulting from the limestone addition, would, at the deposit-metal interface decompose, release sulphur species which could lead to severe attack. Furthermore, practical work on this problem, using in-bed zirconia probes, revealed that the local in-bed environment could fluctuate very rapidly from high reducing, to highly oxidising conditions, within a few seconds.[23] We need to remember this when endeavouring to understand the combustion of solid materials. In the vicinity of burning particles, the local environment can be quite reducing.

In the fluidised bed, these changes from oxidising to reducing conditions, are caused by the passage of so-called dense and light phases. These are respectively, volumes that contain a dense mass of particles, and others that consist mainly of combustion products plus a small amount of solid matter. The impact of the dense phase on the tube surfaces is undoubtedly relevant to the most pressing concern in FBC systems, that is high temperature erosion. Wastage rates, due to wear, in some systems can be up to 25 mm/yr, but are usually very much lower.[22,24]

There is some evidence that oxidation is involved in the wastage process, so that the problem is best described as one of erosion-corrosion, with the erodant consisting of the more abrasive minerals in the coal and coal ash. Adherent oxide scales, being of a similar hardness to silica (quartz) and silicate minerals in coal ashes, can reduce wastage rates under plant conditions, although the exact mechanisms are not well understood. This is partly owing to the lack of understanding of conditions in the plant, and the problems of simulating the situation in the laboratory.

Velocity is a major factor in governing the rate of wear. Laboratory experiments suggest that the rate is proportional to somewhere between the second and fourth power, although sheer speed cannot be the only factor. Pressurisation also has an effect, to the extent that velocities in PFBC systems, at 1–2 m.s^{-1}, are about half those in the AFBC.[25]

As indicated above, there is clear evidence that the wear mechanism involves the collision of the dense phase with the tube material. But it is not just simple impact wear. Hoy has found in fluidised bed experiments, using an oscillating wear target, that the wear involved the sliding of a densified mass or clump of particles over the wearing surface.[26]

Laboratory experiments in which the target has been cooled to below the bed temperature, have shown that as the target temperature increases from near ambient, the wear rate increases quite dramatically, peaking in the case of mild steel at about 200°C. The rate then falls quite markedly, afterwards increasing only very slightly with increased target temperature.[27] This seems to confirm the importance of oxidation in the wear process, since there is no significant change in mechanical properties of steel from room temperature to 300°C. Some observers consider these observations are solely of academic interest, since in-bed tubing generally operates at 250°C and above. However, if oxidation is the cause, it suggests that the impact of hot particles onto the tube surface can stimulate the growth of oxide at temperatures well below those measured under normal conditions.

Practical confirmation of the importance in forming a sufficiently thick adherent oxide scale is shown by the work of Holtzer and Rademakers who stated that there was a significant reduction in corrosion rate when the tube temperature was increased from 260° to 300°C.[27] Hence it would seem that oxide scale effects are important. Furthermore, any model that seeks to explain erosion-corrosion phenomena must account for the temperature/oxidation effects.

The Erosion–Corrosion Maps of Stack, at UMIST, and the thinking behind them, provide an excellent basis for practical design. These show how various combinations of velocity and temperature produce safe and unsafe regions for operation. In principle these maps could be used to develop models covering a wide range of alloys and particulates. Figure 10 shows a typical map for carbon steel.[28]

The wear in commercial boilers is restricted to the in-bed boiler tubes, so there are few creep problems. This low temperature does mean that there are difficulties in forming an oxide with the right characteristics. Alumina would be ideal, but its growth rate would be far too low. There are even problems in forming chromia. Stainless steels do not always perform well. Preoxidation treatments to generate the formation of these hard oxides will only give protection for a short period, suggesting that the constant impact of the hot dense

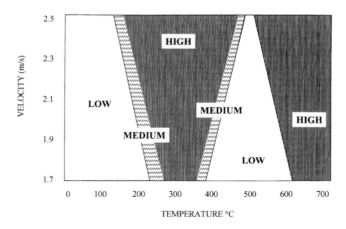

Fig. 10 Laboratory determined corrosion rates for mild steel versus temperature. Note that rate decreases between 350°–600°C, apparently due to formation of a resistant oxide. (Redrawn via M.M. Stack, UMIST)

phase, combined with thermal cycling, leads to the preformed oxide losing its adherence. Hence hard coatings, such as chromised diffusion coatings, nitrided surfaces, and welded or sprayed on metal carbide overlays, are utilised in critical locations. These are used in conjunction with protruding studs or prongs so as to break up the impact of particles onto the metal surface.[22] Modification of the surface contours of tubing to reduce the effects of impact and sliding wear is also under investigation.[29]

3.4   Recuperative Gas Turbine Cycles

3.4.1   Open Cycle Recuperation

The simple cycle gas turbine in its most developed form is now beginning the rival the diesel engine in terms of thermal efficiency. The best aero derived machines have guaranteed efficiencies of 42% when working at the design point. However, unlike the diesel, efficiency falls off quite severely at lower power outputs, even with variable incidence compressor blading.

The inefficiency of the simple cycle gas turbine is reflected in the temperature of the exhaust gases. This, at around 500°C, compares most unfavourably with the 100°–120°C for a typical combined cycle gas turbine or coal fired steam plant. The intention, in the recuperative gas turbine, is to return the heat in the exhaust back into the engine, so reducing the fuel demands and enhancing efficiency.

Figure 11 shows the flows through a recuperative gas turbine as a simple schematic, omitting for clarity, features such as intercooling and reheating. The spent combustion gases from the engine are passed through a large exchanger called a recuperator, before they are released to the atmosphere. The temperature rise for the combustion air would be typically, 200° to 300°C, giving an entry to the combustor of between 500–600°C, before any fuel is burnt.

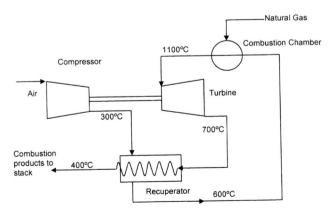

Fig. 11 Schematic of flows and temperatures through a medium scale, open cycle gas turbine. Air after leaving the compressor is preheated in the recuperator before entering the combustion chamber.

A little thought will indicate that the more efficient is a simple cycle machine, the lower the exhaust temperature and the less benefit it will derive from recuperation. Hence recuperative gas turbines need to be specifically designed for this duty. In broad terms there are two distinct approaches:[6,11,12,30]

> To build a fairly complex machine with a relatively high pressure ratio of between 15–25 which uses two or three stage compression with intercooling. This results in a very low *compressor* outlet temperature, enabling the engine to pick up a good deal of heat from the *turbine* exhaust, even though this temperature is itself quite low.
> To build a more simple machine which operates at pressure ratios of between 3–7. Since the pressure ratio is low, the compressor outlet temperature is also low, but the turbine outlet temperature is high. Hence a great deal of heat can be returned to the cycle, without the need for intercooling.

High pressure ratio intercooled variants seem more suited to high output engines for power generation, marine duties and freight locomotives. The Westinghouse–Rolls Royce WR-21 is a 20 MW example of such a machine. Pressure ratio is 16 to 1 at full output, falling to 8 to 1 at 30% load. Development of this machine is being underwritten by the US, UK, and French Navies, where the low fuel consumption that the recuperative gas turbine offers during cruising is a vital feature.[31]

The very low pressure ratio, recuperative gas turbine is best represented by 50–100 kW machines that are being developed for small scale or distributed power. In between these extremes, in output, complexity, and pressure ratio, is the medium sized recuperative gas turbine, as typified by the Solar Mercury 50, offering near diesel engine efficiencies, but operating at a pressure ratio of about 10/1.[32]

It should be noted that given the same turbine inlet temperature, the bigger, more sophisticated, high pressure ratio machines would have significantly lower peak *recuperator* metal temperatures than their smaller, low pressure ratio cousins.

Nevertheless, it can be shown, that in principle, the low pressure ratio design of the recuperative machine is superior, certainly in terms of efficiency. This does suggest that in the longer term, if the concept of distributed power is to come to reality, there will be a need to improve both the design of recuperators and the materials of construction. Recuperator effectiveness, that is the proportion of heat in the turbine exhaust that can be recycled, needs to be pushed up to the mid-nineties. This implies a much bigger recuperator, with all that this means in terms of cost. The rule of thumb is that recuperator size is proportional to the 1.4 power of the effectiveness, so that the size increases at an exponential rate when the effectiveness is over 90%.

High recuperator effectiveness has some advantages, however. It tends to bring down the optimum pressure ratio in the gas turbine cycle, if efficiency rather than specific work is the prime target. This is a real boon in low output machines. Furthermore, due to the reduction of the pressure ratio and pressure difference, the heat transfer rates on the upstream and downstream sides of recuperator are brought into balance, easing the thermal design. However, re-duction in the pressure ratio will tend to increase peak metal temperatures, which again has implications in terms of materials of construction.

The recuperator presents significant difficulties both in terms of heat exchange and mechanical design. The exchanger is of the air to air type, in which very large quantities of heat, in energy terms 1–3 times more than the net power generated, need to be transferred from the hot exhaust, at near atmospheric pressure, to the cold compressor stream at a higher pressure. This has to be done with minimal pressure drop through the exchanger, ideally well under 5%, otherwise any ef-ficiency gain, in terms of heat recovery, will be lost in increased compressor work.

Extremely narrow heat transfer passages are needed to enhance the rate of heat transmission. In the 'prime surface' recuperator, this is accomplished by maximising the area between the cold and hot streams, through which the heat must actually pass, whilst the overall volume of the assembly is kept to a minimum. In early prime surface designs, the recuperator consisted of a nest of narrow diameter tubing, thereby maximising the surface to volume ratio. In industrial machines the tube diameter was about 25 mm, but in low output, ancillary power machines, it was around 5 mm.

In more recent times, prime surface recuperators have been made up of highly convoluted sheets of material that are stacked very closely together. This gives the required combination of high 'primary' surface area, and narrow heat transfer passages. Because of the need to produce such tight corrugations, the sheet material is very thin. As such, the material has to be reasonably strong at temperature and be essentially immune to oxidation or other forms of corrosion.

Oddly enough, in modern primary surface designs, high thermal conductivity from materials of construction is not required. Since the working fluid on both sides of the exchanger is a 'gas', the actual heat transfer coefficients from either air or combustion products to the recuperator sheets are quite modest. Hence it is this 'aerodynamic' factor which determines the rate of heat flow.

The alternative to this type of design is the 'secondary surface' exchanger as represented by the 'plate-fin' configuration. Here fins of various shapes, that is the 'secondary surfaces' are brazed or welded onto what can be regarded as the primary heat transfer surface. Much of the heat flows first into the fins and then along them, until it passes across the 'primary' surface. Obviously, in this case, the conductivity of the fin material is significant, as is the quality of the bond between the base of the fin and the primary surface. These secondary surface exchangers tend to be more strongly built than modern primary surface designs. This, as much as anything, follows from the need to have a reasonably thick section onto which to weld or braze the finning.

Despite these efforts to improve heat transfer and reduce size, the recuperator is always very large, rivalling that of the gas turbine assembly itself. In the bigger engines, the pressure difference between the hot and cold streams is of the order of 15–25 bar, so that pressure loads are significant. Hence plate-fin type exchangers tend to be used in this type of machine, having the necessary strength. The marked difference in the actual volumes of flow and heat transfer coefficients, between the high and low pressure sides is another reason for the use of plate-fin designs in these bigger machines. Fin height and gap can be changed to optimise the two heat transfer coefficients and minimise overall pressure drops. Accordingly, the fin height, if not the gap, tends to be much larger on the turbine outlet side of plate-fin designs, since the volumetric flow is very much larger.

In the smaller size ranges, prime surface heat exchangers have been quantity production for some time for military applications, as exemplified by the AGT 1500 battle tank engine.[30] More than 10 000 units have been produced to date, with over 6 million hours of operation. As mentioned earlier, Solar have recently announced the Mercury 50, a recuperative engine of just over 4 MW output, with a pressure ratio of just over 9 to 1. In the Solar recuperator, as with other prime surface systems of this type, the plates are made to give one another mutual support, so as to resist panting. The plates themselves in the Solar engine are of Type 347 sheet, 0.1 mm thick. Individual pairs of plates are welded together, at the edges and then stacked together to form the recuperator. In addition, tie bars run crossways through the stack of plates, giving additional support, obviating the need for heavy section ends to the recuperator, Fig. 12.

In the mechanical design of recuperators, the most serious concern is that of thermal stress and fatigue, the levels of which, in fins and plates, can be extremely high. Indeed, McDonald, who is perhaps the leading exponent of the recuperative gas turbine, considers that thermal stress, rather than pressure induced stress, to be the governing factor. He concludes that since the allowable stress, for a given material, falls as the operating temperature increases, but thermal stresses increase as the heat transfer rate goes up, there is a trade off between these two functions.[30] This argument may appear to be somewhat simplistic, but the writer broadly supports McDonald's reasoning. In particular, the following factors should be considered:

• The decision to purchase a recuperative gas turbine, rather than an advanced simple cycle machine, implies frequent operation at part load, with all that this means in terms of temperature cycling of the recuperator.
• Depending on the recuperator design there can be marked temperature differences between adjacent hot and cold heat transfer surfaces. Again depending on the design, there can be severe temperature differences between the recuperator proper, and the external pressure 'vessel' or manifolds.

On this basis, it follows that austenitic stainless steels are the least tolerant of materials to thermal cycling. They are relatively weak at temperature. Thermal expansion rates are high, and thermal conductivities are poor. Nickel based alloys are somewhat better. They are stronger, expand less, and conduct heat

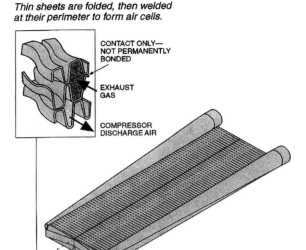

Thin sheets are folded, then welded at their perimeter to form air cells.

CONTACT ONLY—
NOT PERMANENTLY
BONDED

EXHAUST
GAS

COMPRESSOR
DISCHARGE AIR

These air cells are joined by external, high-strength welds into a compact module.

Fig. 12 Schematic of Solar 50 recuperator. Detail shows that the exchanger is of the convoluted primary surface type, and that the plates give mutual support to one another. (Courtesy of Solar Inc.)

more easily. Turning to the ceramics, McDonald claims that silicon carbide is superior to any of the metallic alloys, but indicates that, in terms of thermal stress, carbon-carbon is best of all. This of course is restricted to high temperature environments that are totally inert. McDonald did not cover the martensitic stainless steels in his analysis. Up to about 550°C these would be superior to both austenitic and nickel based alloys, on account of their exceptional strength, high conductivity, and low rate of thermal expansion.

The most serious practical issue, in terms of fabrication, in secondary surface systems at least, is the need to braze or weld the individual plates and fins together. Almost as crucial is the need to join the heat exchanger matrix to the inlet and outlet headers. The connection to the hot end is particularly critical, since this can be exposed to rapid and non-uniform temperature rises during start-up and shutdown.

Much of the technology of fabrication is proprietary. However, the issues involved in brazing of plate-fin designs for automotive gas turbines, have been well explored by Shah.[33] Many of these considerations will be significant in any other fusion joining technique. Shah highlighted, along with the normal problems of brazing a multi surface assembly, the following issues with respect to stainless steel plate-fin exchangers:

- Chromium carbide precipitation, during brazing, can lead to a loss of ductility in the material, and can reduce corrosion resistance.
- Certain types of brazing alloys can lead to liquid metal cracking.
- If vacuum brazing is used, grain growth may occur, and the loss of chromium due evaporation can lead to notch-like grain boundaries.
- Sigma phase can form.
- Guttering of the plates by the molten brazing metal can lead to thinning and premature failure.
- Good set-up of the assembly is critical, so as to maintain joint integrity.

For the present, despite their shortcomings, it would seem that austenitic stainless steels suffice for efficient primary surface designs in low to medium pressure ratio gas turbines. For the advanced intercooled engine of today using secondary surface exchangers, martensitic stainless steels seem to be the preferred option.

Strength, however, is not the only criterion. Since sheet and plate materials, in recuperators, are extremely thin, the resistance to oxidation by air and combustion products needs be excellent. Fortunately, the level of chromium in the martensitics (12–15%) and in the austenitics (16–25%) is adequate for this duty. Under most conditions little more than a tarnish film will develop.

Aqueous forms of corrosion, when recuperative engines operate at very low output are of potential concern. The problem is obviously most serious in a marine engine. Despite filtration, there is always the potential risk from sea borne chloride and sulphate. This could induce corrosion at the dryout point, particularly if there was carry over of water droplets during cleaning or compressor washing. Although not really a high temperature problem, there is risk from acid dewpoint corrosion in the down stream section of the recuperator in the typical marine gas turbine. This would result, primarily from sulphur in the fuel oil. If deposits were to form, due to erratic combustion, the risk increases, since the surface below the deposits would be cooler than normal. High levels of chromium would not be a complete answer to this form of attack, and the author considers that the use of stainless alloys containing high levels of nickel and molybdenum would be advantageous.

For land based engines, the problems of dew point corrosion are greatly eased. Unless situated on the coast or in heavily polluted regions, airborne contaminants will be insignificant. Levels of sulphur in natural gas are normally so low, and excess air levels so high in the recuperative gas turbine, that the risk of acid dew point attack in the machines of today is, in principle, very small. The potential problem will increase, of course, with improvements to recuperator effectiveness and reductions in pressure ratio, so that the situation needs to be watched.

3.4.2   Closed Cycle Recuperation

The closed cycle gas turbine is similar, in principle, to the more sophisticated forms of indirect fired, open cycle machine, but differs in that the compressor and turbine spools are situated inside of a closed loop, through which the working fluid, air or helium, continuously circulates. The closed cycle also differs in needing an efficient cooler, to reject the waste heat from the cycle, and in needing a recuperator of very high effectiveness. The latter, as with the open cycle recuperative gas turbine, returns as much as possible of the waste heat from the gas turbine exhaust, back to the compressor outlet stream.

The main advantage that the closed cycle has over indirect fired open cycle concepts, is that since it is possible to pressurise the equipment above ambient pressure, several times the output can be obtained from the same size of rotating equipment. Furthermore, since part load operation is achieved by reducing the degree of pressurisation in the plant, with no change in operating temperatures, thermal stress problems are largely avoided.

The heat input to a closed cycle can be provided by fossil fuel, in which case the closed cycle loop will also contain an indirect fired heater. This has to operate at full plant pressure, so strength requirements are very exacting. The need for an indirect fired heater is eliminated if nuclear energy is the heat source. Here the reactor is positioned inside of the closed cycle stream, after the recuperator, and before the HP turbine.

To minimise capital costs and maximise output, closed cycle plants are designed to operate at peak pressures of 40–60 bar.[6] For maximum efficiency the gas turbine will operate at very low pressure ratios, in the range 3–5, and for a gas turbine operating at 1100°C, this implies an inlet temperature to the recuperator of around 700–900°C.

The low pressure side of the recuperator will be operating at pressure of around 15–20 bar, the high pressure side at 40–60 bar. In earlier forms of closed cycle plant, the recuperator was built within a cylindrical pressure vessel, as a shell and tube design.[6] To reduce overall volume, the tubes were heavily finned, both externally and internally. Even so, the pressure vessel containing the recuperator was almost as big as the indirect fired heater itself. Because of thermal expansion, such designs are limited to temperatures below 500°C.[6] Alternatives to this are beginning to emerge. However, the combination of very high operating pressures and high inlet temperatures precludes the use of conventional primary and secondary surface designs.

To circumvent the problems of more conventional compact heat exchangers, a recuperator of the printed circuit type was specified, by the author, for a closed cycle demonstrator plant of 1 MW thermal output. In the demonstrator, the printed circuit heat exchanger (PCHE) was designed for operation at 30 bar pressure, and a peak metal temperature of 740°C, although in practice, because of the limitations of the air heater materials, it ran at much lower pressures.

In this particular form of PCHE, the exchanger surfaces are constructed from thin plates of stainless steel. On one side of each plate a zigzag pattern of submillimetre sized heat exchanger channels is etched. Sets of plates are then clamped together and heated in a controlled atmosphere furnace for a number of hours. Diffusion bonding takes place between the protruding channel sides or 'fins', on one plate, with the flat contact surface of the adjoining plate. The channel sides, as well acting as heat transfer surfaces in their own right, hold the plates together, after diffusion bonding, giving an extremely strong assembly. Accordingly, as a heat transfer package, the PCHE has features of both primary and secondary surface exchangers.

In its final form, the heat transfer section of the PCHE resembles a monolithic block of metal, cut through with tiny zigzag channels. Inlet and outlet headers are then welded to each side,[35] Figs 13 and 14.

The outer plates of the PCHE effectively act as the pressure vessel shell. Like the rest of the exchanger this outer skin is very lightly stressed, so that operation at high temperatures is feasible. The elimination of the pressure vessel results in a

Fig. 13 Section taken through a typical Printed Circuit Heat Exchanger matrix. (Courtesy of Heatric Ltd.)

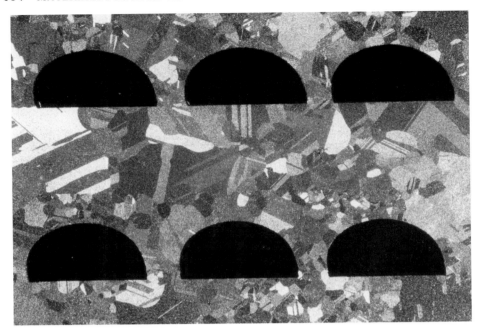

Fig. 14 Micro section taken between pair of PCHE plates showing quality of diffusion bond between the plates. (Courtesy of Heatric Ltd.)

massive reduction in recuperator size. In the aforesaid demonstration plant, a PCHE, about the size and height of a coffee table, gave the same duty as a shell and tube design of about the size of a mobile home.

The material for a PCHE must meet with some additional constraints, besides those normally associated with high temperature alloys. It must not be too resistant to acidic corrosion, since it will impossible to etch in the heat transfer passages. Furthermore, since diffusion bonding implies extended exposure to temperature, the alloy must not embrittle during this form of treatment. It would also seem inadvisable to specify alloys that form alumina, since this would be likely to disrupt the bonding process. Heatric Ltd, who have developed this form of heat exchanger, normally specify Incoloy 800 for higher temperature duties, although it is anticipated that with experience, the range of high temperature alloys that can be diffusion bonded will grow.

3.5   Entrained Bed Gasification

High pressure coal gasification/combined cycle systems, of all the clean coal technologies, have the lowest acid gas and visible smoke emissions. Sulphur compounds and particulates are trapped by the downstream gas treatment plant. The potential efficiency will surpass that of the best pulverised coal/steam cycle systems. Hence the long term prospect for gasification is extremely good.[36]

However, coal gasification-combined cycle is the most costly of all the clean coal technologies, and of the multitude of different processes which have been proposed, only a very small number look set to become truly commercial. Of these, there are basically two classes, fixed bed gasifiers as developed by Lurgi/British Gas, and entrained and circulating bed systems as developed by Shell, Texaco and KRW.[36–40] Every one of these started their existence, many years ago, gasifying lignite or heavy oil, so there is considerable operating experience with all of these systems.

Fixed bed gasifiers have many advantages, not least in minimising the demands on the heat exchange train, which, apart from some low temperature boilers, is virtually non-existent. Despite this, the prevalent view is that fixed bed designs are limited in the range of coals that can be gasified, and hence most interest has focused on entrained bed processes. Here the heat exchanger capability is critical to the economics of the process. Indeed the steam output from syngas heat exchanger is integrated with that from the HRSG in the combined cycle section of the plant. Accordingly, such systems are known by the abbreviation IGCC (integrated gasifier-combined cycle). Unlike most other forms of advanced coal fired generating plant, the dominant issue, in the design of these heat exchangers, is that of corrosion, rather than creep resistance.

The entrained bed gasifier produces a very high temperature '**syn**(thetic)**gas**' stream at about 1500°C, probably moving at between 10–30 m.s^{-1}, and containing molten particles of ash. In the Shell and other processes this **syngas** stream is cooled down to about 800°C by mixing in recirculated low temperature gas from the downstream section of the plant. The molten particles in the gas stream are thereby solidified, before the syngas hits the syngas coolers. The 'coolers' produce high pressure steam for the combined cycle, although a very mild degree of superheating may be practical. The syngas coolers are critical to overall efficiency. Much of the energy in the gas stream in the IGCC plant is present in the form of sensible heat, rather than as chemical energy in the gas stream, Fig. 15.

Syngas is heavily reducing, and depending on the coal type contains significant amounts of H_2S. Levels are currently around 0.5% in the few 'demonstration' systems now operating, but when gasifying high sulphur coals these will climb to 2%. Furthermore, in many cases, when operating commercially, the syngas will also contain up to 0.5% HCl. Although syngas contains significant quantities of steam and carbon dioxide, which should help stainless steels (by enabling them to form protective oxides), in practice these materials cannot withstand the more aggressive species that are present.

In consequence, the steam side temperatures in the syngas cooler are limited to well below 500°C, which is not a good match for modern HRSG equipment. None of the conventional materials of construction offer the prospect of a significant improvement in current operating temperatures. It would seem that if there is to be breakthrough, this will need to come through development of purpose designed materials with greatly enhanced high and low temperature properties. This subject will be dealt with further in Sections 5.2 and 5.3.

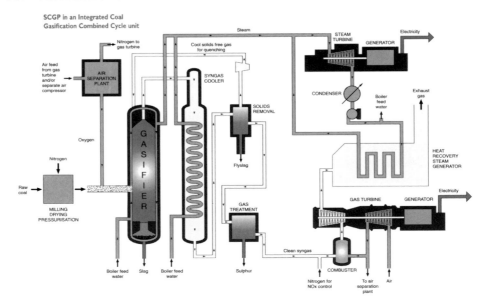

Fig. 15 Schematic of Integrated Gasification Combined Cycle Plant. Note steam and water connection between Syngas Cooler and HRSG. (Courtesy of Shell International BV)

There is one other issue that needs to be addressed, and that is the question of what constitutes an acceptable rate of corrosion in syngas coolers. Most designs of syngas cooler need to be installed within a pressure vessel, and are either of the tube and shell design or consist of helical arrays of tubing. In such circumstances, where the cost of tube replacement, or even the cost of tube cleaning can be prohibitive, it would be sensible to specify materials, which have got more than moderate resistance to attack, and which will not suffer from breakaway corrosion within a ten year time scale. Consequently, it would seem advisable to work well within current limits for metal wastage rates, as applicable for most high temperature corrosion situations, These limits, to the knowledge of the author, are of the order of 0.5 mm/annum.

3.6   Waste Incineration Processes

Waste incineration, combined with power generation, or where viable, District Heating, is now an integral part of our modern way of life. Although it is possible to recycle a good deal of packaging and scrap metal, there is much organic refuse which needs to be disposed of in some other way. Landfill sites, the principal alternative to incineration, result in the evolution of methane, a substance which is reckoned to be thirty times worse than CO_2 as a greenhouse gas. Although there is opposition to waste incineration plant, which to some extent can be traced back to inferior design and materials of construction, the future prospects are good. On average there is a need for one waste incineration plant for every one million of population.

The modern waste incinerator somewhat resembles a 1930s style power plant, using a chain grate stoker. The output would be unlikely to exceed 50 MW, Fig. 16. However, the grate on a waste incinerator is considerably more sophisticated, consisting of separate plates and steps, so as to tumble the waste as it moves along or down the bed. In this way, the bed is continually being broken up, and exposed to the combustion air.[41,42]

Even so, it is difficult to ensure that the waste is properly burnt. Ideally, the waste should remain on the bed, but a considerable proportion is blown up into the furnace chamber where it continues burning, eventually sticking onto the boiler and superheater tubing. These deposits contain up to 10% combustible matter, giving a reducing environment at the tube surface. They are also heavily contaminated with chlorides and sulphates of iron, lead, and zinc.[43]

Temperature and excess air control in the waste incineration furnace chamber is critical. If the off-bed temperature is below 800°C, much of the waste will not burn properly. Levels of CO in excess of 50 ppm imply reducing conditions. If

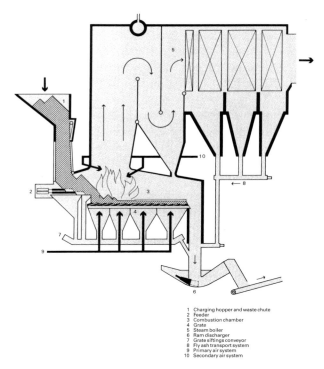

1 Charging hopper and waste chute
2 Feeder
3 Combustion chamber
4 Grate
5 Steam boiler
6 Ram discharger
7 Grate siftings conveyor
8 Fly ash transport system
9 Primary air system
10 Secondary air system

Fig. 16 Schematic of waste incineration plant. There is extensive use of water wall boiler tubing in all sections of the furnace. Waste is fed onto a tumbling type hearth down chute from left. (Courtesy of W + E Umwelttechnik AG)

the temperature is above 1100°C, PVC in the waste will decompose to chlorine rather than HCl, giving a much more aggressive environment.[43]

The actual mechanisms of attack on tubing are not fully understood. The consensus, at present, is that chlorine, as a gas or as HCl, within the deposit, is the main agent or vector of attack. At steam temperatures much above 400°C, the corrosion is believed to involve the formation of iron chloride at the tube surface. This compound is volatile and diffuses away toward the deposit-air interface. Near the surface of the deposit, the iron chloride reacts with oxygen or water vapour, depositing iron oxide. Much of the chlorine or HCl, released as a result of these reactions, migrates back to the deposit-metal interface, reacting with fresh metal. The cycle of events then repeats.[43,44]

Clearly, some chlorine will be lost from the deposit with each 'turn of the cycle', so that there is continual need for a re-supply of this element from the combustion gases and partially burnt deposits. The chlorine can be available in the form of non-equilibrium chlorine in the combustion products, HCl, or metallic chlorides, particularly those of the alkaline metals, sodium and potassium. Adding sulphur to the waste can inhibit the attack. This promotes the formation of iron sulphates in the deposits, and of alkaline sulphates in the combustion gas.

4   DESIGN OF MATERIALS FOR STRESS RUPTURE RESISTANCE

As stated earlier, Fig. 7 shows the stress rupture properties for a variety of austenitic stainless steels and nickel based alloys plotted using a logarithmic stress axis. When drawn in this fashion, the stress versus temperature gradient for most commercial alloys is approximately the same. Indeed an 'average gradient' can be plotted as a straight line that runs right through this family of curves.

This average gradient is governed basically by the climb of dislocations over strengthening precipitates, such as carbides and nitrides, the individual characteristics of which are described in the following part of the paper. The rate of dislocation climb is, of course, governed by diffusion of atoms, within the alloy matrix, and by its elastic modulus. The rate of diffusion is highly dependant on temperature, that of the change in the elastic modulus, very much less so.

Precipitates are then, barriers to the movement of dislocations, and as such, crucial to creep resistance. Without such barriers, the actual *gradient* of stress versus temperature, for a constant life, in high temperature alloys would be little different to that shown in Fig. 7. However, without such precipitates, all the commercial alloys would fall in strength, so that the 'average curve' would be *displaced downwards* to a significant extent.

Conversely, if we could somehow grow a precipitate of optimum size and coherency, and it was absolutely stable, and it did not decompose as the temperature increased, we would again have the same stress versus temperature gradient as in Fig. 7. In this case, the average gradient curve would be displaced upwards, roughly following, but somewhat above the two curves for Inco 617 and H46 M. This would represent a 'wonder material' indeed. The nearest thing to such ideal precipitates, are those in certain of the mechanical alloys, which since they consist of oxides, are almost immune to the effects of temperature. However, even with a wonder material, the fall off in creep properties with temperature is inexorable. The best that can be done to counteract an increase in temperature is to slow up the diffusion rate a little, and increase the elastic modulus of the alloy matrix a trifle, by adding high melting point elements, such as tungsten and molybdenum.

For the more conventional materials, each of the precipitates that we can use is only viable over a relatively small temperature range. All of these precipitates are intermetallic compounds and, like all compounds, can only form when conditions are right. Below the optimum temperature, diffusion is too slow to permit the necessary aggregation of elements that make up that particular intermetallic. Growth of the precipitate is then too slow, hence the alloy is relatively weak. Above the optimum temperature, the intermetallic becomes unstable, the constituent elements redissolving in the matrix. Again there is a loss of strength.

Each of the commercial alloys makes use of one or more of these strengthening precipitates, in conjuction with an appropriate mix of elements in the alloy matrix. This mix of elements should have just the right impact on diffusion rates. At the service temperature the diffusion should be fast enough to permit the precipitate to form, but not so fast that the precipitate overages within the design lifetime. Accordingly every alloy will have a temperature range over which the alloy will perform at its best. Above or below this range, the creep resistance, relative to the average curve, tends to fall away.

Even when the increase in mechanical properties appears to be due to a 'solid solution' hardening effect, as with additions of nitrogen and carbon over certain temperature ranges, the strengthening process probably involves the interaction of a number of different types of atom with a dislocation line. This being a complex, diffusion dominated process, it too will only be really effective at one temperature.

Figure 17 shows schematically the useable temperature ranges for a variety of common strengthening 'precipitates' in the heat exchanger alloys under review. For the reasons given above, as the temperature increases the creep rupture strength falls. Note, however, that this is no reflection upon the capabilities of the individual strengthening mechanisms. The fall off is governed by the fundamental characteristics of the alloy, which leads to the increasing ability of dislocations to climb over, or free themselves from, the various obstacles as the temperature rises.

4.1   ODS Fecraloys

The ODS (oxide dispersion strengthened) materials represent the ultimate in high temperature performance for metallic heat exchanger materials. The ferritic versions of these were developed from the Fecraloy type of alloy, currently in production as a catalyst support for automotive exhaust systems. In this form,

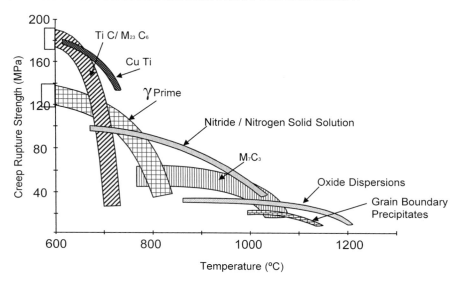

Fig. 17 Temperature ranges over which specific precipitation and solid solution hardening mechanisms are considered to operate. Note that stress rupture strengths are shown schematically only.

sans the nanometre sized oxide particles, such materials are amazingly weak, even at moderate temperatures. The strong ODS variants are produced by mechanical alloying, so as to incorporate the particles of yttria. Commercial compositions are in the range Fe–(16–20)Cr–(4.5–6.0)Al–(0.3–1.0)Y_2O_3, and are represented by such alloys as MA 956, PM 2000, and ODM 751. All are alumina formers, suffering very little oxidation when exposed to 'clean' air environments at temperatures in the 1100°C range. Providing that the alloys are preoxidised in clean air, they can be expected to withstand carburising atmospheres for long periods.

Because of the potential of this class of materials, Dourmetal ODM 75I was used in the construction of a natural gas fired harp type heater in the demonstration closed cycle gas turbine plant mentioned earlier. The heater was designed to operate at 8 bar, with an outlet temperature of 1100°C. The experience gained in this programme is reflected in this section of the paper. Some of the results of this work have been described at the COST 1994 Conference.[45] Further information was given at the San Diago NACE Meeting in 1998.[46] However, the key issues will be outlined in this paper.

Mechanical alloying has been well described by Elliot and Hack, and is essentially a sophisticated powder metallurgy process in which a mixture of the matrix alloy and yttria is first pelletised in ball mills before being sintered.[47] The ball milling breaks up the oxide into sub micron sized particles, around 20 nm in diameter, which become more or less uniformly distributed throughout the pellets of matrix alloy. In this way, the material receives a dispersion that is not too dissimilar to that produced by gamma prime in the early Superalloys. However, unlike gamma prime, the oxide particles are not coherent with the matrix, so that their strengthening effect is not so dramatic. Furthermore, the interparticle distance and concentration is much less uniform. Because of these factors, and the basic weakness of ferritic matrix, at temperatures below 900°C, the Fecraloy ODS alloys are inferior to conventional, austenitic type, high temperature exchanger materials. Above this, since the oxide dispersion is essentially highly stable, the ODS alloys can be significantly better.

The ODS ferritics also differ from conventional materials in a number of important ways. These are:

- Acceptable strength is only obtained with a very coarse grain size, since most of the creep occurs at grain boundaries, rather than in the matrix.
- The alloys have highly directional mechanical properties, due to the formation of grains with a high, three dimensional, aspect ratio (i.e. the long direction is significantly bigger than the other two directions).
- High aspect ratio coarse grains are only obtained after a carefully controlled working process and heat treatment. Tube fabrication processes, for example, of Mannesmann type, which produce varying amounts of strain across a billet, are unsuitable.
- Conventional welding breaks up the microstructure, resulting in weak welds.
- Cold working and subsequent recrystallisation is likely to destroy the original coarse grained structure.

Materials for exchanger tubing are initially produced by extrusion at 1100°C. In this condition the grains are very fine and elongated, resulting in extremely poor creep properties. Accordingly, the tubing is then lightly cold drawn, and afterward recrystallised at temperatures in the region of 1300°C, *for many hours*. This gives coarse, high aspect ratio grains, resulting in excellent longitudinal properties. The degree of cold working and recrystallisation temperature are very critical.

The strength in the hoop direction does not gain as much from this processing route, as would be desirable. In small diameter tubing it is impossible to get high aspect ratio grains and a very large grain size *in the circumferential direction* when using simple extrusion as the initial fabrication step. Furthermore, current ball milling and extrusion equipment limits the size of tubing to less than 3 cm OD. Fortunately, this is around the optimum diameter for high value heat exchanger materials, where it is necessary to provide the maximum heat transfer area for a given tube weight.

Because of the presence of the oxide dispersion, ODM 751 exhibits a moderately high stress sensitivity of around 6 in the hoop direction (in the longitudinal direction, stress sensitivity and strength are probably very much higher). It follows that even minor improvements to the material, whilst not increasing the allowable hoop stresses very much, can have remarkable effect on the life of tubing. Conversely, high stress sensitivity implies the need to avoid stress concentrations, in terms of abrupt changes in section, and fabrication defects, such as laps or gouges.

At present, tubing of around 2.5 cm diameter and 3 mm wall thickness can be used for pressures of up to 10 bar at 1150°C, corresponding to a stress level of around 4–5 MPa. Although low, it is roughly twice that of the best cast stainless alloys. The oxidation resistance of the ODS alloys is greatly superior, since the wastage rate, of a typical cast stainless, at this sort of temperature, is around a few millimetres a year, and is just about acceptable. The rate of oxidation of the best of the ODS materials is roughly two orders of magnitude slower than this, as one might expect from the difference between a chromia and an alumina former.

Because of the problems in the fusion welding of ODS alloys, less conventional techniques are required. For the furnace tubing in the demonstrator heater, the welds were made by first explosively welding an ODS tube to a short stub pipe of cast stainless steel. The free end of the stub pipe was then welded to the header manifold. These explosive joints proved to be quite sound for short-term use, the principle problem being that of interdiffusion between the stub material and the ODS alloy, particularly if the stub consists of a nickel rich alloy. In this situation the aluminium migrates into the stub material, forming nickel aluminides. The depletion of aluminium then leads to breakaway type oxidation in the ODS alloy and the formation of gross porosity.

An objective view of the present status of ODS ferritic alloys would be that although clearly superior to the cast stainless steels, the relatively low hoop strength is a drawback. Nevertheless, these materials, even in their present form, have potential for use in ethylene furnaces and in heaters for an open cycle indirect fired gas turbine system. Ideally for the indirect fired gas turbine, tubing would need to operate at around 20 bar pressure and 1225°–1250°C, and to be available in longer lengths at an economic cost.

This does seem possible, given the necessary investment into R&D and in the fabrication equipment. Tube wall thicknesses could be increased without too much difficulty, thereby helping to reduce wall stresses. In terms of increased mechanical strength, the greatest improvement would be obtained by producing a tube in which the grains were arranged in helical fashion around the circumference. If this was practicable, we could expect tubing to sustain pressures of 40 bar or more at over 1200°C.

More prosaic methods could bring significant gains. The addition of higher melting point elements should have some effect on temperature capability. The effect of these may be indirect, lowering the diffusion rate of vacancies and thereby preventing grain boundary sliding and vacancy coalescence. Indeed, if near term improvements are to be made to these alloys, it will come from closer attention to grain boundary phenomena. Some form of pinning, by precipitates, at grain boundaries, is desirable, this having been shown to be beneficial in a very advanced Japanese alloy (see Section 4.2.4). This technique could create problems during the recrystallisation procedure, which is intended to actually encourage grain growth. A more subtle means of strengthening would be to change the morphology of the grain boundaries. A number of workers have informally suggested that when grain boundaries are serrated, they interlock, thereby increasing creep resistance.[48] Initial work at Dourmetal suggests that there are good prospects of being able to do this in a controlled fashion.

4.2   Precipitation Hardening

In the *intermediate* temperature range, providing that an alloy has a fully stable matrix, and it is one in which diffusion rates are slow, many forms of precipitation hardening can be used with good effect, so that long term stress rupture properties in excess of 25 MPa at around 800°C can be obtained. Values are correspondingly higher or lower at temperatures below or above this level.

4.2.1   Gamma Prime

The oldest forms of precipitation hardening in high temperature heat exchanger alloys were from gamma prime, Ni_3 (Al,Ti) and from the chromium based carbides. Nevertheless, in the early high temperature alloys, as represented by Incoloy 800, there appears to have been little awareness that either of these precipitates had much to offer. Hence the aluminium, titanium, and carbon contents of this alloy were fairly loosely drawn. There is now much greater recognition of the importance of these precipitates.

Gamma prime acts in the classical way, forming a coherent precipitate and thereby impeding dislocation movement. However, the density of precipitates is significantly lower than that in modern superalloys. Gamma prime itself is relatively unstable above 800°C. Such materials are at their best in the intermediate temperature ranges.

In modern iron based super stainless alloys, the Al + Ti content is in the range 0.85–1.3 as is shown by alloys such as Inco 800HT. The minimum acceptable level of chromium in all high temperature alloys is around 25%. In combination with the aluminium, this element will give protection to above 1000°C in clean atmospheres and reasonable resistance to fireside corrosion in the 750°C range.[49]

Higher levels of aluminium and titanium, in conjunction with the chromium, would lead to sigma phase embrittlement. In the latest variant of the '800' range, Inco 803, the level of nickel has been increased to counter this effect. There can be other drawbacks in attempting to use a greater degree of gamma prime strengthening. In the Type 300 stainless steels if Al + Ti exceeds 3.0%, there is a danger that the BCC phase Ni_2(Al Ti) may form. This rapidly overages.[50,51] It is noteworthy that RA85H, which can be regarded as a highly modified 18/8, is limited to 1% aluminium.

4.2.2   Carbide Precipitation

For higher temperatures, above which gamma prime will dissolve, older alloys relied on carbide precipitation, and this approach has gradually become more sophisticated with experience.[49,52]

In the wrought alloys, such as RA330, RA333, Inco 800HT, Inco 803 and Inco 617, carbon levels are currently, in general, restricted to 0.1% or less. Some early work on Haynes 230 suggested that there was no benefit in going above 0.05% in this material, particularly since it then had an adverse effect on creep ductility.[53] Matsuo *et al.* suggest that maximum strengthening occurs when the carbon content is just supersaturated at the operating temperature.[65] At higher levels than this, rapid overageing of the precipitate will occur. Since the solubility of carbon falls with increasing nickel content, this may explain why for example, materials such as Inco 617 and Haynes 230 have relatively low carbon contents in comparison to high temperature iron-based alloys.

In contradiction to this, the recently developed nickel based alloy, Nicrofer 602CA contains 0.2% carbon. However, in this case, the carbon combines with zirconium and titanium to form a fine dispersion of carbides, giving this alloy extremely good properties. In consequence the manufacturers quote stress rupture data up to 1200°C. Micron sized particles of chromium carbide are also present, their main purpose being to restrict grain growth.[54] It is noteworthy that in his review of high temperature heat treatments, Klarstrom states that titanium and zirconium carbides, are destabilised by the presence of tungsten and molybdenum, and Inco 602 CA is free from these elements, unlike some other high temperature alloys.[55] Essentially then, 602 CA can be regarded as a nickel rich, wrought analogue of the microalloyed cast stainless steels.

As was mentioned in the introductory passages, high levels of carbon transformed the prospects for the cast stainless steels. Strength in these materials is due to the presence of interdendritic grain boundary carbides formed during casting, and the subsequent precipitation of $M_{23}C_6$ carbides during service. Creep rupture properties, in these materials as stated earlier, increase very rapidly as the carbon level rises from 0.2% to 0.45%, then fall off above this value, Fig. 2. The increase is due to the fact that the austenite matrix is capable of dissolving up to around 0.4% carbon at very high temperatures. Above this level, the decrease in properties is thought to be due to the formation of a continuous network of carbide at the grain boundaries, during casting, resulting in easy crack propagation. The consensus view on the grain boundary carbide is that it prevents excessive grain growth. It could also be argued that by acting as a reservoir of carbon, it prevents dissolution of the intergranular carbides at temperature.

Since the original developments, as represented by alloys of the HK group, (Fe–25Cr–20Ni–0.4C), there have been a number of innovations, the background to which has been outlined, respectively, by Atkinson and Jones.[56,57] To increase strength and oxidation resistance, and to reduce the tendency to sigma phase embrittlement, the HP grades came into wider use (Fe–35Ni–25Cr–0.45C). An alternative approach was pioneered by Inco, with IN 519. Here the carbon levels in HK-like alloys were dropped slightly to increase ductility, and the nickel increased to reduce sigma phase formation. However the main trend has been the use of increasing amounts of niobium and tungsten to promote the formation of more stable carbides. This technique was adopted for the HP grades, giving us HP45 Mod Nb (Fe–35Ni–25Cr–1Nb–0.4C). The latest group of alloys uses the 'microalloy concept' in which combinations of zirconium and titanium are used in conjunction with a rare earth element. In this form the high temperature ductility of these alloys is improved considerably.

4.2.3 Nitrides and Nitrogen

A number of high strength wrought alloys which rely on nitride precipitation have now reached the market, and can be regarded as the most recent commercial innovations in terms of alloy design. These include the 153, 253, and 353 MA austenitics from Sweden, NF709 and HR3C from Japan, and a range of nickel based compositions such as Haynes 120, 160 and 556 from the USA.[58–61]

The nitrides appear to be more stable than the carbides, but there are other advantages. Nitrogen stabilises austenite. In consequence sigma is suppressed, as shown by the stability, of 253MA for example, in the critical 650°–800°C range,[62] Fig. 18. Nitrogen additions give us the option to increase levels of solid solution strengtheners, which are usually ferrite formers. Alternatively we can choose to increase the resistance to oxidation by additions of chromium and silicon, elements which also promote the formation of sigma.

The effect of nitrogen may be very complex. In a Sumitomo study on HR3C (Fe–25Cr–20Ni–0.4Nb–0.25N–0.01C), transgranular precipitates of CrNbN were considered to be responsible for at least some of the strength increase at temperature. There is a suggestion that much of the nitrogen remained in solid solution, and this too contributed to the increased resistance to stress.[58]

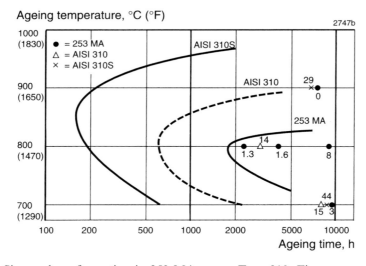

Fig. 18 Sigma phase formation in 253 MA versus Type 310. Figures on graph are percent sigma. (Courtesy of AB Sandvik Steel)

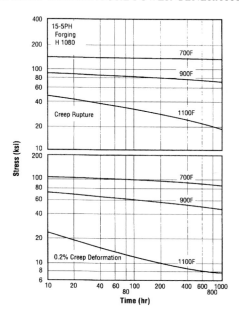

Fig. 19 Medium term rupture strength of 15–5 PH precipitation hardened martensitic stainless. Note rapid drop in properties with time at c.480°C (1100°F). For a 100 000 life at similar stresses, temperatures would need to be about 70°C lower.[66]

Investigations on HR-120, by workers at Haynes International, also suggested that the strengthening in this alloy was due to CrNbN. This precipitate did not go into solution below 1175°C, although there were even more stable phases present.[60] Earlier studies on Haynes 556 indicated that a small amount of carbon is needed in nitrogen strengthened alloys for optimum properties. Creep rupture strength peaked at around 0.1% carbon although there was considerable scatter. Nevertheless it is noteworthy that both HR-120 and 253 MA have between 0.05–0.10 carbon as well as the nitrogen addition.

So far, there are no reports of any adverse effect of nitrogen in the current round of commercially produced alloys. However, as with all additions, it has its limitations. Nitrogen cannot be used in alumina forming alloys, since the aluminium would react selectively with this element. This would almost certainly affect alloy strength and probably the oxidation resistance as well.

4.2.4 *Novel Precipitation Hardened Alloys from Japan*
A number of Japanese alloys seem worth keeping under review, since at least two seem reasonably close to production. All of these materials make use of novel intermetallics, solid solution, or grain boundary strengthening mechanisms. The alloys are:

- *Tempaloy-A1: Fe–18Cr–10Ni–3Cu–0.27Nb–0.16Ti–0.1C (also known as St3Cu)* This material relies on the usual combination of $M_{23}C_6$ carbide, and (Nb,Ti)C to ensure good strength in the 650°–750°C range. However, the titanium, niobium and carbon ratios and the solution temperature are all arranged to minimise grain boundary sensitisation, and to control grain growth during annealing.[63]

  A major innovation is the use of copper to promote the formation of a copper rich phase. Copper has the advantage of promoting austenite stability, being about twice as good as manganese and roughly equivalent to that of nickel. Hence its presence reduces the tendency to the formation of sigma.

- *Tempaloy 30A: Ni–30Cr–19Fe–2Mo–0.27Nb–0.2Si–0.2Ti–0.1Al–0.06C–0.02Zr–0.004B* This is an alloy containing around 40% nickel, in which an alpha chromium rich phase is precipitated to form a transgranular Widmanstätten structure. The chromium addition was intended to give fireside corrosion resistance rather than strength. It may be that its ability to enhance mechanical properties is fortuitous. In Europe and the USA, the presence of alpha chromium is normally regarded as being quite detrimental. Clearly, with such a microstructure, there must be concern about the possibility of embrittlement, but both creep ductilities and post exposure Charpy tests indicate that the alloy appears to have an acceptable performance.[64]

- *SSS113MA: Ni–23Cr–18W–0.48Ti–0.03C–0.035Zr* This is a nickel based alloy, developed at the National Research Institute for Materials in Japan, and was intended for very high temperature, gas cooled reactors. It was designed to be superior to Hasteloy X and Inconel 617, the principal off-the-shelf candidates for such a system.[65]

  The basic matrix itself, because of the tungsten addition is extremely strong at temperature. Above about 15% of this element, a grain boundary precipitate of 'alpha tungsten' forms during service, which adds considerably to the strength. Here again, we have an anomalous observation. The view of the spun cast alloy producers is that heavy grain boundary precipitates (that is chromium carbides in their materials), do little other than to reduce grain growth. It would appear, then, that this tungsten phase has some very unusual properties, unless our understanding of grain boundary phenomena is less complete than we think.

  The Japanese studies on this material are very comprehensive. Additions of titanium, manganese and silicon, at a given level of tungsten, have the effect of reducing the solid solubility of tungsten and give increased grain boundary coverage due to precipitation of the alpha tungsten phase. A balance of these elements is therefore needed to maximise the strengthening effect of the tungsten in the matrix, whilst obtaining the optimum grain boundary coverage of the secondary phase. The Japanese also consider that, in this alloy and some others, carbon itself can have a solid solution effect at very high temperatures.

4.3   Corrosion Resistant Martensitic and Bainitic Alloys

Alloys of this type are normally covered in discussions about LP steam turbine blading, where ultrahigh strength and moderate resistance to temperature are required. Here we are dealing with materials for secondary surface gas turbine recuperators, where the key demands are thermal stability up to about 500°C, high strength so as to avoid thermal ratchetting effects, good conductivity, and the lowest possible thermal expansion coefficient.

Allied Signal Inc., one of the leading fabricators of recuperative systems in the USA, makes use of a proprietary alloy which is called 14–4Cu. Few details have been published about this material, but it is probably a precipitation hardened stainless steel of the 17–4PH, or 15–5PH variety. These materials are basically of the air hardening martensitic type, but they are subject to an additional precipitation heat treatment between 480°C and 620°C.[66] In such materials the UTS rises from around 1100 MPa in the simple martensitic condition, rising to 1450 MPa when fully aged, dropping to around 950 MPa when overaged. Overageing is used primarily to inhibit hydrogen embrittlement. As with the Japanese austenitics, a copper rich phase is responsible for the rise in hardness in the standard 17–4PH and 15–5PH materials, and it is more than likely the same phenomena occurs with the Allied Signal alloy. The standard alloys are quite good up to about 400°C, with a design strength in the region of 200 MPa for long term service. There is a deterioration above this temperature, and it may be this is why new alloys are under development, Fig. 21.

Are there any other potential candidates for secondary surface recuperators? A 12Cr nitrogen containing bainitic, developed in the COST programme, could be a candidate if it can be produced in the form of thin weldable plate.[67] Yield strength in this alloy does not fall below 900 MPa up to 550°C, suggesting that this material is superior to the conventional precipitation hardened stainless steels. Some modification would be necessary, however. Somewhat higher levels of chromium, and perhaps molybdenum, will be needed to give the necessary corrosion resistance. To compensate for these changes, so as to provide an austenite phase field at solution temperature, additional levels of austenitic stabilising elements will be needed.

5   HIGH TEMPERATURE CORROSION

Just as with resistance to creep, in which no practical strengthening mechanism is effective over the entire temperature range, no single combination of alloying elements will give resistance to all forms of high temperature corrosion. Fortunately there is now a good understanding as to why this is so, and why certain alloying elements are beneficial under some conditions, but not in others.

The need for such awareness is particularly critical in the advanced heat exchanger field which, as we have seen, is under continuing development, and is taking us into new areas of high temperature corrosion. Figures 20 and 21 indicate schematically the range of temperatures and conditions to which ad-

Temperature Domains for Industrially Significant Corrosion Mechanisms

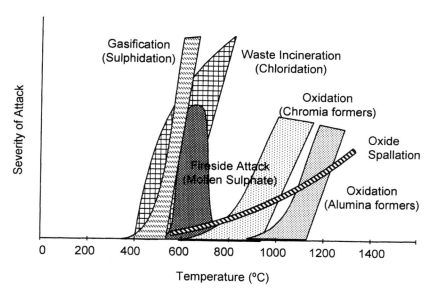

Fig. 20 Temperature ranges of corrosion mechanisms as experienced in modern energy conversion systems. Note that in general attack rate increases dramatically with temperature, apart from fireside corrosion, where the rate falls off above 750°C.

vanced heat exchangers in the energy conversion field are subject, and indicate the types of alloy which are needed to give an acceptable life. The following sections deal with the mechanisms of attack that can be encountered in the advanced systems of today.

5.1   Clean Environment Oxidation Resistance

A variety of technologically important environments can be considered to be clean, that is, essentially free from *aggressive* high temperature attack. These environments include the *air side* of combustion air preheaters, the *furnace side* of natural gas fired heat exchangers, and the interiors of high temperature steam pipework. The corrosion mechanism under such conditions is that of simple oxidation, by oxygen from the air or combined oxygen in water vapour or steam. At temperatures below 650°C, for most of the advanced alloys under review, corrosion under these conditions is not a real issue, the material being protected by a thin solid layer of impervious, well adherent oxide.

At temperatures just above this, steam side attack in advanced pulverised fuel plant is becoming an issue, particularly with steels that are modifications of the 18Cr–8Ni stainless range.[68–70] Such alloys are used for superheaters, and when untreated, form iron-chrome spinels rather than a true pure chromia oxide.

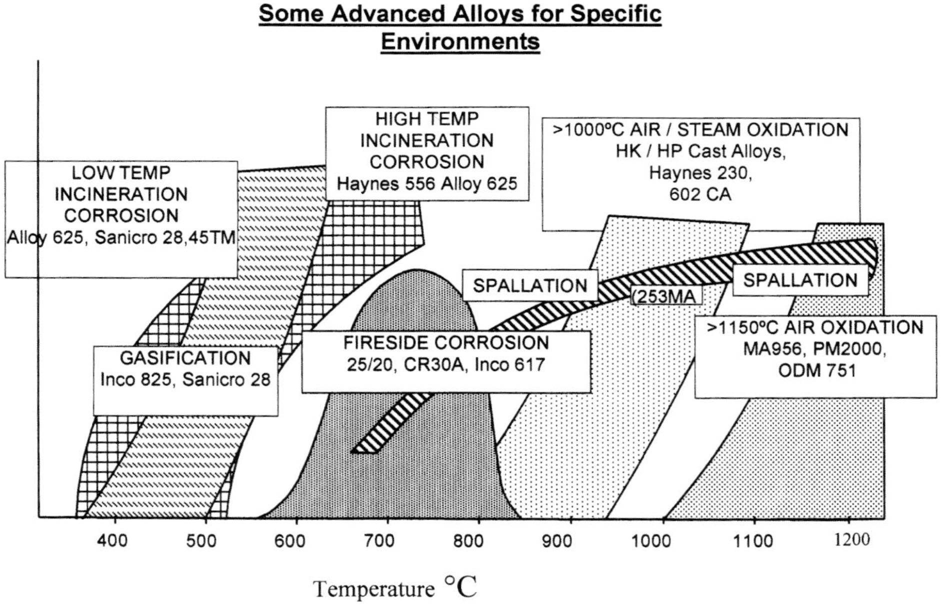

Some Advanced Alloys for Specific Environments

Fig. 21 Alloy types capable of giving reasonable resistance to the forms of attack shown in Fig. 20. All materials encounter an upper useable temperature limit.

Spinels are relatively fast growing oxides, which, like most scales, once they reach a critical thickness will spall, causing erosion of steam turbine blading and blocking of superheater pendants. Thick scales will also reduce overall tube conductivity, significantly reducing superheater life. To suppress the formation of the spinel, and to promote the formation of chromia, materials such as Tempaloy A1 (St3Cu) will need to have their internal surfaces shotblasted.[63]

The standard alloys for high temperature air preheaters are the Type 410 and 430 ferritic stainless for lower temperatures, and Type 310 austenitic stainless steels for more extreme conditions. Reference 71 describes the design and development of a preheater, using Type 310, for a high temperature furnace, the practical peak metal temperature of this being around 1000°C when running on natural gas.

Alloys with over 20% chromium are described as true chromia formers. Even here, a detailed examination will show the scales to be complex and multi-layered, but the overriding feature is that in air, after moderate lengths of time, the principal layer consists of Cr_2O_3. There are a wide range of alloys of this type, of which the higher alloyed stainless steels, both cast and wrought, are typical examples.

Chromia, as the experience with the natural gas air preheater described above shows, is not particularly good at very high temperatures, suffering from an excessive growth rate, leading to spallation and alloy depletion effects. Another problem, volatilisation at above 900°C in air, through the formation of CrO_3 has long been recognised to be a limiting factor in governing the life of high temperature pipework using conventional stainless alloys. More recently, thermodynamic studies, have indicated that volatilisation of chromium hydroxides, in pure steam, could be a significant problem at higher temperatures, and in steam-air environments in the 700°C region.[72,73] This may in part account for the higher rate of attack of stainless steels in clean combustion products, compared to dry air. Kofstad, however, favours a more direct effect, in which hydrogen protons are incorporated into the lattice, speeding up diffusion and oxide growth.[74] Kofstad also suggests that silica and alumina formers may also suffer from the presence of water vapour, although long term tests at British Gas on ODS ferritic alloys showed that there was no significant difference between tests in air and in the products of natural gas combustion.[75]

To slow down the growth rate of chromia, elements which form thin subscales are added, at a level corresponding to around 0.5–1.0 *atomic percent*. Older alloys used silicon as an addition, so that a thin glassy (i.e. amorphous) sub layer of silica forms which impedes diffusion of oxygen or chromium. Above 900–1000°C a thicker layer of quartz, rather than amorphous silica, will form and, during cooling, differential expansion stresses result in the debonding of this layer with the spallation of the entire oxide scale.[76,77] Due to this shortcoming, the tendency is to use aluminium as an addition in modern alloys, although the actual growth rate of alumina is somewhat higher than that of silica. Alloys making use of the aluminium as an addition would be older Inco 601 and more recent Haynes 556.[54,78,79] It should be noted that these materials are quite different to the alumina formers described subsequently. In materials like Inco 601, the alumina forms a sub-layer, so that the oxide growth rate is still basically that of a chromia former.

As mentioned above, at temperatures in excess of 1000°C, the volatility of chromia becomes a critical factor, so that extremely high levels of chromium are required to maintain a protective layer. Even so, at 1150°C, the reduction in wall thickness for a typical 35/45 cast stainless is 2.0–2.5 mm/annum.[80] There can be significant denudation of the sub-surface layers of chromium in these circumstances, such is rate of volatilisation. Furthermore, at these temperatures, despite the presence of silicon, which should slow up subsurface effects, grain boundary oxidation can be significant. In cast stainless steels, this limits the temperature of HP Nb modified grades to 1075°C. The tungsten containing spun cast materials are said be less susceptible to this form of attack.[81]

It follows that at temperatures in excess of 1100°C, true alumina forming alloys are essential. These offer a reduction in the rate of attack of between one and two orders of magnitude over the chromia formers. Nevertheless, all the *commercial* 'alumina forming' alloys contain significant amounts of chromium. This element is needed to suppress the formation of rapidly growing iron or nickel oxides in the initial stages of oxidation, and allows the more slowly growing alumina to form, initially as a sub scale. Given an adequate level of chromium, 12–25%, somewhere between 2 and 4% aluminium is needed to form a truly protective scale in the longer term. Materials like 602CA, with 25% chromium and 2% aluminium, are on the borderline. Brill and Argarwal, indicate that at high temperatures, with this alloy, small areas of chromic oxide break through the alumina layer.[54] These extrusions quickly reheal, but the effect is such that the rate of attack, at temperatures in excess of 1150°C, is significantly higher than that of true alumina formers, such as MA956 and ODM 751(iron based) or Haynes 214 (nickel based).

The prospects for the alumina forming alloys have been transformed by the use of active elements, such as yttrium and zirconium, which provide greatly increased scale adherence. Spallation, indeed, was a major impediment, in all applications, to the use of alumina forming alloys, confining them to the manufacture of high temperature heating elements of the Kanthal A type. Although chromia formers are rather less susceptible to spallation than alumina formers, active elements have been added to the more modern alloys and are represented by such materials as 253MA and Haynes 230.

There are a variety of opinions on the mechanism(s) by which active elements key the scale, and a number of major conferences have been devoted to this issue.[82,83] At this point of time the views of Smialek are attracting considerable support. His contention, now well supported by experimental evidence, is that the active elements tie up sulphur, which would otherwise interfere with the oxide-substrate bond.[84] However, one should not neglect the earlier observations of Meier, who pointed out the marked effect on oxide plasticity of various trace elements.[85]

Despite the excellent performance of the ferritic ODS alloys, in clean environments, oxidation can be a limiting factor in governing the life of this type of exchanger tubing. In cycling duties, spallation is likely once the oxide reaches a thickness corresponding to a weight change of 2.5 mg/cm. Re-growth and progressive spallation of the oxide can eventually lead to the depletion of aluminium in the matrix. Once this falls to below 1.4%, the alumina is no longer able to reform, and breakaway corrosion begins.[86] Essentially, the material begins to behave as a 12 Cr ferritic stainless steel, forming a rapidly growing, non-protective, black iron-chrome spinel, Fig. 22.

The time to breakaway depends on the basic oxidation rate, the critical weight at which spallation takes place, the aluminium content of the alloy, and the tube wall thickness. Here the tube wall is acting as a reservoir for aluminium, and it is found that providing the tube wall thickness is in excess of 2.5 mm, a 100 000 hour life at 1150°C should be obtainable with the better alloys, Fig 23. Quadakkers has formulated a number of equations, which involve the factors that permit the time to breakaway to be predicted with a reasonable degree of accuracy.[87]

In this context, later work has shown that the level of yttria can influence the basic oxidation rate of alumina formers. High levels of yttria are detrimental, tending to inhibit grain growth *in the alumina*. In slowly growing oxides such as

Fig. 22 Formation of massive nodules of mixed iron-chromium oxide on a ferritic alumina forming ODS alloy tube following commencement of breakaway corrosion. Failure occurred after 6000 hours at 1200°C. Tube OD 30 mm.

this, it is the grain boundaries, rather than the lattice, which act as channels for oxygen or oxygen ions. A small grain size in the alumina therefore gives a high channel-to-volume ratio, so increasing the rate of attack. Quadakkers considers the optimum level of yttria in the ODS alloys to be around 0.2–0.3% to give the best balance between spallation resistance and oxide growth. Titanium additions are also critical.[87]

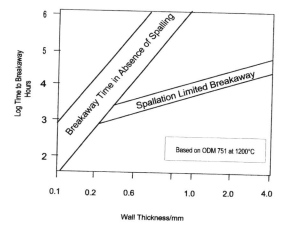

Fig. 23 Simplified time to breakaway diagram (after Quadakkers) showing effect of component wall thickness.

5.2   Sulphidation

Many forms of high temperature corrosion can be regarded as modifications of simple oxidation in clean atmospheres. Under such conditions an otherwise protective scale is undermined by the attack of a secondary oxidant. Of the secondary oxidants, historically, sulphur has given us most concern. It accelerates wastage rates by roughly two orders of magnitude, compared to clean conditions, bringing down operating temperatures by 200°–400°C with virtually every standard alloy.[88]

Since sulphur is present in coal and heavy oils, work in this area is of vital economic importance. As Section 5.3 will show, sulphur is the principle factor in causing fireside corrosion, which, as we have seen, is a major challenge to the development of advanced conventional power generation. In this Section, the emphasis is on the effects of sulphur in reducing atmospheres, typical of those encountered in oil refinery practice, and in coal and heavy oil gasification processes.

In extremely reducing atmospheres, for example, in the thermal or catalytic cracking of oil, sulphur compounds will react directly with the tube material, to form iron, nickel, and chromium sulphides. In this case the sulphur is acting as a *primary* oxidant, since the conditions are far too reducing for *oxygen based* oxides to be stable. Sulphides, because of their high growth rate, limit useable temperatures to under 550°C with the conventional high temperature alloys.

This high rate of attack is governed by factors of a fundamental nature. Sulphides are bulky in comparison to the original metal, and are liable to spall off and crack. More significant is the high rate of self-diffusion of metal ions in the sulphide lattice. This is thought to be a result of the mismatch between the ionic sizes of the anion (sulphur) and cations (iron, nickel, cobalt, and chromium). Sulphides of these elements also melt at a comparatively low temperature. It follows that there can be little prospect of improvement with alloys, which use chromium, and nothing else, for protection.

As mentioned at the very start of this paper, the practical implications of this have long been known from the early days of oil refining. The effect of sulphur in oils in thermal cracking processes, and H_2S in hydrocracking, is now quantified with the use of McConomy, and Couper Gorman diagrams, Fig. 24. Hence, the oil industry has largely come to terms with the sulphur problem, and where there is a risk of attack, limits temperatures to an appropriate level.[89,90]

Currently the most pressing issue, involving sulphur induced attack, here acting as a *secondary* oxidant, is that of the development of materials for syngas coolers for IGCC plant. As such there have been a number of major conferences on coal gasification materials, notably those in 1987, 1993 and 1996. Sadly none of these meetings have pointed towards any commercial breakthroughs in alloy development.[91,92,93] Conventional materials of construction are strictly limited to comparatively low temperatures. Even so, where good on–plant and laboratory based performance is reported with such alloys, this is invariably with gases of a modest sulphur content. Although far better materials have been evaluated un-

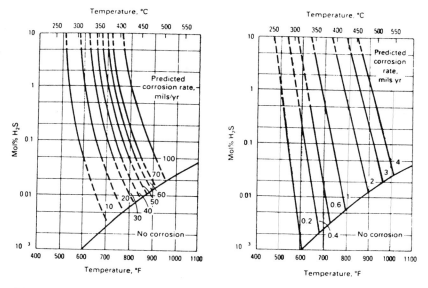

Fig. 24 Couper Gorman curves for corrosion of carbon and stainless steels in a carburising H₂H₂S environment. Corrosion rates are given in US mils/year (0.025 mm/annum). (Courtesy of NACE International)

der laboratory conditions, these in general would be expensive to produce, and present strength, ductility, and welding problems.

But the fact is that the IGCC does need such a breakthrough if it is to have a commercial future, in the near term. As we have seen, this implies the need for syngas cooler materials which are essentially immune to attack by gases containing between 1–2% H₂S, at metal temperatures of around 650°C. Is there anything which has been done that gives us the prospect of reaching this sort of target?

At an earlier stage, in the efforts to gasify coal, the aim was to run heat exchangers even hotter. These were gas to gas exchangers with syngas on one side and a hydrogen rich gas on the other. During the 1970s, the emphasis was on the production of a substitute or synthetic natural gas, rich in methane, for distribution to consumers. Steam raising was not an option with such processes. It was vital to recycle the waste heat from the gasification stage back into the process at the highest possible temperature.

It is well worth looking at the history of this work and interpreting this in the light of modern thinking, since it has implications for the development of the syngas materials of today. In this earlier period there was some hope that the presence of substantial amounts of oxidising gases, such as CO, CO₂ and H₂O, in gasification environments would inhibit gross sulphidation. Thermochemical stability diagrams, and the calculation of equilibrium gas compositions appeared to support this hypothesis, Fig. 25 shows that although the pO₂ (the partial

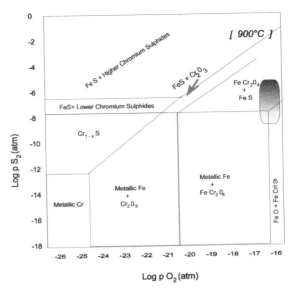

Fig. 25 Oxygen–Sulphur–Metal thermochemical stability diagram for iron chromium 'alloy' at 900°C. O_2 and S_2 partial pressures of typical US Gasifier are shown in shaded area. Surface forming chromium based oxides are stable and will, in many cases, form an initially protective layer, particularly at lower pS_2 pressures. (Redrawn from Ref. 133)

pressure of oxygen) in gasification atmospheres is very low, it is still high enough to enable chromia and alumina to be stable.

In this context the pO_2 is a method of defining the oxidising and reducing quality of an atmosphere, and is basically determined by a combination of the temperature, and the ratios between the oxidising and reducing gases in the environment. It can actually be equated to the number of oxygen molecules which are physically present in the gas mixture. In defining the pO_2, the key ratios, in gasification environments, if they are at equilibrium, are those between H_2 and H_2O, and CO and CO_2. In view of the fact that even when the gas composition is fixed, the pO_2 will increase with temperature, it is virtually meaningless to quote the pO_2 in the absence of a full gas composition. Unfortunately this is done all too often. The same remarks can be made about the use similar thermodynamic functions, involving sulphur, carbon, nitrogen, and chlorine compounds.

These early tests showed that at temperatures above 750°C, which was the original area of interest, many chromia forming alloys behaved well in the early stages. They formed slowly growing oxides, particularly if the test was started up in an oxidising gas, free from H_2S. Most materials, after an induction period, which might last between a few hundred and few thousand hours, then went into breakaway corrosion, forming molten sulphides or oxysulphides.[94] Fig. 26.

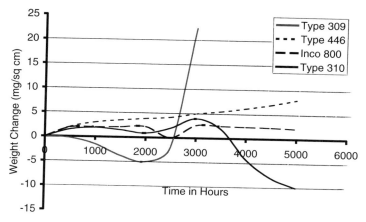

Fig. 26 Breakaway corrosion of Type 310 stainless steel and other alloys in a coal gasification environment containing 0.5% H_2S at c.890°C (1650°F). Note parabolic behaviour up to 500–1000 hours. Subsequent erratic behaviour is due to sulphidation and spallation. (Redrawn from Ref. 94)

We now believe that thermochemical stability diagrams were being used in too simplistic a fashion. They can however be used to clarify the mechanism of breakaway, and, in so doing, reveal the problems in attempting to use conventional alloys in coal gasifiers, and in many other complex environments. At the oxide-metal interface, the pO_2 is by definition, at the decomposition pressure of the oxide. For chromia, it is around 10^{-23} bar at 750°C. In essence, the oxide is barely stable at the interface, so there is the possibility that other compounds could form instead.

Given time, H_2S will diffuse along grain boundaries or micro cracks within the chromia to the interface with the metal. Once a sufficient concentration has built up, the H_2S will react to form a stable sulphide, this corresponding to the relevant phase in the thermochemical stability diagram, for the requisite pO_2 and pS_2 (partial pressure of sulphur) at the interface. To put things in a simple form, at the interface, the metal is exposed to a very reducing atmosphere, containing sulphur compounds, rather than the mildly reducing atmosphere it would 'see' at the surface. The newly formed sulphide will begin to cause spallation and gross cracking of the chromia layer. Once the chromia layer has begun to be disrupted, the corrosion rate will begin to increase, heralding the onset of breakaway.

The time to breakaway will be determined by a number of factors, the most important being temperature, H_2S concentration, the concentration of oxidising gases in the environment, and possibly, the 'solubility' and diffusivity of sulphur in the underlying alloy. Natesan has shown that the pO_2/pS_2 ratio can be used to draw 'kinetic boundary' stability diagrams. These lines are determined empirically, and invariably are to the right of the true thermodynamic stability line between the oxide and sulphide. These sets of lines indicate, as might be expected, that the higher the pO_2/pS_2 ratio, the longer it will be before an alloy goes into breakaway. In principal it should be possible to determine the position of the kinetic boundaries theoretically, or at least show how they may be used to predict the time to breakaway from relatively short term experiments.[95]

Today, most of the work in gasification atmospheres is concentrated on temperatures between 400°–650°C. Even this is a range over which it might be difficult to form a truly protective chromic oxide scale, unless the chromium level is extremely high. The same is true for aluminium and silicon containing alloys.

But is it essential to form a protective oxide, at these relatively low temperatures, in view of the experience in the petroleum refining industry? Bakker, indeed, suggests that at temperatures of around 540°C, typical of peak metal temperatures in current designs of syngas cooler, the growth rate of chromium sulphide may acceptable, *if it can be made to form in preference to the oxide.* Nevertheless, he shows that the presence of oxidising gases can be significant. The environments of entrained bed gasifiers, operating with coal-water slurry mixtures, are found to be marginally more protective than those operating with a so-called dry feed, due to the presence of greater quantities of steam. At intermediate steam conditions, Bakker concludes that there is an increase in the rate of attack. This is ascribed to the formation of a chromic oxide layer which is only partially protective, but disrupts the formation of a continuous layer of sulphide. At even lower levels of steam, the attack rate falls again, presumably due to the disappearance of the oxide.[96,97]

Nevertheless, in the opinion of the author, conventional materials in gasification plant are limited to temperatures of around 500°C. This due to the rapid growth of chromium sulphide, and to the relative instability of the oxide. We need to look back at those earlier studies that were aimed at producing alloys which would withstand temperatures in excess of 750°C.

Information on alloys that are in service is proprietary, but Inconel 800, Inconel 825, and Sanicro 28 are known to have been used. These may be utilised in a co-extruded form, particularly with Sanicro 28, which suffers from embrittlement at around 500°C.[98] Super stainless alloys of this type should be capable of resisting so-called 'downtime corrosion'. This form of corrosion is aqueous in nature, and is due to the interaction of moisture with sulphur and chlorine species in deposits and corrosion scales.[99]

Despite the difficulties, there is good prospect that it may be possible to develop truly resistant materials for coal gasifiers. Work is going on in the USA on iron aluminide intermetallic compounds. Medium term tests at 650°C show good performance, particularly with the FeAl type compositions, even in the presence of both H_2S and HCl. The good performance is due to the formation of an alumina scale, which is much more stable, thermodynamically, than chromia. Very high levels of aluminium are needed, in the alloy, to enable the alumina to begin to grow, within a short time of exposure to the environment, otherwise non-protective iron sulphides will form in preference. Natesan suggests a minimum of at least 12% aluminium.[100] Reactive elements are required to prevent spallation of the alumina. These intermetallics have been tested on the laboratory scale as overlay coatings, with apparent success.[101]

On the Continent, good work has been carried out on the effects of silicon additions to 12% chromium steels, levels in excess of 4% having been shown to be effective. Temperatures were however, restricted to 450°C.[102]

In the early eighties, British Gas sponsored a programme, at Newcastle Polytechnic, (now The University of Northumbria) which the author helped to initiate, that made use of the observation that refractory metals, such as Mo, W, and Nb have excellent resistance to H_2S. The growth rate of sulphides on these elements, in H_2S, is similar to that of chromic oxide on stainless steels in air at the same temperature.[103–105]

There was therefore good prospects of being able to use the refractory metals themselves, at temperatures up to 1000°C, providing reducing conditions were maintained, Fig. 27. This would not be too difficult, since the principal oxides of the refractory metals are far more easily reduced than those of chromium and aluminium. In consequence, in gasification atmospheres, the refractory metals tend to be covered with a protective layer of sulphide.

The original view was that the good properties of the refractory material sulphides were related to their thermodynamic stability. This no longer seems to be the whole truth; good performance is probably due to the fit in the sulphide

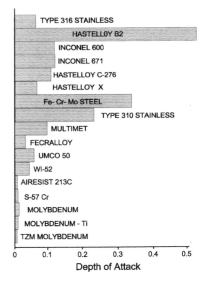

Fig. 27 Seven day corrosion rates of materials in an extremely reducing $H_2/CH_4/$ H_2S/H_2O environment at 750°C. Note good performance of molybdenum alloys and AiResist 213b, a Cocrally alloy containing refractory metals. (Redrawn from Ref. 105)

lattice between cations and anions, which slows up diffusion rates, although the details still need to be confirmed.

At one stage, tubing made from the refractory elements themselves was envisaged for the construction of a high temperature gas to gas exchanger. Molybdenum in particular, it was understood, had been considered for liquid sodium coolers in fast breeder reactor systems. There was concern, however, about the mechanical properties of fabricated components, and corrosion rates of the refractory metals during start-up and shut down, when the gasifier environment could be significantly more oxidising.

An alternative was incorporate a significant percentage of refractory metals into a Fecraloy or Cocraloy type base composition. Here, an alumina scale was to give the basic protection, and to seal off the underlying metal as far as possible from the ingress of sulphur compounds. The addition of refractory metal, niobium for example, was intended to postpone the onset of breakaway sulphidation. Any H_2S that found its way down micro cracks in the alumina was supposed to react preferentially with the niobium, forming a protective plug of niobium sulphide at the base of the microcrack, preventing sulphidation of the iron or chromium components in the alloy. In the parlance that was used, the niobium 'gettered' the H_2S.

Good results in laboratory tests were obtained, particularly when the alloys were preoxidised. This was even at 900°C, in atmospheres containing up to 2.8%

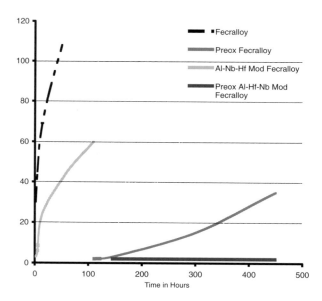

Fig. 28 Corrosion rate of Fecralloy and experimental high aluminium, niobium/hafnium Fecrally in a $H_2/H_2S/H_2O$ environment. Note excellent performance in preoxidised condition. (Redrawn from Ref. 106)

H_2S, Fig. 28. One should also add, that because of their unique composition, these alloys throw new light on the mechanisms mixed oxidant corrosion, in a way that tests on conventional alloys could not.[106,107]

Fabrication investigations showed that, although it would be possible to produce certain of these alloys in tube form, a number of them were brittle at room temperature, but superplastic above 800°C. Nevertheless, if there are reservations about mechanical and fabrication properties, it should be possible to use these materials in the form of coatings for the syngas cooler tubing, particularly since some of these alloys did so well at temperatures well above the critical 650°C range.

5.3   Molten Salt Corrosion

In both coal and oil fired generating plant, 'molten salt' attack is the principle mechanism of degradation in fireside corrosion. This has been for, many years, of critical importance in the UK. Fireside attack has not received quite so much attention elsewhere. The set of papers issued following the 'Workshop on Fireside Corrosion in Electrical Utility Boilers at Karlsruhe' in 1988 exemplifies the situation.[108] Steam temperatures in German, Dutch and Danish plants at that time were limited to 540°C, whereas in the UK the standard was 565°C, with two units operating at 593°C.[109,110]

These temperatures brought UK plants well into the so-called 'Bell Shaped Curve' for fireside corrosion in superheaters, that is the temperature regime between 575°–750°C, over which molten alkaline trisulphates are stable,[64,111] Fig. 5. Above this range the trisulphate begins to decompose into solid monosulphate and base metal oxide, so that this form of molten salt attack ceases. At still higher temperatures, circa 900°C, the alkaline mono-sulphate itself becomes is molten and can cause enhanced attack, although this is far beyond normal operating conditions in generating plants.[112]

In one sense then, fireside corrosion is an old problem, certainly in the UK, and is not one that fits well into a paper on advanced heat exchangers and materials. However, it is clear that it is a new issue to many, and with the development of fossil fuel power stations with steam temperatures in the 600°C regime, it is a subject which needs to be addressed.

Despite over forty years work in the UK and USA there is no final consensus about the mechanism of fireside corrosion, except the recognition that it is of the molten salt type, involving sulphates and SO_3. The most recent view is that the deposition of ash on the leading face of superheater tubing gives an appropriate medium in which the SO_2 in the combustion products is catalysed to SO_3. The trioxide then reacts with alkaline materials in the ash and elemental iron in the tube material, thereby producing a molten layer of ferric iron rich alkali metal-trisulphate.[113] The older view, that it was the level of sulphur trioxide in the combustion products that determined the degree of attack, is no longer considered viable, although the basic thermodynamic argument, that the presence of the trioxide gives rise to the molten trisulphate, is still accepted.

There is considerable disagreement about the details of the mechanism of hot corrosion itself. Some have the view that trisulphates react directly with the iron or chromium in the tube material, forming sulphides at the metal/molten salt interface and have provided powerful quantitative arguments in support.[114] Others consider that the trisulphate acts as a carrier for oxidants such as SO_3, which then form molten sulphates with fresh metal. In an opinion, not dissimilar to this, Flatley et al. suggest that the decomposition of the trisulphate near the outer surface of the melt, has the effect of liberating SO_3 which is then recycled to the corrosion front.[115]

Decomposition of the trisulphate undoubtedly does occur, and accounts for the formation of a red layer, rich in base metal oxides, which lies between the outer ash layer and the trisulphate. It follows from this argument, as is found in practice, that as the combustion product temperature increases so does the rate of attack. The equilibrium thickness of the trisulphate layer is reduced, thereby increasing concentration gradients and mass transfer rates. This mode of attack can be followed by plotting a reaction path using a thermochemical stability diagram of the appropriate type.

However, there is indirect evidence that the corrosion is of the acid fluxing or Type II form of attack, as experienced in gas turbines.[116] This electrochemical mode of corrosion is said to account for three critical observations. These are:

- The attack is very often of the pitting type, and is concentrated at the edge of the deposit where iron sulphides are formed. Furthermore, if a chromised diffusion coating is used to protect the underlying metal, when it fails, it results in accelerated pitting, presumably due to galvanic coupling.[117]
- The attack is at a minimum under the thickest layer of ash, suggesting some type of 'differential aeration' processes.[115]
- Very large concentrations of protective elements are needed to inhibit corrosion.[116]

Empirically, it has been found that levels of chromium in excess of 25% are needed to give protection to both iron and nickel based alloys in moderately severe environments.[111,112,113,118] Conversely there is some evidence that percentage levels of nickel can be somewhat detrimental, so that the 12% Cr ferritics are almost as good as the Type 300 stainless steels. A more conservative view would be that all of the new set of 9–12 Cr superheater steels, which have been under development for so many years, will need protection against fireside attack, and this is an area needing more research. The same is true of standard austenitic stainless steels, some of which are being proposed for $600 + °C$ steam temperatures. In the opinion of the author, chromium contents are too low in these materials. Alloys such as Tempalloy 30A, and 35A, 45TM, will be required.

All of this would be in line with UK experience. In this country, 25Cr–20Ni stainless alloys were used to produce co-extruded tubing using both ferritic and austenitic materials as the inner layer. To give a higher degree of resistance,

50Cr–50Ni alloy was applied as a powder spray or weld overlay. Flatley, in 1989, gave a good review of the status of the various technologies used in the UK.[119] However, the 'Rolls Royce solution' had been developed in the USA some years previously, with the fabrication of an Inco 671/800H composite tube.[120]

5.4 Chloridation Attack

Of the various forms of high temperature corrosion, attack by chlorine bearing species is the most insidious. It is normally an adjunct to other forms of degradation and often leaves no direct evidence of its presence, although overall metal wastage rates will be inexplicably higher. The effects of chlorine are becoming more important in the corrosion of heat exchangers, particularly with the increasing use of waste and agricultural biomass as fuels in power generation. Major reviews of corrosion in waste incineration plant were carried out at the NACE Corrosion 87, 89 and 91 symposia, the papers to which are available in a bound volume.[121]

In the UK, there has long been an appreciation of the importance of chlorine, because of our coal types, which contain up to 0.4% of this element. Chlorine was thought to affect both superheater and furnace wall corrosion, and this subject was covered in a Conference entitled 'Chlorine in Coal'.[122] Unpublished studies by the author and his colleagues at British Gas on corrosion in gasification atmospheres in the 750°–1000°C range also showed the dramatic effect of HCl, under reducing conditions. More recently, the NPL have shown that HCl has considerable impact on syngas cooler corrosion, both during operation and when the plant is shutdown.[99]

The direct effect of chlorine is two-fold. In simple oxidation, chlorine increases the rate of oxide spallation, apparently due to attack on the bond between the oxide and metal. It is not clear whether the presence of active elements helps or hinders in this respect. The disruption of the bonding is probably due to the formation of minute quantities of low melting point or gaseous metal chlorides, which result from a reaction of small amounts of chlorine bearing species with the metal itself. Under reducing conditions, however, for example within deposits that contain carbonaceous materials, or in coal gasification environments, tube materials can be attacked directly, with the formation of large amounts of non protective molten or gaseous metal chlorides. It would appear that chlorine gas is more aggressive than HCl under such conditions.[123,124]

It is particularly unfortunate that iron, which forms the principal constituent of the less costly heat exchanger alloys, reacts with chlorine or with chlorine bearing species to form a series of low melting point chlorides or oxy-chlorides, all of which have a relatively high vapour pressure. Although nickel and cobalt will also form such compounds, those of iron are more likely to form, being more stable thermodynamically. In consequence the chances of attack are greater, at a given pCl_2 pressure, on for example, stainless steel as against nickel based alloys. In this respect Grabke has given a good basic review the thermodynamics of the situation.[125] One needs to emphasise, in addition, that since we are dealing with volatile species, a series of thermochemical stability diagrams is needed to fully evaluate the thermodynamic of the situation. These diagrams show that as the partial pressure of the volatile species falls, the range of stability of these compounds increases, at the expense of the metal itself or of protective oxides, Fig. 29.

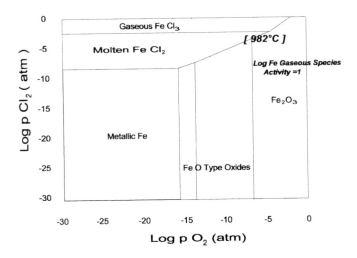

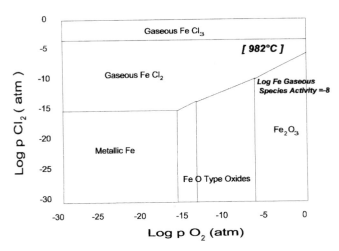

Fig. 29 Thermochemical stability diagrams for Fe–O–Cl, at pressures of 1 and 10^{-8} bar of the gaseous phases. Note increased area of stability gaseous chlorides at expense of metallic iron and protective oxides at low vapour pressures.

It also follows that vanishingly small amounts of HCl or chlorine can have an untoward effect on wastage rates, particularly where the local gas velocity is high.

As has been noted, *corrosion products* containing chlorine do not in general form a major component of scales and deposits, although other chlorine bearing species, vectored in from the environment, may be present. If detected at all, chlorine based corrosion products are likely to be found at the metal/scale-deposit interface. This is at a point where the oxide or sulphide have become thermodynamically unstable, in comparison to chlorides.

If such chlorides are present in quantity in deposits, their effect will be dramatic. Their visible presence implies that chlorine compounds have been reacting with the metal, probably forming a molten sub-layer at the deposit-metal interface. Since these compounds have a high vapour pressure, the corrosion products will migrate away from the metal surface by gaseous diffusion processes. Hence, unlike solid deposits, they give no protection to the underlying metal. Furthermore, as these gaseous chlorine compounds diffuse towards the outer surface of the deposit, they will, in most practical environments, come into contact with water vapour or oxygen, reacting with these species. The oxides produced, as a result of these reactions, are simply dumped, typically forming an iron rich non-protective layer, close to the surface of the deposit. Some of the resulting chlorine or HCl is then free to migrate back to the metal surface and cause further degradation.

This recycling effect will gradually increases the level of chlorine in the deposit, both in gasifiers and waste incineration plants. In the latter, the chlorine content of deposits has been found to range between 1 and 20%, depending on the temperature and chlorine loading in the gaseous environment.[126,127] Not all of this is due to the recycling effect, however. In waste incineration, as the combustion gas and tube temperatures fall, various chlorine compounds condense out of the environment. Each will have a characteristic temperature range of deposition, depending principally on the vapour pressure of the individual compound. Such deposits will obviously increase the corrosivity at high temperatures, and during shut down conditions the deposits, being deliquescent, will induce high rates of chloride pitting.

Chlorine bearing compounds can influence high temperature corrosion in other ways. In fireside corrosion, Flatley suggests that it will increase the reactivity of alkaline trisulphates.[115] In waste incineration, the combustion of PVC, at high temperature, promotes the formation of alkaline chlorides through reaction of HCl with alkaline species in the combustion products. These chlorides, theoretically unstable once the combustion temperature falls, can be carried straight onto boiler or superheater tubes where they will induce scale cracking.[126,127]

It is in waste incineration where the problem of chlorine is universally recognised, and there are extensive programmes to develop more resistant alloys. ERA Technology gives a fuller account of this in a review, which is now available

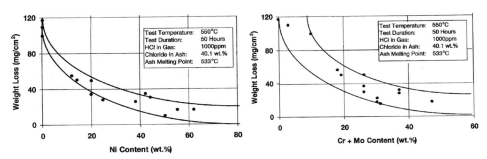

Fig. 30 Beneficial effect of Ni and Cr + Mo in simulated waste environment. Note that high Ni implies low Fe.[134]

as a European Union document.[43] Waste incineration processes are characterised by low metal temperatures and it is unlikely that the presence of chromium will have a very significant effect, unless present in large quantities. Of greater importance is the nickel to iron ratio. Whitlow *et al.* have shown, in practical tests on the Gallatin Incinerator, that with the exception of Inco 825, there is a direct correlation between the iron content of the alloy and the corrosion rate,[128] Fig. 30.

Accordingly, a number of high nickel stainless and nickel based alloys, including Sanicro 28, 45TM and Inco 625 have proved to be rather better than the Type 300 stainless steels. In addition, the silicon content of Nicrofer 45 TM is thought to contribute to the good resistance of this alloy.[129] In the case of Inco 625, although the main factor must be the very high nickel content, the presence of 9% molybdenum is undoubtedly important, although we cannot really explain its effect.

The indirect evidence suggests that, as with conventional fireside attack in pulverised fuel plants, corrosion in waste incineration processes, at lower temperatures at least, involves an electrochemical aspect.[130] It is possible to explain the beneficial effects of nickel, and to a lesser extent, cobalt and silicon, using simple gas-metal reaction arguments. These cannot account for molybdenum, since this element forms liquid phase chlorides of high volatility very easily. Further experiments to assess the importance of electrochemical effects are desirable.

As mentioned above the importance of chlorine in coal gasification is attracting wider attention. Work at the University of Northumbria, at temperatures in excess of 750°C showed that under some conditions, at high temperatures, selective grain boundary attack could occur, leaving deep, stress raising trenches.[131] The mechanism involved the rapid transport of iron to the grain boundaries where it was carried off as volatile chlorides. Once again, nickel based materials behaved better, but here we have a real quandary, in that high nickel alloys in reducing atmospheres can sulphidise extremely rapidly. On thermodynamic grounds, cobalt based, alumina formers seem the best compromise, if the cost can be tolerated. The trenching phenomenon needs to be evaluated for its effect on creep resistance.

At lower temperatures, in entrained bed gasification environments, Bakker *et al.* have shown that the attack by HCl will result in the rapid removal of Fe and Ni from the metal.[132] This leaves a micro-porous surface that allows deep penetration by other oxidants, leading to rapid rates of corrosion. High levels of chromium are needed to prevent corrosion, but attack on aluminised materials can be very rapid owing to the formation of aluminium chloride. Fortunately, the Oak Ridge, iron aluminide type materials seem reasonably tolerant of HCl.[101]

6 DISCUSSION AND CONCLUSIONS

This paper has highlighted how developments in the advanced heat exchanger field are imposing new demands on the materials of construction, in particular in terms of temperature and in the corrosivity of the environment. The paper has also provided opportunities to highlight some of the more innovative concepts in heat exchanger design, particularly those involving indirect fired and re-cuperative gas turbine cycles. Examples of modern recuperative gas turbines are now close to commercialisation. They are feasible because of the development of highly compact heat exchange designs that can dispense with external pressure vessels. Large scale indirect fired gas turbines are some way off, but we may see their introduction into the biomass field much sooner. The thermal fatigue aspects of these new structures will need close analysis.

Many of the advanced materials discussed in this paper are more costly than the standard types of alloys, and they will be used under rather novel conditions. Is there not a case for reviewing existing design standards and procedures, and using modern life assessment techniques, at the detailed design stage, in an effort to cut down on the cost of construction?

An underlying theme of this paper has been the continuing competition between the wrought stainless steels, the high nickel alloys, and their spun cast 'stainless' analogues, and to show, that in a historical context, this is not a new trend. At a time when Europe looks towards ultra supercritical 700°C power plant, we need to be aware that there is more than one way of producing tube. We need to recognise that the higher the creep strength of an alloy, the more difficult it is to produce tubing in the wrought form, due to the reduction of the hot working range.

Indeed, although there are some striking innovations with high temperature wrought alloys, there is much practical experience with spun cast materials in very high temperature plant. As the introduction to Section 4 suggested, we seem to be very close to the practical limits with high strength wrought tubing. We need to acknowledge this when evaluating advanced materials for power generation systems, whether for advanced plant or in more conventional equipment. Spun cast tubing gives us that little bit extra in terms of creep resistance, and, possibly, in corrosion performance. But can we accept the ductility problems associated with the use of such materials?

The need for higher corrosion resistance severely compromises alloy design for creep strength, although the Japanese, with materials such as St3Cu and Tempaloy 30A show what can be done, even with wrought alloys. But rather than attempt the difficult task of producing a really strong alloy with good corrosion resistance, protection of creep resistant tube by claddings or thick fusion welded overlays is the alternative. The UK has considerable experience in this area. An alternative approach is to use spun casting to produce a two layers, or even triple layer tube, so as to give high strength with good fireside and steam side properties.

There are some areas in which there is a need for some truly innovative research. If IGCC is to progress in terms of steam temperatures, specific output, and efficiency, it will be vital to increase the corrosion resistance of materials. Within an acceptable time horizon, this implies the development of improved tube coatings, perhaps relying on high chrome-silicon, or aluminide type compositions, in conjunction with gettering elements. The increased importance of chlorine in governing corrosion resistance needs better understanding, and there is a real need to investigate how dominant are electrochemical effects in the 300°–750°C range. Downtime and under-deposit corrosion aspects should not be neglected.

If spun cast material is to be used more generally, greater knowledge of the effect that primary or interdendritic carbides have on high strain thermal fatigue, after long-term exposure is essential. If these alloys are to be used in aggressive fireside corrosion situations, how significant will be selective attack on the carbides? Will this type of attack lead to reduced creep or thermal fatigue resistance? Spun cast material, if it is to be used for power generation, will need good steam side oxidation resistance, and there is undoubtedly a need for basic work in this area. Little seems to be known, for example, on whether active elements can help prevent the spallation of oxides that contain silica sublayers.

Others will have their own personal list of preferences for commercially oriented R&D and demonstration work in this area. It is to be hoped that this paper will have stimulated new thoughts amongst workers in the field and within appropriate funding agencies. The experience of the author would suggest, however, that with the development of new coatings and alloys, the successful development of new materials depends upon recognising that there is more than one route to success. In addition, although any programme must have the support of the alloy producers, fabricators, plant designers and operators, it is vital to have among the team leaders, experienced and innovative R&D organisations. It is they, who in the early stages, should be prepared to do the critical thinking about alloy development programmes. This thinking should be driven by the needs of the market, but be of a fundamental and far reaching character.

OLDER HIGH TEMPERATURE AUSTENITIC AND NICKEL BASED ALLOYS

Alloy	Fe	Ni	Co	Cr	Si	C	N	Al	Ti	Nb	Ta	W	Mo	Active elements	Mn	Other
304	Bal	9.0	—	18	0.5	0.06	—	—	—	—	—	—	—	—	<2.0	—
321	Bal	10.0	—	18	0.8	0.05	—	—	0.8	—	—	—	—	—	<2.0	—
347	Bal	10.0	—	18	0.5	0.05	—	—	—	0.9	—	—	—	—	<2.0	—
316	Bal	11.0	—	17	0.7	0.03	—	—	—	—	—	—	—	2.0	—	<1.0	—
310	Bal	20.0	—	25	0.7	0.02	—	—	—	—	—	—	—	—	—	<1.7	—
Alloy 800	Bal	32.0	—	21	0.2	<0.1	—	—	0.15 to 0.60	0.15 to 0.60	—	—	—	—	—	<0.9	—
Alloy 800H	Bal	32.0	—	20	0.2	0.05 to 0.1	—	—	0.15 to 0.60	0.15 to 0.60	—	—	—	—	—	<0.9	Note Al + Ti 0.30–1.2
Alloy 825	30.0	Bal	<2.0	22	0.3	0.05	—	—	0.1	1.0	—	—	—	3.0	—	1.0	—
RA330	Bal	35.0	—	18.5	1.0	0.08	—	—	—	—	1.0	—	—	—	—	<2.0	3.0
RA333	15.0	Bal	2.5	26.0	1.4	0.05	—	—	—	—	—	—	2.7	3.8	—	1.5	—
Alloy 600	7.0	Bal	—	16	0.2	0.05	—	—	—	—	—	—	—	—	—	0.2	—
Alloy 601	16.0	Bal	—	22.5	0.2	0.05	—	1.2	—	—	—	—	—	—	—	0.2	—
Alloy 617	—	Bal	12.5	22.0	—	0.07	—	1.0	—	—	—	—	—	8.0	—	—	—
Alloy 671	1.0	Bal	—	50.0	0.4	0.06	—	—	—	—	—	—	—	—	—	0.06	—
Haynes 188	1.2	22.0	Bal	22.0	0.4	0.1	—	—	—	—	—	—	14.5	0.6	0.07 La	0.05	—
Essette 1250	Bal	10	—	15	0.6	0.1	—	0.2	—	—	—	—	—	1.0	—	6.0	0.03V, 0.006B

NEWER HIGH TEMPERATURE AUSTENITIC AND NICKEL BASED ALLOYS

ALLOY	Fe	Ni	Co	Cr	Si	C	N	Al	Ti	Nb	Ta	W	Mo	Active elements	Mn	Other
153MA	Bal	10	—	18	1.6	0.05	0.15	—	—	—	—	—	—	0.05 Ce	1.0	—
253MA	Bal	10	—	21	1.6	0.1	0.20	—	—	—	—	—	—	0.05 Ce	<0.8	—
Inco 800HT	Bal	31	—	20	0.2	0.06 to 0.1	—	0.15 to 0.60	0.15 to 0.60	—	—	—	—	—	0.9	Note Al+Ti 0.85–1.20
Inco 803	Bal	34	—	25	0.7	0.07	—	0.5	0.6	—	—	—	3.0	0.1 Ce	0.7	—
Sanicro 28	Bal	31	—	28	<0.7	<0.02	—	—	—	—	—	—	8.5	—	<2.0	1.0
Sanicro 65	8	Bal	—	21	<0.5	<0.025	—	—	—	<0.5	—	—	2.0	—	<1.0	<0.015 B
Haynes 230	<3.0	Bal	<5.0	22	0.4	0.1	—	0.3	—	—	—	<2.5	2.0	0.02 La	0.7	—
Haynes 120	Bal	37	<3.0	25	0.6	0.05	0.2	0.1	—	0.7	—	<1.0	<2.5	—	0.5	0.004B
Haynes 160	<3.5	Bal	30.0	28	2.8	<0.15	—	—	0.6	<1.0	—	—	<1.0	—	1.0	—
Haynes 556	Bal	21	18.0	22	0.4	0.1	0.2	0.2	—	0.1	0.6	2.5	3.0	0.02 La, 0.02 Zr	1.0	—
AC66	Bal	32	—	27	<0.3	0.08	—	—	—	1.0	—	—	—	0.1 Ce	—	—
Alloy 602CA	10	Bal	—	25	<0.5	0.2	—	2.1	0.15	—	—	—	—	0.8 Y, 0.05 Zr	<0.1	—
45TM	23	Bal	—	27	2.8	0.1	—	—	—	√	—	—	—	RE	0.5	—
Alloy 625	<5.0	Bal	<1.0	21	<0.5	<0.1	—	<0.4	<0.4	√	√	—	9.0	<0.5	<0.5	Note Nb+Ta 3.5

ADVANCED JAPANESE WROUGHT ALLOYS

ALLOY	Fe	Ni	Co	Cr	Si	C	N	Al	Ti	Nb	Ta	W	Mo	Active elements	Mn	Other
St 3Cu	Bal	9	–	18	0.2	0.1	0.1	–	–	0.4	–	–	–	–	0.8	3 Cu
NF 709	Bal	25	–	20	0.5	0.15	–	–	0.1	0.2	–	–	1.5	–	1.0	–
HRC3	Bal	20	–	25	0.4	0.06	0.2	–	–	0.45	–	–	–	–	1.2	–
NF707	Bal	35	–	22	0.5	0.08	–	–	0.1	0.2	–	–	1.5	–	1.0	–
Tempaloy A–3	Bal	15	–	22	0.4	0.05	0.5	–	–	0.7	–	–	–	–	1.5	0.002B
CR 30A	Bal	50	–	30	0.3	0.06	–	–	0.2	0.2	–	–	2.0	0.03 Zr	0.2	–
CR 35A	Bal	45	–	35	–	0.06	–	–	–	0.5	–	–	–	–	0.03	–
HR 6W	Bal	43	–	23	0.4	0.08	–	–	0.08	0.2	–	6.0	–	–	1.2	0.003B
SSS 113MA	–	Bal	–	23	–	0.03	–	–	0.48	–	–	18	1	0.04 Zr	–	–

FERRITIC OXIDE DISPERSION STRENGTHENED ALLOYS

	Fe	Ni	Co	Cr	Si	C	N	Al	Ti	Nb	Ta	W	Mo	Active elements	Mn	Other
MA 956	Bal	–	–	20	–	0.05	–	4.5	0.5	–	–	–	–	0.5 Y_2O_3	–	–
PM 2000	Bal	–	–	20	–	<0.04	–	5.5	0.5	–	–	–	–	0.5 Y_2O_3	–	–
PM 3000	Bal	–	–	20	–	0.05	–	6.0	–	–	–	3.5	2.0	1.1 Y_2O_3	–	–
ODM 751	Bal	–	–	20	0.1	0.01	<0.1	4.5	0.6	–	–	–	1.0	0.4 Y_2O_3 (0.15 Zr)	0.1	0.01B

SPUN CAST HIGH TEMPERATURE AUSTENITIC ALLOYS

ALLOY	Fe	Ni	Co	Cr	Si	C	N	Al	Ti	Nb	Ta	W	Mo	Active elements	Mn	Other
HU/HT (18/37)	Bal	37	–	18	<2.5	0.4	–	–	–	–	–	–	–	–	<2.0	–
HK–40 (25–20)	Bal	20	–	25	<2.0	0.4	–	–	–	–	–	–	–	–	<2.0	–
HP45 Mod Nb	Bal	35	–	25	1.5	0.45	–	–	–	1.0	–	–	–	–	<2.0	Micro Zr
HP45 Mod Nb M.A	Bal	35	–	25	1.5	0.45	–	–	–	1.0	–	–	–	–	<2.0	–
35/45	Bal	45	–	35	1.8	0.45	–	–	–	1.2	–	–	–	–	<2.0	–
MO–RE–2	Bal	50	–	32	0.3	0.2	–	–	–	–	–	16	–	–	<2.0	Micro Zr
Paralloy H4 ST 20–32	Bal	48	–	28	1.5	0.5	–	–	–	–	–	–	–	–	<2.0	–

ELEVATED TEMPERATURE PRECIPITATION HARDENING MARTENSITIC SHEET ALLOYS

ALLOY	Fe	Ni	Co	Cr	Si	C	N	Al	Ti	Nb	Ta	W	Mo	Active elements	Mn	Other
15–5 PH	Bal	5	–	15	<0.6	0.04	<0.04	–	–	0.3	–	–	–	–	0.5	3.2 Cu

ACKNOWLEDGEMENTS

As is usual in these cases, with a paper of this length, it was written at home, in the author's 'spare' time. He would therefore wish to acknowledge the forbearance and support of his family over the past fifteen months.

More formally, the author wishes to acknowledge the encouragement of Dr Roger Townsend, formerly of ERA Technology Ltd, in proposing to the IOM that he should write the paper. He also wishes to thank Mr Andrew Strang, as general editor, for giving him the latitude to cover the subject of the paper in reasonable detail.

The author also wishes to thank ERA Technology Ltd for its support in enabling him to present a shortened version of the paper at the IOM Materials Congress in 1998.

REFERENCES

1. I. Turner: 'Chapter 9, The Air Used in the Blast Furnace', *The Metallurgy of Iron*, Griffith Scientific Text Books, 1890.
2. Symposium on the Effect of Temperature Metals, ASTM and ASM, 1932.
3. R.B. Cooper: 'Introduction to the Use of Cast Alloy Material for Hydrocarbon Processing in the USA', *Materials Technology for Steam Reforming Processes*, C. Edeleanu ed., Pergammon, 1966.
4. *Report on the Shortages of Gas Suppliers in the West Midlands During the Winter of 1965–66*, HMSO, 1966.
5. *Alloy 800 Proc. Petten International Conf. 1978*, W. Betteridge et al. ed., North Holland, 1978.
6. 'Chapter 6, Closed Cycle Gas Turbines', *Total Energy*, R.M.E. Diamant ed., Pergammon, 1970.
7. D.B. Meadowcroft: 'An Introduction to Fireside Experience in the Central Electricity Generating Board', *Werkstoffe und Korrosion*, 1988, **39**(2).
8. D.R. Holmes: 'New Corrosion Resistant High Temperature Heat Exchanger Materials', *Corrosion Science*, 1968, **8**, 608–622.
9. C.J. Middleton, R. Timmins and R.D. Townsend: 'The Integrity of Materials in High Temperature Components; Performance and Life Assessment, *Int. J Pressure Vessel and Piping*, 1995, **66**, 33–67.
10. I.G. Wright and J. Stringer: 'Materials Issues for High Temperature Components in Indirectly Fired Cycles', *ASME Gas Turbine Conference*, Orlando, 1977.
11. 'Chapter 5, Open Cycle Gas Turbines', *Total Energy*, R.M.E. Diamant ed., Pergammon, 1970.
12. D.G. Wilson: 'Chapter 3, The Thermodynamics of Gas Turbine Power Cycles', *The Design of High Efficiency Turbomachinery and Gas Turbines*, MIT Press, 1984.

13. J.L. Blough and S. Kihara: 'Coal-Ash Corrosion in Superheaters and Re-heaters', *Paper 189, Corrosion 88*, St Louis.
14. M. Tamura, N. Yamanouchi, M. Tanimura and Y. Minami: 'High Temperature Performance of 35Cr-45Ni-Fe Alloy', *High Temperature Corrosion in Energy Systems*, M. Rothman ed., AIME.
15. R. Blum: 'Materials Developments for Power Plants with Advanced Steam Condition; Utility Point of View', *Conference on Materials for Advanced Power Engineering*, Coutsaradis *et al.* eds, Kluwer, 1994.
16. N. Birks and C.A. Smith: 'Factors in the Selection of Advanced Alloys for Advanced Heat Exchange Applications', *Materials Issues in Heat Exchangers and Boilers Conference*, Institute of Materials, 1997.
17. Y. Nakabayashi, S. Ikeda, T. Yashioka, A. Hizume, K. Shimomura, N. Yamada and T. Fujikawa: 'Japanese Developments in High Temperature Steam Cycles', *Conference on High Temperature Materials for Power Engineering*, Bachelet *et al.* eds, Kluwer, 1990.
18. E. Metcalf, W.T. Bakker, R. Blum, R.P. Bygate, T.B. Gibbons, H. Hald, F. Masuyama, H. Naoi and Y. Sawaragi: 'New Steels for Advanced Coal Fired Plant up to 620°C', *IMechE Conference Transactions*, 1997.
19. D.H. Scott: 'Emerging Coal Fired Power Generation Techniques', *Ibid*.
20. H. Jericha: 'High Temperature Reactor Heat Conversion by Novel High Efficiency Steam Cycle', ASME, 1988, COGEN-TURBO, ASME, 1988.
21. J. Stringer: 'Practical Experience of Wastage at Elevated Temperatures in Coal Combustion Systems', *Wear*, Elsevier, 1995, 186–87.
22. I.G. Wright: 'A Review of Experience of Wastage in Fluidised Bed Boilers', *Mat at High Temperatures*, 1997, **14**(3).
23. T.B. Lindemer and R.L. Pearson: 'Chemical Thermodynamic Analysis of the Interaction of 304 Stainless Steel with the Gases and the $CaSO_r$ Deposit in the Fluidised Bed Combustor', *High Temperatures, High Pressures*, 1982, **14**(5).
24. J.W. Slusser, A.D. Bixler, S.B. Bartlet: 'Materials Experience in the Stockton CFBC', *Proc Workshop on CFBCs*, J.W. Stallings ed, EPRI Report No. GS-6747, 1990.
25. A. Hansson and S.O. Ostman: 'Status of PFBC and Experience with Materials Performance During 75000 of Operation in ABB's PFBC Plants', *Mat at High Temperatures*, 1997, **14**(2).
26. P.Y. Hou, S. Macadom, H. Zhang and J. Stringer: 'Summary of Results from the Berkeley In-Bed Tube Erosion Simulator', *Ibid*.
27. P. Holtzer and P. Rademakers: 'Studies on 90 MW(Th) AKZO and MW(Th) TNO FBC Show Excellent Corrosion Results', *Proc 1991 Int. Nat Con on Fluidised Bed Combustion*, E.J. Anthony ed., ASME, 1991.
28. M.M. Stack: 'Optimising the Performance of Materials in FBC Conditions Using Erosion-Corrosion Wastage and Materials Performance Maps', *Mat at High Temperatures*, 1997, **14**(2).

29. T. Wolzenburg, D. Koneke and P.M. Weinspach: 'Lifetime Enhancement of Erosively Laden Components by Modelling the Outline Changes Under Erosive Wear Conditions', VGB PowerTech, Jan 1998.

30. C.F. Macdonald: 'Gas Turbine Recuperator Advancements', *Materials Issues in Heat Exchangers and Boilers Conference*, Inst. of Materials, 1997.

31. E.R. Watson, M.L. Parker and D.A. Branch: 'Development Testing and Validation of the WR-21; An Intercooled and Recuperated Marine Gas Turbine', Paper 19, *INEC 96*, Inst. of Marine Engineers.

32. 'Solar Mercury 50: A Fresh Approach to Efficiency', *Turbomachinery International*, Nov/Dec 1997.

33. R.K. Shah: 'Brazing of Compact Heat Exchangers', *Compact Heat Exchangers*, Hemisphere Pub., 1990.

34. F. Starr, A.R. White, B. Kazimierzak: 'Pressurised Heat Exchangers for 1100°C Operation Using ODS Alloys', *Advanced Power Engineering*, Coutsaradis *et al.* eds, Kluwer, 1994.

35. N.M. Johnson: *Development of Applications of Printed Circuit Heat Exchangers*, c/o Heatric Ltd, Poole, Dorset, England, c1996.

36. *Prospects for Advanced Coal Fuelled Power Plant and the Requirements for Materials Research*, c/o Quo-Tech Ltd, England, 1996.

37. S.A. Posthuma and P.L. Zuideveld: 'Gasification for Power Generation', *Power Technology Int.*, Spring 1997.

38. J. van Liere: 'Present Status of Advanced Coal-Fired Plants', *Mat at High Temperatures*, 1997, **14**(2).

39. W. Schellberg, N. Ullrich, W.T. Bakker and R.G.I. Keferink: 'The PRE-NFLO Gasification Process; Design and Materials Experience, *Ibid*.

40. S.A. Posthuma, E.E. Vlaswinkel, P.I. Zuideveld: 'Shell Gasifiers in Operation', *Conf on Gasification Technology in Practice, IChemE*, Milan, 1997.

41. L. Barniske: 'Waste Incineration in the Federal Republic of Germany', *Conference on Fireside Problems While Incinerating Municipal and Industrial Waste*, Florida, 1989.

42. M. Kunzell and E.W. Haltzinger: 'The Incinerator Grate's Central Role in Thermal Waste Disposal', ABB Review, Oct 1990.

43. F. Starr: 'Fireside Corrosion in Waste Incineration', *ERA Report 97-0296*, Leatherhead, ERA Technology Ltd, 1997.

44. H.H. Krause: 'Corrosion by Chlorine in Waste Incinerators', *Conference on Fireside Problems While Incinerating Municipal and Industrial Waste*, Florida, 1989.

45. 'Section 3.2, Fe-Base ODS Alloys', *Materials for Advanced Power Engineering 1994*, Coutsaradis *et al.* eds, Kluwer, 1994.

46. F. Starr: 'Corrosion Aspects in the Design and Operation of ODS Fired Heat Exchangers', *NACE Corrosion 98*, San Diego.

47. I.C. Elliot and G.A.J. Hack: 'MA Alloys for Aerospace Applications', *Structural Applications of Mechanical Alloying*, Froes and deBarbadillo eds, ASM 1990.

48. D.M. Jager and A.R. Jones: 'The Development of Grain Shape in Iron Based ODS Alloys', *Materials for Power Engineering 1994*, Coutsaradis *et al.* eds, Kluwer, 1994.

49. P. Ganesan, J.A. Plybum and C.S. Tassen: 'Incoloy Alloy 803, a Cost Effective Alloy for High Temperature Service', *Heat Resistant Alloys II*, Natesan *et al.* eds, ASM.

50. R.F. Decker and E.G. Richards: 'Metallurgical Aspects of Nickel Alloys for Superheater and Steam Pipe Applications', *Int. Power Conf*, Lausanne, 1968.

51. B.R. Clark and R.B. Pickering: 'Precipitation Effects in Austenitic Stainless Steels Containing Titanium and Aluminium Additions', *Journal of the Iron and Steel Institute*, Jan 1967.

52. M.R. Diglio, H. Straube, K. Spiradek and H. Degischer: 'Improving the Creep Resistance of Alloy 800H', *High Temperature Materials for Power Engineering 1990*, Bachelet *et al.* eds, Kluwer, 1990.

53. *Preliminary Properties of Vacuum Investment Cast Test Bars for Haynes Alloy 230*, c/o Haynes International, Kokomo, Ia, USA.

54. U. Brill and D.C. Agarwal: 'Alloy 602 CA – A new Alloy for the Furnace Industry', *Heat Resistant Alloys II*, K. Nateson *et al.* eds, ASM, 1995.

55. D. Klarstrom: 'Heat Treatment Property Relationships for Solid-Solution Strengthened Heat Resistant Alloys', *Heat Resistant Alloys*, ASM, 1991.

56. R.F. Atkinson: 'The Development of Heat Resistant Alloys for Reformer and Ethylene Furnaces', *Materials Issues in Heat Exchangers and Boilers Conference*, Inst. of Materials.

57. J.J. Jones: 'Developments in Heat Resistant Alloys for Petrochemical Plant', *Research and Development of High Temperature Materials for Industry*, Bullock *et al.* eds, Elsevier, 1989.

58. *Characteristics of New Steel Tube (HR3C) with Elevated Temperature Strength and High Corrosion Resistance for Boiler* (sic), c/o Sumitomo Metal Industries, Japan, May 1988.

59. *Avesta High Temperature Steels Within the Steel and Metals Industries*, c/o Avesta, Sweden.

60. S.C. Ernst and G.Y. Lai: 'A New High Strength Fe-Ni-Cr-Nb-N Alloy for Elevated Temperature Applications', *First Int. Conf Heat Resistant Materials*, Fontana, USA, ASM, 1991.

61. *Quality and Mechanical Properties of NF 709 for Power Plant Boilers*, c/o Nippon Steel Corporation, May 1993.

62. T. Andersson and T. Odelstam: *Sandvik 253 MA (UNS S30815) – The Problem Solver for High Temperature Applications*, c/o Sandvik, Sandviksand, Sweden.

63. A. Tohyama, Y. Minami and M. Miyauchi: 'Development of 18 Cr-10Ni-3Cu-Ti-Nb Stainless Steel for Ultra Supercritical Boilers', *Materials for Advanced Power Engineering 1994*, Coutsaradis *et al.* eds, Kluwer, 1994.

64. M. Tamura, N. Yamanouchi, T. Shimada, Y. Kuriki and Y.Minami: 'Corrosion Performance and Mechanical Properties of the Austenitic Stainless Alloys with Very High Chromium Content', *First Int. Conf Heat Resistant Alloys*, Fontana, USA, ASM, 1991.

65. T. Matsuo, M. Klkuchi and M. Takeyama: 'Strengthening Mechanisms of Ni-Cr-W Based Superalloys for Very High Temperature Gas Cooled Reactors', *First Int. Conf on Heat Resistant Alloys*, ASM, 1991.

66. J.L. Shannon: '15-5PH Ferrous Alloys', *Aerospace Structural Metals Handbook*.

67. G. Stein and J. Menzel: 'Nitrogen Alloyed Material for Advanced Blading and Turbine Designs', *High Temperature Materials for Power Engineering 1990*, Bachelet *et al.* eds, Kluwer, 1990.

68. M. Thiele, H. Teichmann, W. Schwarz and W.J. Quadakkers: 'Corrosion Behaviour of Ferritic and Austenitic Steels in the Simulated Combustion Gases of Power Plants Fired with Hard Coal and Brown Coal', *VGB Kraftwerktechnik*, 1997, **77**(2).

69. N. Otsuka and H. Fuijikawa: 'Scaling of Austenitic Stainless Steels and Nickel-Based Alloys in High-Temperature Steam at 973K', *Corrosion*, 1991, **47**(4).

70. 'Corrosion Problems in Coal-Fired Boiler Superheater and Reheater Tubes', *Stream-Side Oxidation and Exfoliation: EPRI Report CS-1811*, April 1981.

71. 'High Temperature Metallic Recuperator', *Gas Research Institute Report PB84-207869* (obtainable through USA Department of Commerce NT15), c1983.

72. D.P. Fleetwood and J.E. Whittle: 'The Cyclic Oxidation of Nickel Based Alloys and Heat Resistant Steels', *British Corrosion Journal*, May 1970.

73. B.B. Ebbinhaus: 'Thermodynamics of Gas Phase Chromium Species: The Chromium Oxides, the Chromium Oxyhydroxides and Volatility Calculations in Waste Incineration Processes', *Combustion and Flame*, 1993, **13**, 119.

74. P. Kofstad: 'Fundamental Aspects of Corrosion by Hot Gases', *High Temperature Corrosion 2*, Streiff *et al.* eds, Elsevier, 1988.

75. N. Wood, Q. Mabutt, J. Wonsowski, and F. Starr 'The Long Term Oxidation Behaviour of Iron Based ODS Alloys', *13th Int. Plansee Seminar*, Bildstein and Eck eds, Metallwerke Plansee Reutte, 1993.

76. M.J. Bennett: 'Beneficial and Detrimental Effects of Silica in the High Temperature Oxidation of 20Cr/25Ni/Nb Stainless Steel', *10th Int. Congress on High Temperature Corrosion*, India, 1989.

77. H.E. Evans: 'Modelling Oxide Scale Spallation', *Materials at High Temperatures*, 1994, **12**(2–3).

78. R.B. Herchenschroeder: 'Major Effects of Minor Alloying Elements on an Oxidation Resistant Fe-Ni-Cr-Co Alloy (Haynes Alloy No. 556)'.

79. M.A. Harper, J.E. Barnes, G.Y. Lai: 'Long-Term Oxidation Behaviour of Selected High Temperature Alloys', *Paper 97132*, NACE Corrosion 97.

80. *Paralloy H46M* (*35/45 Cr/Ni + Nb Microalloy*), c/o Paralloy, Billingham UK.
81. *Improve Ethylene Furnace Efficiency with Better Alloys*, c/o Paralloy, Billingham, UK.
82. 'Oxide/Metal Interface and Adherence', *Special Issue of Materials and Technology*, 1988, **4**(5).
83. *The Role of Active Elements on the Oxidation Behaviour of High Temperature Metals and Alloys*, Lang ed., Elsevier, 1989.
84. J.A. Smialek: 'Sulphur Impurities and the Microstructure of Alumina Scales', *Microscopy of Oxidation 3*, Inst. of Materials, 1997, 127–139.
85. G.H. Meier: 'A Review of Recent Advances in High Temperature Corrosion', *High Temperature Corrosion 2*, Streiff *et al.* eds, Elsevier, 1989.
86. W.J. Quadakkers and M.J. Bennett: 'Oxidation Induced Lifetime Limits of Thin Walled, Iron Based, Alumina Forming, Oxide Dispersion Strengthened Alloy Components', *Materials Science and Technology*, Feb 1994.
87. W.J. Quadakkers, D. Clemens and M.J. Bennett: 'Measures to Improve the Oxidation Limited Service Life', *Microscopy of Oxidation 3*, Inst. of Materials, 1997, 195–206.
88. S. Mrowec: 'The Problem of Sulphur in High Temperature Corrosion', *Oxidation of Metals*, 1995, **44**(1/2).
89. J. Gutzeit: 'High Temperature Sulphidic Corrosion of Steels', *Process Industries Corrosion*, NACE, 1986.
90. G. Sorrell, M.J. Humphries and J.E. McLaughlin: 'Alloy Performance in High Temperature Oil Refining Environments', *Heat Resistance Materials II*, Natesan *et al.* eds, ASM, 1991.
91. *Materials for Coal Gasification*, Bakker *et al.* eds, ASM, 1988.
92. 'Materials for Coal Gasification Power Plant', *Materials at High Temperatures*, 1993, **11**(1–4).
93. 'Corrosion in Advanced Power Plants', *Materials at High Temperatures*, 1997, **14**(2–3).
94. M.A. Howes: 'Corrosion of Alloys in Simulated Coal Gasification Environments at Elevated Temperatures', *Materials for Coal Gasification*, Bakker *et al.* eds, ASM, 1988.
95. K. Natesan: 'High Temperature Alloy Corrosion in Coal Conversion Environments', *High Temperature Corrosion*, R.A. Rapp ed., NACE, 1981.
96. W.T. Bakker and J.H.A. Bon Vallet: 'The Corrosion of Stainless Steels on the Wrong Side of the Kinetic Boundary', *Heat Resistant Materials II*, Natesan *et al.* eds, ASM, 1995.
97. F.H. Stott and J.F. Norton: 'Laboratory Studies Involving Corrosion in Complex, Multicomponent Gaseous Environments at Elevated Temperature', *Materials at High Temperatures*, 1997, **14**(2).
98. *Sandvik Sanicro 28 – Composite Tube for Steam Boiler Applications*, c/o Sandvik, Sandviken, Sweden.

99. S.R. Saunders, D.D. Gohil and S. Osgerby: 'The Combined Effect of Downtime Corrosion and Sulphidation on the Degradation of Commercial Alloys', *Materials at High Temperatures*, 1997, **14**(3).

100. K. Natesan: 'Corrosion Performance of Fe-Cr-Al and Fe Aluminide Alloys in Complex Gas Environments', *Heat Resistant Materials II*, ASM, 1995.

101. P.F. Tortorelli, G.M. Goodwin: 'Weld Overlay Iron-Aluminide Coatings for Use in High Temperature Oxidising/Sulfidising Environments', *Ibid*.

102. J.F. Norton, M. Maier and W.T. Bakker: 'Corrosion of 12% Cr Alloys with Varying Si Contents in a Simulated Dry-Feed Entrained Slagging Gasifier Environment', *Materials at High Temperatures*, 1997, **14**(2).

103. K.N. Strafford and D. Jenkinson: 'The High Temperature Degradation of Some Refractory Metals in Hydrogen-Sulphur Atmospheres', *Corrosion Resistant Materials for Coal Conversion Systems*, Meadowcroft *et al.* eds, Applied Science, 1983.

104 K.N. Strafford, A.F. Hampton and D. Jenkinson: 'The Sulphidation Behaviour of Vanadium, Niobium, and Molybdenum', *High Temperature Alloys; Their Exploitable Potential*, Elsevier, 1985.

105. F. Starr and S.G. Denner: 'High Temperature Corrosion Aspects of Thermal Hydrogenation Processes', *Environmental Degradation of Materials*, Inst. of Materials, 1980.

106. K.N. Strafford, P.K. Datta, A.R. Cooper and G. Forster: 'Composition Optimisation of Coating Materials to Inhibit Breakaway Corrosion in Complex Gas Atmospheres at Elevated Temperatures', *Surface Engineering Practice*, Ellis Horwood, 1990.

107. K.N. Strafford: 'The Influence of Refractory Elements on the Sulphidation Behaviour of Cobalt-Based Alloys', *Corrosion Science*, 1989, **29**(6).

108. 'Issue Dealing with Workshop on Fireside Corrosion in Electric Utility Boilers', *Werkstoffe und Korrosion*, 1988, **32**(2).

109. R.D. Townsend: 'Effect of Legislation, Regulation, Privatisation and Metallurgical Issues on the Electricity Supply Industry', *this publication*.

110. W. Schoch: '100 Years of Power Plant Engineering (with emphasis on materials technology)', *VGB Kraftwerkstechnik*, 1983, **63**(7).

111. J.L. Blough and S. Kihara: 'Coal-Ash Corrosion in Superheaters and Reheaters', *NACE Corrosion 88*, St Louis.

112. S. van Wheele, S.L. Blough and J.H. de Van: 'Attack of Superheater Tube Alloys, Coatings and Claddings by Coal-Ash Corrosion', *NACE Corrosion 94*.

113. S. KIhara, A. Ohtomo, I. Kajigaya and F. Kishimoto: 'Recent Plant Experience and Research into Fireside Corrosion in Japan', *Werkstoffe und Korrosion*, 1988, **39**(2).

114. N. Bolt and J.T.W. Pastoors: 'Fireside Corrosion in Boilers of the Dutch Electricity Undertakings', *Ibid*.

115. T. Flatley, E.P. Lathem and C.W. Morris: 'Mechanistic Features of Molten Salt Corrosion in Coal Fired Boilers', *Ibid.*

116. 'Section 6, Corrosion and Sulphidation Mechanisms Related to Fireside Corrosion', *Corrosion Problems in Coal Fired Superheater and Reheater Tubes*, EPRI Report CS-1653, Nov 1980.

117. E.C. Lewis and A.L. Plumley: 'Chromizing for Combating Fireside Corrosion', *Advances in Material Technology for Fossil Power Plants*, Viswanathan *et al.* eds, ASM, 1987.

118. J.L. Blough, G.J. Stanko, M. Krawchuk and W. Wolowodiuk: 'In-Situ Coal-Ash Corrosion Testing', *Heat Resistant Materials II*, Natesan *et al.* eds, ASM, 1995.

119. T. Flatley and C.W. Morris: 'Claddings and Co-extruded Tubes', *Research and Development of High Temperature Materials for Industry*, Bullock *et al.* eds, Elsevier, 1989.

120. D.W. Rahoi and J.F. Delong: 'Evaluation of Incoclad 671/800H Tubing After Nine Years of Reheater Service in a Coal Fired Utility Boiler', at ASME Winter Meeting, Chicago, 1980.

121. *Materials Performance in Waste Incineration Systems*, Lai *et al.* eds, NACE, 1992.

122. *Chlorine in Coal*, Stringer *et al.* eds, Elsevier, 1991.

123. M.J. McNallan: 'Gaseous Environments', *Ibid.*

124. U. Brill and E. Alpeter: 'Long Time Corrosion Behaviour of a Wide Range of Commercial Heat Resistant Alloys in a Chlorine Bearing Environment', *Heat Resistant Alloys*, ASM, 1990.

125. H.J. Grabke: 'Fundamental Mechanisms of Attack of Cl, HCl and Chlorides on Steels and High Temperature Alloys in the Range 400–900°C', *Conference on Fireside Problems While Incinerating Municipal and Industrial Waste*, Florida, USA, 1989.

126. P.L. Daniel, L.D. Paul and J. Barns: 'Fireside Corrosion in Refuse-Fired Boilers', *Materials Performance in Waste Incinerators*, Lai *et al.* eds, NACE, 1992.

127. H.H. Krause: 'Effects of Flue-Gas Temperature and Composition on Corrosion from Refuse Firing', *Ibid.*

128. G.A. Whitlow, P.J. Gallagher and S.Y. Lee: 'Combustor and Superheater Materials Performance in a Municipal Solid Waste Incinerator', *Materials Performance in Waste Incinerators*, NACE, 1992.

129. G. Sorrell: 'Materials of Construction for Incinerator Heat Transfer Equipment', *NACE Corrosion 93*, New Orleans.

130. G.A. Whitlow, W.Y. Mok, W.M. Cox, P.J. Gallagher, S.Y. Lee and P. Elliot: 'On-line Materials Surveillance for Improving the Reliability in Power Generation Systems', *Materials Performance in Waste Incinerators*, NACE, 1992.

131. K.N. Strafford, P.K. Datta and G. Forster: 'On the Design of Coating Materials to Resist High Temperature Chloridation', *Surface Engineering Practice*, Ellis Horwood, 1990.

132. W.T. Bakker: 'The Effect of Chlorine on Mixed Oxidant Corrosion', *Materials at High Temperatures*, 1997, **14**(3).
133. G.J. Yurek, M.H. La Branch and Y.K. Kim: 'Oxidation of Cr and Fe-25Cr Alloy in H_2-H_2O- H_2S Mixtures at 900°C'. *High Temperature Corrosion in Energy Systems*, M Rothman ed., and AIME.
134. G. Sorrell: 'The Role of Chlorine in High Temperature Waste to Energy Plants', *Materials at High Temperatures*, 1997, **14**(3).

Welding and Fabrication of High Temperature Components for Advanced Power Plant

A. M. BARNES, R. L. JONES, D. J. ABSON and T. G. GOOCH

TWI, Granta Park, Great Abington, Cambridge CB1 6AL, UK

ABSTRACT

Within the power generation industry, in the UK and worldwide, many changes have been brought about, partly as a result of widespread deregulation, but primarily because of the drive for improved thermal efficiency and reduced emissions. This has led to material developments, to meet the demands of higher operating temperatures and pressures, while ensuring the continued and safe opertaion of plant throughout its life.

While a wide range of materials is employed within power generation plant, Cr–Mo steels play an essential role in high temperature components. This paper therefore concentrates on the family of Cr–Mo steels, and provides an overview of the metallurgy, and the fabrication and service problems associated with these alloys. It also outlines the welding processes traditionally employed in the construction of high temperature plant. Current and future developments in materials for high temperature power plant applications are considered, together with the changes envisaged in welding techniques and processes.

The paper concludes with the identification of a number of specific areas, in terms of welding behaviour, weldment properties, and service performance, where further work is required.

INTRODUCTION

There are ever increasing demands on designers, fabricators and operators, for both economic and environmental reasons, to increase the thermal efficiency of power plant. This has been further promoted by the widespread deregulation within the power generation industry, which has given rise to significant competition amongst the generators. The net effect is an increase in operating temperatures and pressures, which in turn places higher demands on material and weldment performance to ensure the safe and reliable operation of the plant throughout its life. It is to meet such demands that we see the moves to combined cycle gas turbine plant and 'clean coal' technology. Many of the independent power producers favour gas turbine based plant in new construction, as it is both cost-effective and can be built relatively quickly, allowing them to compete with the established utilities. Combined heat and power installations are also now an attractive proposition. However, it seems hard for many of us living in the western world to recognise that there is still a very high percentage of the world's population that do not have access to electricity. It is no surprise that there are huge plant construction programmes, either planned or indeed in progress, to meet the increasing supply demands, particularly in countries such as Korea, China and India. Clearly such programmes are constrained by environmental factors, particularly the need to minimise emissions to slow down global warming, but by far the most significant driving factor is cost. Therefore, at least in the short/medium term, efforts in these geographic areas are concentrating on the existing technology of pulverised fuel plant.

A large number of materials, section thicknesses and product forms is used in the construction of power plant, requiring a variety of welding techniques, and each accompanied by different fabrication and service problems. Further, an inevitability of the diversity of materials used within power plant is the frequent need for dissimilar metal joints, involving either ferritic to austenitic combinations or different ferritic alloys.

It is clearly impractical, in a single paper, to cover all material and joining issues related to high temperature plant, so the principal emphasis will be on the family of chromium–molybdenum (Cr–Mo) steels. This review is intended to provide a brief historical overview of materials and welding issues associated with both the fabrication and service operation of power plant, through to the present day. In light of problems that have arisen in the past, and the changes occurring in the operating environment, a projected view of future developments in both fabrication practice and the materials employed will be presented. The paper will broadly be divided into two sections , the first addressing many of the materials issues, and the second covering fabrication practice and welding processes.

METALLURGY OF LOW ALLOY CREEP RESISTING STEELS

The low alloy creep resisting steels typically contain between 0.5 and 9% chromium, to give good corrosion/oxidation resistance, rupture ductility and resistance to graphitisation. This, together with carbon and 0.5–1.0% molybdenum, contribute to the creep strength through secondary hardening. In addition, they contain small additions of carbide-forming elements such as Nb, V and Ti to achieve grain refinement and precipitation strengthening. Such materials are supplied in a number of heat treatment conditions, for example quenched and tempered, normalised and tempered, and annealed. The microstructure will be determined by both the composition and the heat treatment condition; it can be a ferrite/pearlite or ferrite/bainite mix, fully bainitic, or martensitic, and the mechanical properties will vary accordingly. Ultimately it is the required balance of toughness and strength for a particular application that will govern the heat treatment and particularly the tempering temperature employed. Broadly this family of steels can be divided into three categories, viz the Cr–Mo steels, the Cr–Mo–V steels and the modified Cr–Mo steels.

APPLICATION OF WELDING TO CR–MO STEELS FOR HIGH TEMPERATURE APPLICATIONS

In conventional, fossil-fired power plant in the UK., low alloy steels such as 0.5Cr–0.5Mo–0.25V, 1.25Cr–0.5Mo, 2.25Cr–1Mo, to name but a few, have been used in a variety of applications, for example header pipes, steam pipework, boiler tubes, and turbine casings. The Cr–Mo–V and Ni–Cr–Mo–V steels have been employed in the manufacture of rotors, although high temperature rotors, bolting, and the turbine blades have generally been made from 12%Cr stainless steels, with various Mo, W and V modifications.

The low alloy steels have traditionally been welded using the conventional arc welding processes. Although these materials, with their potentially high hardenability, are susceptible to hydrogen induced cracking, procedures have been developed incorporating appropriate preheat levels and often the use of temper bead techniques, to minimise the risk of such cracking. The occurrence of solidification cracking has generally not been a major problem, although isolated instances have been encountered.

In the following sections, some of the principal fabrication and service problems associated with power plant components will be highlighted. Where appropriate, the remedial action taken will be indicated, in terms of design changes or the use of alternative materials.

Reheat Cracking

When a weld is reheated during stress relief or high temperature service, intergranular 'reheat' or 'stress relief' cracking can occur if either the weld metal or coarse grained HAZ have insufficient creep ductility, Fig. 1. As illustrated by Middleton et al.,[1] in such instances the relaxation of welding residual stresses occurs through either grain boundary cavitation, or intergranular cracking, as opposed to plastic flow. The susceptibility to cracking is dependent on the level of restraint, and hence joint geometry, as well as alloy composition (for example 0.5Cr–0.5Mo–0.25V showed a higher susceptibility than 2.25Cr–1Mo steel).

Historically there have been a number of widely reported instances of reheat cracking. A detailed review of reheat cracking was undertaken in 1987 by Dhooge and Vinckier.[2] Their review considered the details of the cracking mechanism and factors affecting the cracking susceptibility. In the 1950s, significant cracking problems were encountered, particularly in high temperature steam piping made from type 347 stainless steel.[3] The 1960s brought cracking in 2CrMo and CrMoV steam pipework, and later constructional low alloy steels, including the quenched and tempered grades used in pressure vessel fabrication, were found to be potentially susceptible. Since the 1970s, the problem has also been encountered in the form of 'underclad cracking' in nuclear pressure vessels.[4,5] The reheat cracking problems in low alloy steels largely resulted from the use of weld procedures that led to the development of a high proportion of coarse grained microstructure in the heat-affected zone (HAZ) with little re-

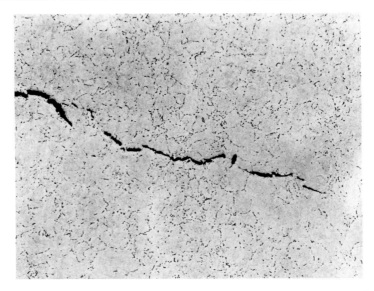

Fig. 1 Optical micrograph showing reheat cracking in the coarse grained heat affected zone of Cr Mo.V steel.

finement. The development of controlled deposition procedures, giving rise to significant weld metal and HAZ refinement,[6-10] together with tighter control of impurity levels, and optimisation of the postweld heat treatment procedures, have virtually eliminated the problem.

Temper Embrittlement

Temper embrittlement is a phenomenon that has plagued industry for many decades, and is primarily associated with tempered commercial alloy steels. A comprehensive and interpretive review of studies on temper embrittlement was carried out by McMahon.[11] Temper embrittlement is encountered particularly with the higher alloying levels required to achieve through-section hardening at large section thickness; the larger ingots involved give rise to greater segregation, and require slower cooling through the embrittlement temperature range of ~375–575°C. Further, embrittlement can also occur during long-term service within this temperature range. Temper embrittlement only occurs when certain impurity elements are present within the steel[12] and segregate to the prior-austenite grain boundaries. The principal embrittling elements (in decreasing order of importance) are antimony, phosphorus, tin and arsenic, in trace quantities. In addition, larger quantities of Si and Mn have been found to have an embrittling effect. The susceptibility increases with increased levels of these elements, and exhibits a C-curve time-temperature dependence. Temper embrittlement is a reversible phenomenon; the toughness of an embrittled material can be restored by heat treatment at a temperature above 600°C. The embrittlement is manifest in a number of ways, viz. the ductile/brittle transition temperature is raised, the fracture path is intergranular with respect to the prior austenite grains, and the prior austenite boundaries may show preferential etching. Whilst there have been many studies carried out to determine the details of the mechanism associated with temper embrittlement, this has yet to be established definitively. A number of compositional parameters have been defined, for both base material and weld metal, to allow an assessment to be made of the susceptibility to temper embrittlement. Examples of these are the Bruscato factor, X,[13] and the 'J' factor proposed by Watanabe.[14]

Ligament Cracking in Headers

The superheater headers of a boiler are used to transport high pressure steam from the boiler to the steam turbine. The temperatures encountered are typically in the region of 540°C, which is within the creep regime for the materials that were traditionally employed, namely 2.25Cr–1Mo and 1.25Cr–1Mo steel. As reported by Viswanathan et al.,[15] and Middleton et al.,[1] the headers were designed for base loading and not for the cyclic operation that was frequently encountered; the latter led to thermal stresses and cyclic loading. This, in combination with the high temperature operation, gave rise to premature cracking in the headers by a creep-fatigue mechanism widely referred to as 'ligament cracking'. The cracks typically occurred around tube bore holes, and often extended, linking adjacent holes. The cracking generally initiated on the inside of the header, and propagated radially into the ligament between holes, and axially into the tube-hole. A thick oxide layer has been frequently associated with the ligament cracks, arising from localised temperatures above those of the bulk material. The occurrence of ligament cracking was reportedly independent of material type or the age of the header, and depended primarily on the thermal/stress history.

When cracking was discovered in a header, the remedial action depended largely on the operating conditions and the extent of the cracking. In some instances, the header was deemed fit for continued service. For more severe cracking, weld repair was carried out or, in extreme cases, the header was replaced. Many of the low alloy steel headers have now been replaced with modified 9Cr1Mo (grade 91) headers. The use of this alloy allows considerable reduction in section thickness by virtue of its higher creep strength. This in turn leads to a decrease in the thermal gradients, reducing the cyclic loading experienced, and thereby reducing the propensity for ligament cracking.

In-service Cracking

For the most part, the in-service problems encountered with the the ferritic alloys widely adopted in high temperature plant have been few. However, many of these alloys, for example 0.5Cr–Mo–V, 2.25Cr–1Mo and modified 9Cr–1Mo (grade 91), are susceptible to Type IV cracking. This cracking occurs at the edge of the heat affected zone adjacent to parent material.[16] The occurrence of cracking is attributed to the development of a localised 'soft zone' in this region, from the weld thermal cycle and postweld heat treatment, giving rise to localised creep deformation under the action of bending stresses. Type IV cracking can result in the weldment creep performance being significantly poorer than that of the base steel. It has been shown[17] that some improvement in performance and reduced propensity for Type IV cracking can be achieved through careful heat treatment procedures, whereby the base material is supplied after only a partial tempering treatment, and the full component subjected to a post-weld stress relieving treatment. This is clearly an approach of limited practical application. An alternative approach is to incorporate a greater safety margin in terms of stress into the design, but this carries an economic penalty.

Dissimilar Metal Joints

In view of the wide variety of materials employed in power plant construction, the need for dissimilar metal joints is inevitable, for example between low alloy steels and modified 9Cr1Mo steel, or between ferritic and stainless steels. However, there is uncertainty in the choice of weld procedure, consumable, and heat treatment schedule to optimise the mechanical properties of the joint. The recommendation given in AWS D1.1[18] is that the filler material should be selected to match the low alloy material and the joint heat treated as for the higher alloy material. However, because of the differences in the Cr content of the materials, carbon diffuses from the low Cr material into the high Cr material or weld metal, giving rise to the formation of a carbon-depleted zone in the low Cr steel (and a carbon-enriched region in the neighbouring material), as shown in Fig. 2. The presence of such regions can result in premature failure due to strain concentration in service. The extent of the carbon migration will depend on the heat treatment temperature and time, and can really only be eliminated by the use of nickel-based weld metal. However, in principle the extent of carbon migration can be reduced by the introduction of a buttering layer onto the higher alloy material, to allow a more gradual compositional gradient. For the welding of modified 9Cr–1Mo (grade 91) to 2.25Cr–1Mo steel, a commonly adopted procedure involves the buttering of the modified 9Cr–1Mo steel with conventional 9Cr weld metal, intermediate PWHT (to temper the hard, potentially brittle high alloy heat affected zone), completion of the main fill, again using conventional 9Cr weld metal, followed by a lower temperature PWHT.

MATERIAL DEVELOPMENTS

With the drive for higher operating temperatures and pressures to improve the thermal efficiency of new plant, as well as the bid to extend the life of existing plant, there have been significant changes in the materials employed. For operation up to 620°C, a new generation of ferritic steels with 9–13%Cr has been developed, containing additions of tungsten (1–2%) to give improved high temperature properties over the traditional modified 9Cr grades. These materials,

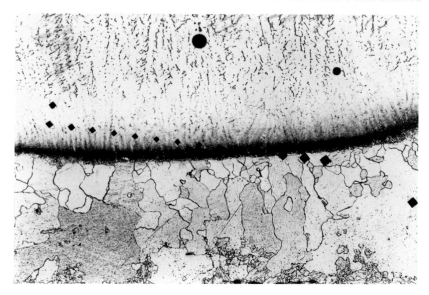

Fig. 2 Optical micrograph of the interface between 2.25Cr 1Mo base material and stainless steel weld metal (309LNb/347) showing clearly the carbon depleted zone in the Cr–Mo HAZ. Vickers hardness indents show hardening at the fusion line and softening in the decarburised region.

like grade 91, are predominantly martensitic, with varying amounts of retained delta-ferrite. Such materials have been the subject of extensive research programmes, for example EPRI project 1403–50[19] and COST 501. A number of these materials now have ASME code case approval, and indeed have now been introduced into actual plant for full-scale trials. However, further long term trials are required and, in view of the difficulty of matching the properties of the cast and PWHT weld metal microstructure with those of the base material, additional welding consumable development is needed before we can expect to see the widespread application of these grades.

The principal grades that have evolved to date are:

E911–as studied in COST 501
NF616–produced by Nippon Steel [A213 T92 / A335 P92]
HCM12A–produced by Sumitomo Metal Industries [A213 T122 / A335 P122]
TB12M–produced by Forgemasters Steel and Engineering Ltd

The compositional requirements of these grades, and typical creep strength values are detailed in Table 1.

These materials offer considerable advantages over conventional grade 91. The use of NF616 (code case 2179), for example, may allow a ~35% increase in allowable stress at 600°C. This in turn permits a decrease in section thickness,

Table 1 Chemical composition of 'new' 9–13%Cr steels.

Element		Grade 91	NF616	HCM 12A	TB 12M	E911
C		0.08–0.12	0.07–0.13	0.07–014	0.10–0.15	0.10–0.13
Mn		0.30–0.60	0.30–0.60	≤0.70	0.40–0.60	0.30–0.60
Si		0.20–0.50	≤0.50	≤0.50	0.50 max	0.10–0.30
S		0.010 max	0.010 max	≤0.010	0.010 max	0.010 max
P		0.020 max	≤0.020	≤0.020	0.020 max	0.020 max
Cr		8.00–9.50	8.50–9.50	10.00–12.50	11.0–11.30	8.50–9.50
Mo		0.85–1.05	0.30–0.60	0.25–0.60	0.40–0.60	0.90–1.10
W		–	1.50–2.00	1.50–2.50	1.60–1.90	0.90–1.10
Ni		0.40 max	≤0.40	≤0.50	0.70–1.0	0.20–0.40
Cu		–	–	0.30–1.70	–	–
V		0.18–0.25	0.15–0.25	0.15–0.30	0.15–0.25	0.15–0.25
Nb		0.06–0.10	0.04–0.09	0.09–0.10	0.04–0.09	0.06–0.10
N		0.030–0.070	0.030–0.070	0.040–0.100	0.04–0.09	0.050–0.080
Al		0.04 max	≤0.040	≤0.040	0.010 max	–
B		–	0.001–0.006	≤0.005	–	–
Sn		–	–	–	0.010 max	–
As		–	–	–	0.010 max	–
Sb		–	–	–	0.005 max	–
Creep strength	600°	94	(115)	(115)	(150†)	(115)
in 10^5 hrs at	650°	50	(60)	(60)	(80†)	(65)
ref20						

* Ref. 20 Orr *et al.*
† 10 000 hr
( ) estimated

and thereby a reduction in weight. Consequently, the through wall temperature gradients will be lowered, giving a reduction in the thermal fatigue loading experienced.

The diagram given in Fig. 3 compares the relative wall thickness for a pipe of 290 mm internal diameter for operation at 557°C, 20 MPa for grade 22, grade 91 and grade 122, although the benefits of the new advanced ferritic steel are even more pronounced at higher temperatures and pressures.[21]

On-going steel developments are now looking at non-austenitic steels that are suitable for service up to 650°C to give further improvement to the thermal efficiency of ultra supercritical power plant.[22] One of the new steels, NF12, designed for boiler application, contains ~12%Cr, ~2.5%W and ~2.5%Co, the addition of cobalt preventing the retention of delta-ferrite in the microstructure.[22,23] A rotor steel, HR1200, has also been developed, intended for use in ultra supercritical turbine rotors for service at temperatures of 620 and 650°C. This steel contains alloying additions of W, V, Nb, Co, and B and a low N content of ~200 ppm. Data generated to date on the material have indicated that it exhibits excellent creep rupture strength, corresponding to that of the precipitation hardened austenitic alloy A286, but with a more favourable (lower) coefficient of thermal expansion. A bolting and blading material has been also

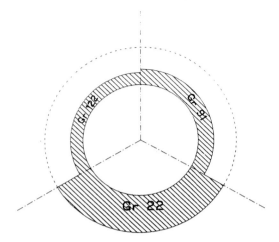

Fig. 3 Schematic showing the variation in wall thickness with material grade (with an internal diameter of 290 mm) for service at 557°C 20 MPa.[21]

been developed[24] TAF650, which possess extremely good high temperature properties, significantly above those of AISI 422. Further work is required to optimise these alloys and their heat treatment, and to develop the associated welding consumables.

WELDING PROCESSES FOR HIGH TEMPERATURE POWER PLANT

Current Position

The welding processes employed in the manufacture of both fossil-fired and nuclear power plant are similar, despite differences in system design and plant components. A common feature, particularly for steam generation and turbine generators, is that a large number of welds, often repetitive in nature, must be manufactured to high quality levels and rigorous quality assurance regimes. For this reason, the welding processes employed have traditionally been based on conventional fusion techniques, predominantly arc welding. An accurate knowledge of the process techniques and the weld properties achievable, particularly at high temperatures, has enabled the required welds to be made consistently, and to an acceptable level of quality. In recent years, welding process developments have been aimed at increasing welding productivity and automation, in order to reduce manufacturing costs and to improve plant reliability. However, progress in their practical exploitation for the manufacture of power plant has been slow. This is due partly to the inherent conservatism that exists within the power generation industry and the reluctance to adopt new technology. Furthermore, the fragmentation of the UK power generation industry has created a commercial climate which has restricted longer-term capital invest-

Table 2 Welding processes for high temperature power plant.

Components	Joint type	Current welding process technology	Possible evolution over next 5–10 years
Boiler			
Boiler panel (small bore tubing)	Fin to tube	Dedicated machine using mechanised/robotic SA and/or high deposition MIG. Resistance butt welding	Higher level of automated arc welding. Possible use of CO_2 and/or YAG laser. Possible use of integral fin/tube extrusion to eliminate fin to tube weld
	Fin to fin	Mechanised SA and/or MIG/FCAW	Higher level of automated arc welding. Possible use of CO_2 and/or YAG laser
	Fin to fin	MMA	Increased use of Orbital TIG including A-TIG. Possible use of YAG Laser. Increased use of Orbital TIG including A-TIG
	Tube to tube	Manual TIG and MMA. Fixed head/Orbital TIG. Flash Butt	
	Site welding	Manual TIG and MMA. Manual/Orbital TIG and MMA	As above
Superheater/reheater/economiser (small bore tubing)	Tube to tube	Fixed/Orbital TIG. Flash butt. Manual TIG and MMA	Possible use of YAG laser
	Spacers & attachments	Manual TIG and MMA. Robotic MIG	Increased use of Orbital TIG, including A-TIG
	Site welding	Manual TIG and MMA. Orbital TIG	
Steam pipework and headers	Butt welds	TIG, MMA and SA Ng welding for thicker sections	Increased use of mechanised arc welding. Possible use of EB, High Vacuum and/or reduced pressure. Possible use of YAG laser in combination with Ng TIG
	Stub to header butt welds	Manual TIG/MMA. Mechanised TIG/MIG	Increased use of mechanised arc welding. Possible use of YAG laser
	Site welding	Manual TIG and MMA. Orbital TIG	Increased use of Orbital and/or Ng-TIG. Manual/Mechanised FCAW

Table 2 (Continued)

Components	Joint types	Current welding process technology	Possible evolution over next 5–10 years
Pressure vessels, e.g. steam drums, mud drum, steam generators etc.	Butt welding	TIG, MMA and SA	Increased use of mechanised welding including narrow gap techniques
Steam Turbines Valve chests	Butt welds	Mainly TIG, MMA and limited SA where facilities exist	Use of Ng, TIG or SA Possible use of EB High Vacuum and/or Reduced Pressure Possible use of YAG laser in combination with Ng TIG
Loop pipe work	Butt welds Site welding	As above Manual TIG and MMA	As above Increased use of Orbital Ng TIG and/or manual/mechanised FCAW
Diaphragms	Butt welds	Manual TIG/MMA and SA	Use of Ng-TIG or SA Possible use of EB High Vacuum Possibly fabricated by Reduced Pressure EB
Rotors		Mainly one piece forgings In some cases intermediate and low pressure rotors may be fabricated using TIG and SA	
Casings		Currently one piece castings	Possibly fabricated by Reduced Pressure EB

ments, such as are generally necessary to implement new welding methods. The current state of welding process technologies for boiler and turbine generator plant, together with their projected future developments, are summarised in Table 2.

Boiler Plant Components

The panels in a utility boiler form the walls of the combustion area and the containment for the products of combustion. The panel itself consists of alternate tubes and flat fin bar fillet welded together on the tube centreline. The fin to tube weld is fabricated using a dedicated automatic welding machine, commonly based on the submerged arc (SA) process. Typical equipment comprises a large frame that carries sets of rolls, which align the tube and fin components. These are driven through the rolls, with welding taking place between the first two sets of rolls and simultaneously on several adjacent joints. Panels are then welded together to form larger widths, by using gantry-mounted or tractor-driven SA machines.[25]

Tube to tube welds in both the boiler panels and superheater/reheater elements are commonly welded by the Tungsten inert gas (TIG) or flash butt processes. The latter welds are less easily inspected by radiography, and in some cases may be precluded from use for this reason. With regard to TIG welding, where practical, girth welds are made by rotating the tube beneath a stationary TIG welding head; the welding process is generally controlled automatically. In cases where the fabricated tube cannot be rotated, the orbital TIG method may be used for thin walled tubes, if adequate clearance for the welding head exists and sufficient numbers of welds are required to justify the initial set-up costs involved. Alternatively, these welds may be completed using manual TIG, and, for thicker walled tubes, manual metal arc (MMA) welding.

Thick section steam pipework and headers have very similar dimensions, and generally employ the same materials, so the manufacturing methods are common to both. The use of TIG root runs and MMA fill runs is typical practice. Where the component can be rotated, and the pipe diameter permits (i.e. > 200 mm OD), the joint can be filled with the SA process. Narrow gap techniques have found increasing use for such applications; the reduced weld volume provides benefits in terms of faster joint completion rates, reduced consumption of consumables and less distortion. The narrow gap TIG process in particular has been found to be suitable for this application, and the low heat input associated with this process is an advantage for welding Cr–Mo–V materials which may be susceptible to a degradation of mechanical properties with the higher heat input processes.

Turbine Generator Components

Several components, including steam chests, diaphragms, and loop pipework, require the welding of very thick section materials. Welding procedures are commonly based on TIG root runs followed by MMA fill passes, but, where practical, and if facilities are available, SA may also be used for the deposition of the fill passes. Whilst most rotors are manufactured as one-piece components, some intermediate and low pressure rotors may be fabricated using the SA process.

Site Welding

Despite the prefabrication of large assemblies in the form of large panels of welded tube wall, with tube/header packs for superheaters, reheaters and economisers, site erection of the boiler still requires extensive in-situ welding, predominantly of tubes. Similarly, installation of turbine generators requires in-situ welding of the high temperature pipework system. Site welding is often required in areas of restricted access and in all positions, so that the preferred processes are normally restricted to manual TIG and/or MMA.

Repair Welding Without Postweld Heat Treatment

In a wide variety of structures, vessels and pipework, defects may be found on final inspection after postweld heat treatment has been carried out. The presence of defects of a sufficient size to warrant repair raises the question of whether it is necessary, after completion of the repair welding, to carry out a further PWHT operation. Relevant fabrication codes may not make provision for repair welding without PWHT. Hence, its omission will normally require the agreement of all interested parties (which is likely to include the certifying authority, insurers, operators, and owners) and that a fitness for purpose analysis be carried out, with data generated in one or more procedure qualification tests. However, recent studies have shown that repair welding without subsequent PWHT will be viable in many situations.

Welding Procedures for Repair without PWHT

In principle, two alternative approaches may be adopted, the half bead technique and the temper bead technique. Each requires the careful placement of a regular arrangement of uniform layers of beads, with each bead laid down so as to overlap the previous bead. The intent is to produce a smooth overall profile for the inner and outer surfaces of the layer, with the substantial overlap of beads giving a high proportion of grain-refined HAZ microstructure, together with a measure of tempering and softening. The extent of bead overlap is important, with an overlap amounting to 50% typically being required. As weld bead shape is influenced by deposition technique[26] and by welding parameters, welding procedures must be devised with care. A second layer is required, to replace any grain-coarsened HAZ with grain-refined HAZ.

In order to permit adequate penetration of a second layer into the first, the half bead technique[27,28] requires the removal, by grinding, of half of the depth of each layer, before the next layer is deposited. These procedures specify particular preheating and post-heating requirements. Largely because of the difficulty in determining, controlling and monitoring the depth of metal removed, this technique now finds little, if any application, having been superseded by the temper bead technique.

The temper bead technique was devised to generate a fine-grained HAZ in low alloy (Cr–Mo–V) steels which suffered temper embrittlement[10,29,30] and reheat cracking sensitivity. The principle of this technique is that a first layer of over-lapped beads is deposited, essentially as for the half bead technique; appropriate penetration of a second layer into the first is achieved by employing a higher arc energy, which is typically 1.5 to 2.5 times that of the first layer.[6,31,32] The in-tention is that a second layer deposited in this way will generate a region of coarse-grained reheated weld metal which is contained within the first layer, with the corresponding grain-refined region re-austenitising any coarse-grained HAZ in the underlying parent steel. In hardenable steels, further layers may be de-posited to temper the grain-refined HAZ. The MMA process is commonly em-ployed, but MAG welding[32] and an automated TIG repair procedure[33] have also been employed. Where ferritic consumables are used, preheating is commonly required when welding thicker sections, particularly in the more hardenable steels. However, nickel-base electrodes have been used without preheat.[34]

Residual Stresses

While it is commonly assumed that yield magnitude residual stresses are present in the weld and HAZ of an as-welded repair, the actual levels are not necessarily always of this magnitude.[35] Nevertheless, consideration must be given to the consequences of any residual stresses for the defect tolerance of the newly-re-paired fabrication, particularly when it is first subjected to its service loading. Where the fabrication operates at elevated temperature, and where C–Mo or Cr–Mo steel consumables are employed, some relaxation of residual stresses may occur during early service, particularly if the lower carbon variants are chosen. Clearly, the extent of any residual stress reduction will depend on the service temperature and on the parent steels and weld metals employed.[35] It is beneficial to use consumables which overmatch (the yield strength of the parent steels) by as little as possible.[32] Hrivnak *et al.*[36] described the relief of residual stresses in repair welds by over-stressing, with the optimum temperature for the pressure cycle being determined by evaluating the effect of test conditions on fracture toughness.

Code Requirements

Several National standards and procedures devised by the major utilities now permit post-weld heat treatment to be omitted, provided that certain criteria are met. Such National codes include BS 1113,[37] BS 2633,[38] ASME VIII,[39] and ASME XI.[40] In addition to restrictions on the parent material types and wall thickness, the limitations imposed include, variously, that the fabrication was previously given a PWHT, that the repair does not exceed a particular depth, and that particular preheat and post-heat requirements are met. For example, in the UK, post-weld heat treatment is not necessary in new pipe-work in C and C–Mn steels, and 2.25Cr–1Mo steel,[38] provided that certain criteria are met. As noted above, for the half-bead technique, ASME XI[27] specified particular preheat and post-heat requirements. However, this code now allows repair by the temper bead technique,[40] followed by post-heating. In the National Board Inspection Codes (NBIC) in the USA, covering repair after service exposure, the 1977 issue was the first to include weld repair procedures for which PWHT could be omitted, for C, C–Mn, C–Mn–Si, C–0.5Mo and 0.5Cr–0.5Mo steels. As discussed by Doty,[41] in the light of data provided by several research programmes, the 1995 issue extended the list of steels to include ASME P4 and P5 (Cr–Mo) steels. In view of the high cost, and in some cases impracticability, of carrying out PWHT after repair, it is likely that similar changes will be adopted more widely, as adequate supporting data are generated.

DEVELOPMENTS IN WELDING PROCESSES

Arc Welding

The most likely trend in manufacturing methods for power generation plant over the next decade will be the increased use of mechanised and automated welding based on the more conventional process, i.e. TIG, MIG/FCAW and SA welding. The driving force for this evolution will be the increasing need to reduce manufacturing costs and to increase plant reliability. A further factor will be the declining availability of skilled and fully qualified welders. There is presently a major shortage of such personnel, and the current welder training schemes which are aimed at giving the individual specific skills and not the broad skill range required of competent pressure vessel welders, will not make up this shortfall.

In addition to increased mechanisation, it is envisaged that developments in process techniques and consumables will also be exploited.

The introduction of vacuum-packed MMA low alloy steel electrodes with moisture-resistant electrode coatings has imparted significant benefits, especially for site welding. Very low weld metal hydrogen contents can be guaranteed, without the need for electrode baking before welding, thereby alleviating significantly the quality control requirements.

A significant development of the TIG process, originating from the Paton Welding Institute, Ukraine, is the use of activated fluxes which result in a significant enhancement of weld penetration.[42] Active fluxes are available commercially for a range of carbon and stainless steels. The process utilises a square edge preparation and no filler wire addition is necessary. The active flux is applied in the form of a paste which is painted onto the joint surface or by means of an aerosol spray.

During welding, the active flux causes a constriction of the arc, and the resultant concentrated heat source gives a 2 to 3 fold increase in penetration. In practical terms, welding of up to 10 mm thick steel in the flat position can be achieved in a single pass, and up to 6 mm thick is possible in-situ. Furthermore, the process appears to be unaffected by compositional variations which can arise between parent materials of the same specification.

The major application of the A-TIG process for power plant is tube to tube welding, for which significant improvements in welding productivity is envisaged. A-TIG welding can be carried out in a single pass on up to 6 mm thick material in all positions. In comparison, conventional TIG welding requires the use of a prepared bevel preparation, and up to six weld passes to complete the joint. Furthermore, the process is simple to apply, and requires no specialised equipment.

The gas-shielded flux-cored arc welding process is currently being developed for the welding of low alloy steel steam pipework as a more productive alternative to MMA welding.[43,44] A key requirement is to validate the long-term high temperature performance of FCA weldments. An extensive programme of creep rupture testing is currently underway amongst the principal UK power generators, to characterise their performance in comparison with MMA welds. Rutile flux-cored wires are preferred because of their superior positional welding characteristics. Further, preliminary studies on C–Mn structural steels at TWI[45] involving controlled thermal severity (CTS) testing have indicated that, for a given weld hydrogen level, rutile flux cored wires, may show a reduced sensitivity to HAZ hydrogen cracking compared to baisc and metal cored wires. The application of rutile flux cored wires to thick section pipe welds appears to be most attractive where the metal deposition rate advantages can be fully exploited.

The increased use of narrow gap welding, particularly using TIG or SA for thick-sectioned butt joints, is anticipated. Improved welding procedures, together with more reliable welding equipment, make it possible to consistently manufacture high quality welds at higher production rates compared to conventional welding. Even unsophisticated joint tracking systems can give good assurance of sidewall fusion. The simple joint geometry involves the repeated deposition of similar weld passes, making it a relatively easy task to automate. Narrow gap orbital TIG welding systems have also been developed, which are also suited to site applications. However, the undoubted advantages in terms of reduced welding time must be offset against the equipment set-up and more accurate requirements for pipe alignment, which can be difficult and time consuming to achieve in site conditions.

A number of sensor technologies have been developed for welding applications, in order to facilitate automated process control. Whilst the majority of these suffer from insufficient resolution and lack of robustness, the use of arc-based and vision-based systems has become well-established for joint tracking purposes. The latter approach, in the form of either optical or infra-red cameras, is being further developed to achieve top face weld penetration control. Changes in the weld pool geometry, and associated thermal profiles, can be used to monitor changes in weld penetration. Real time adjustment in welding parameters can then be made to maintain constant weld geometries. The introduction of such control systems is expected to have major impact on the reliability of automated welding systems. Future welding systems may be able to minimise the incidence of weld defects via the intelligent control of welding parameters, thereby reducing the need for traditional post-weld non-destructive examination.

Laser Welding

Lasers are attractive manufacturing tools for a range of welding, cutting and heat treatment tasks, due to their associated reduced manufacturing costs and increased flexibility. For welding applications, the primary advantages are controlled and predictable distortion, high joint completion rates and consistent, reliable weld quality, although there will be significant capital outlay. At present, two main laser types are available for materials processing. These are the CO_2 gas laser and the Nd–YAG solid state laser. CO_2 lasers with powers up to 15 kW and Nd–YAG lasers with powers up to 4 kW are currently being used in production. Although power is limited, the Nd–YAG laser beam has the advantage of being able to be delivered to the workpiece by flexible optical fibres, at least up to 200 m long, giving substantial production advantages over CO_2 laser beams which can only be manipulated via moving mirror systems. The Nd–YAG laser is thus suited to difficult access or remote applications. Furthermore, the fibre optic can also be easily attached to robotic systems for three-dimensional manipulation.

Recent work has resulted in the commercial availability of a 5 kW Nd–YAG laser source, with even higher power systems under development. With this laser, welds in 10 mm thick steel can be made, as well as high speed processing of thin materials. This new high power laser technology, and expected future developments, will considerably expand the opportunities for laser exploitation in the manufacture of power generation plant.

For boiler plant fabrication, the fin to tube weld is well suited to laser welding. Low distortion, high speed welding should be possible. Laser welding, in combination with narrow gap-TIG for example, could also offer production advantages for thick section weldments, including pipework, headers, and turbine generator components. However, its main applications in the power generation industry, at present, is for remote repair operations.

Electron Beam Welding

Electron beam welding is a mature technology which has been used extensively in industry for more than 40 years. Conventionally, the process is operated at a vacuum pressure of better than 5×10^{-3} mbar, which dictates that the component to be welded must be wholly contained in a vacuum chamber. To an extent this has precluded the application of the process for the fabrication of large structures, although many components in the power generation industry, including rotors, turbine diaphragms and steam generating plant are routinely welded using electron beam at high vacuum outside of the UK. Attempts to use local sealing at high vacuum levels have been largely unsuccessful in the past, due to difficulties with leaks in the local sealing causing fluctuations in the vacuum pressure, and subsequent poor weld quality.

Recent developments at TWI in electron beam welding equipment technology have resulted in a system which can be operated at a pressure in the range 0.1–10 mbar. In consequence, this system can be employed for welding large structures by using local vacuum sealing and a simple pumping arrangement. To date, the equipment at TWI has been used extensively for welding copper of over 70 mm thickness for sealing nuclear waste containers. Furthermore, it has been demonstrated that steel of 150 mm thickness and aluminum alloys of 70 mm thickness can be welded reliably. Nevertheless, the experience of welding alloy steels, and in particular stainless steels, is limited and the full performance capability of the process has yet to be proved.

Clearly there are many power plant applications which could benefit from the use of a low heat input, low distortion welding process which is capable of producing narrow parallel-sided welds of high integrity in a single pass. In particular thick section pipework, and header and steam chest components are viewed as candidate applications for this process. Furthermore, in the future, rotors and casings, which are currently manufactured as one-piece forgings, could be fabricated by utilising electron beam welding, with minimal requirements for subsequent machining.

In common with laser welding systems, the application of electron beam welding will require significant capital investment. However, the present demands to achieve short term payback on such investments, together with the uncertainties regarding the volume requirements for power plant over the next decade or so, are likely to be major obstacle to the practical exploitation of power beam technologies.

AREAS FOR FUTURE RESEARCH

Welding Behaviour

It is axiomatic that the application of advanced welding procedures, as considered above, will require definition of conditions under which sound joints can be reliably and economically produced. In the general sense, the various forms of cracking which occur as a result of welding are well-defined, but it should be recognised that appropriate preventative measures may limit the welding procedures applied. To exemplify the point, weld metal solidification cracking is likely to occur if high travel speed is employed, because the resultant elongated weld pool will promote solidification directly towards the centreline, with enhanced segregation of impurity elements in the region last to solidify. Further, it is known that modified 9Cr steel weld metal is sensitive to the formation of small 'hot cracks'. The problem has been encountered in deposits produced by both flux and gas-shielded processes, under a range of welding conditions. The effect is apparently a consequence of the specific material compositions employed, rather than the solidification structures developed, and unpublished work at TWI on T91 material and higher chromium alloys has indicated a link with the presence of niobium. However, further study is required to identify more clearly the elements primarily responsible so that joint completion rate can be maximised for high temperature alloys.

Weldment Properties

While creep resistance represents the sine qua non of advanced ferritic steels for high temperature power plant, it is essential also that materials display sufficient toughness at normal ambient temperature for pressure testing to be safely carried out. In large part, this has not been a particular problem with low alloy grades, but the situation is a little different with, for example, the modified 9Cr alloys. Most fabrication codes require toughness to be assessed by an impact test method, and considerable consumable development has been necessary to obtain weld metal composition such that typical code requirements can be met. In general, impact toughness is not seen as a particular problem. However, it has become apparent from fracture mechanics tests on pre-cracked samples that modified 9Cr weld metals can display an appreciable tendency to 'pop-in' behaviour, with initiation and arrest of local brittle cracks during testing. The behaviour is influenced by material analysis, and further study is required to clarify the optimum consumable composition. Work to date indicates that changes such as reduced niobium or increased manganese relative to base steel are of benefit in respect of toughness, but such changes are not necessarily consistent with achieving optimum creep strength. It is not known how far similar trends will be observed with other more recently developed alloys, and appropriate fracture mechanics testing is required.

Post-weld heat treatment procedures for high chromium creep-resisting steels were originally developed on the basis of achieving a reasonable degree of weld area softening from the as-welded condition, with the formation of a carbide distribution expected to be conducive to good creep properties. There is no doubt that the precise thermal cycle employed for post-weld heat treatment can appreciably influence the weld area toughness, yet there has been surprisingly little attention paid to optimising heat treatment conditions.

In essence, design of high temperature plant is based on control of the operative stress level to achieve the required life under creep conditions. Depending on the fabrication code involved, a factor may be applied to recognise the presence of welded joints. Creep ductility is not explicitly recognised. Nevertheless, it is desirable that weld metals display a reasonable capacity for accepting strain, for example, so that incipient failure from upset conditions or other cause can be identified before complete wall penetration takes place. Work on austenitic stainless steels has shown that inclusions in arc weld metal can appreciably reduce the rupture ductility, but there has been negligible analogous study on high temperature ferritic grades.

Service Behaviour

The creep resisting steels are normally supplied and welded in the tempered condition. The welding operation induces an additional thermal cycle, and can promote additional tempering in the subcritical area beyond the transformed heat affected zone. Even further tempering can take place on post-weld heat treatment, leading to a region on the periphery of the weld that may be of appreciably lower hardness and creep strength than other regions. The consequences of this are manifest as 'Type IV' cracking in service, as discussed earlier. Further study of all factors contributing to this cause of premature service failure is required. However, the effect is clearly associated with the presence of local soft zones in welded joints prior to entering service, and particular attention is needed to the significance of welding conditions and the resultant thermal cycles, and to the specific post-weld heat treatment procedure employed.

In completed plant, modified 9Cr and similar steels will be utilised only for the high temperature regions, with lower alloy grades employed elsewhere for reasons of economy. It will therefore be necessary to make a dissimilar metal joint between the two classes of steel at some point in the unit. When steels of varying chromium content are welded and exposed to elevated temperatures either during post-weld heat treatment or in service, carbon migration takes place towards the higher alloy side in consequence of the lower carbon activity. The immediate consequence of this is that a decarburised zone is formed in the lower alloy material, with potentially much reduced properties relative to the unaffected base metal. The phenomenon of carbon migration in dissimilar metal joints has been well studied for welds between austenitic and ferritic steels, but much less information is available for welds between ferritic/martensitic alloys of different composition. It has been shown that development of a decarburised zone does not follow a direct parabolic relationship with time,[46] as would be expected if diffusion alone were controlling, but the rate of decarburisation tends to diminish with increased time. This is a consequence of carbide precipitation on the high alloy side and depletion of the matrix alloy content and hence the driving force for carbon diffusion. However, at present a comprehensive model has not been developed, and it is therefore difficult to predict the extent and consequences of carbon migration at dissimilar metal joints between steels of similar metallurgical structure.

CONCLUDING REMARKS

The family of Cr–Mo steels assume an essential role in power plant construction. A summary has been presented of the metallurgy of these steels, and some of the fabrication and service problems that have been encountered. Historically, power plant fabrication has favoured the use of conventional arc welding techniques, with considerable success, although there is an emerging trend for increased automation and the use of flexible, high deposition rate processes. With the drive for improved thermal efficiency, reduced emissions and cost control, we are clearly seeing the start of many exciting changes in the fields of both materials and welding technology. Indeed, we have already seen the widespread introduction of modified 9Cr 1Mo steel (grade 91) to replace 2.25Cr1Mo components during both the upgrading of existing plant and in new plant construction. Such changes, and the advent of further new high temperature alloys are opening up further new challenges and research and development opportunities to be met by materials scientists and welding engineers.

REFERENCES

1. C.J. Middleton, R. Timmins and R.D. Townsend: 'The integrity of materials in high temperature components; Performance and life assessment', *Int. J. of Pres. Ves. and Piping*, 1996, **66**, 33–57.

2. A. Dhooge and A. Vinckier: 'Reheat cracking – a review of recent studies', *Int. J. of Pres. Ves. and Piping*, 1987, **27**(4), 239–269.

3. C.F. Meitzner: 'Stress relief cracking in steel weldments', *Welding Research Council Bulletin*, November 1975, 1–17.

4. A. Dhooge *et al.*, 'Review of work related to reheat cracking in nuclear reactor pressure vessel steels', *Int. J. of Pres. Ves. and Piping*, 1978, **6**, 329–409.

5. A.G. Vinckier and A.W. Pense: 'A review of underclad cracking in pressure vessel components', *Welding Research Council Bulletin 197*, August 1974, 1–35.

6. L.M. Friedman: 'EWI/TWI controlled deposition repair welding procedure for 1.25%Cr-0.5%Mo and 2.5%Cr-1%Mo steels', *Welding Research Council Bulletin 412*, June 1996, 27–34.

7. W.H.S. Lawson, P.Y.Y. Maak and M.J. Tinkler: 'In-situ repair of chrome-moly-vanadium turbine cylinder and valve chest cracks', *Ontario Hydro Research Division Report CEA G170*, Canadian Electrical Association, Montreal, August 1982.

8. J. Myers and J.N. Clark: 'Influence of welding procedure on stress-relief cracking. Part 2: Repair evaluation in Cr-Mo-V steels', *Metals Tech.*, 1981, **8**(10), 389–394.

9. B. Chew and P. Harris: 'HAZ refinement in CrMoV steel', *Metal Con.*, 1979, **11**(5), 229–234.

10. P.J. Alberry, J. Myers and B. Chew: 'An improved welding technique for HAZ refinement', *Welding and Metal Fab.*, 1997, **45**(9), 549–553.

11. C.J. McMahon Jr: 'Temper Embrittlement - An Interpretive Review', Temper Embrittlement in Steel, ASTM STP 407 American Society for Testing and Materials 1968, 127–167.

12. W. Steven and K. Balajiva: 'The influence of minor elements on the iso-thermal embrittlement of steels', *J. Iron and Steel Inst.*, 1959, **93**, 141–147.

13. R. Bruscato: 'Temper embrittlement and creep embrittlement of 2¼Cr-1Mo shielded metal arc deposits', *Weld. J. Res. Suppl.*, 1970, **49**(4), 148s–156s.

14. Y. Murakami, T. Nomura and J. Watanabe: 'Heavy section 2¼Cr-1Mo steel for hydrogenation reactors' Applications of 2¼Cr-1Mo steel for thick wall pressure vessels', *ASTM Special Technical Publication 755*, 1982, 383–417.

15. R. Viswanathan, M. Berasi, J. Tanzosh and T. Thaxton: 'Ligament cracking and the use of modified 9Cr1Mo alloy steel (P91) for boiler headers', Proc 1990 Pressure Vessels and Piping Conf. on *New alloys for Pressure Vessels and Piping*, (PVP vol 201, MPC vol 31), 97–109.

16. R.T. Townsend: 'CEGB experience and UK developments in materials for Advanced fossil plants' Proc. Conf. on *Advances in materials technology for fossil fuel power plants*, Chicago, Illinois, 1-3 September 1987, R. Viswanathan and R.L. Jaffee, Eds, ASM 1987, 11–20.

17. B.W. Roberts and D.A. Canonico: 'Candidate uses for modified 9Cr1Mo steel in an improved coal fired plant', Int. Conf. on *Improved coal fired power plants*, EPRI '88, 5–203.

18. AWS D1.1-96: *Structural welding code - Steel*, American Welding Society, 1996.

19. 'New Steels for Advanced Power Plant up to 620°C', Proc EPRI/National Power Conference, May 1995.

20. J. Orr, L. Woollard and H. Everson: 'The development and properties of a European high strength 9CrMoNbVWN steel - E911', Proc Int. Conf. on *Advanced Steam Plant*, IMechE, Supplement paper 3, London, 21–22 May 1997.

21. Z. Yang, M.A. Fong and T.B. Gibbons: 'Steels for thick section parts: comparison of economics of usage in a typical design', Proc EPRI/National Power Conference, London, 11 May 1995, 174–183.

22. T. Fujita: 'Future ferritic steels for high temperature service', Proc EPRI/National Power Conference, London, 11 May 199, 190–200.

23. H. Naoi *et al.*: 'Mechanical properties of 12Cr-W-Co ferritic steels with high temperature creep rupture strength', *Proc. 5th Int. Conf. on Materials for Advanced Power Engineering*, Liege, October 1994.

24. T. Fujita: 'Heat resistant steels for advanced power plant', *Advanced materials and processes*, 1992, **141**(4) 42–47.

25. G. Mathers: 'Recent development in tube and boiler panel welding'. Paper 27 in The Welding Institute Conf. *Advanced Welding Systems*, November 1985.

26. D.J. Allen and C. Earl: 'The effect of overlap and deposition technique on weld bead shape', *Metals Technology*, 1984, **11**(6), 242–248.

27. ASME Boiler and pressure vessel code, Section XI "Rules for service inspection of nuclear power plant components", paragraph IWA-4513 Repair Welding Procedures A-SMA, The American Society of Mechanical Engineers, New York, July 1992.

28. M. Higuchi, H. Sakamoto and S. Tanioka: 'A study on weld repair through half bead method', *IHI Engineering Review*, 1980, **13**(2),14–19.

29. P.J. Alberry and T. Rowley: 'Repair of Ince "B" power station reheat pipework', *Central Electricity Generating Board Report R/M/N 982*, 1978.

30. P.J. Alberry and K.E. Jones: 'Two-layer refinement techniques for pipe welding', Second Int. Conf. on *Pipe Welding*, The Welding Institute, London, November 1979.

31. R.L. Jones: 'Development of two-layer deposition techniques for the manual metal arc repair welding of thick C-Mn steel plate without post-weld heat treatment', *The Welding Institute Research Report 335/1987*, April 1987.

32. N. Bailey: 'Prospects for repair of thick sectioned ferritic steel without post-weld heat treatment', Proc. 5th Int. Symp. of the Japan Welding Society on *Advanced technology in welding materials processing and evaluation*, The Japan Welding Society, Tokyo, 17–19 April 1990.

33. G.W. Gandy, S.J. Findlan and W.J. Childs: 'Repair welding of SA-508 Class 2 steel utilising the 3-layer temperbead approach', Proc. Conf. on *Fatigue, fracture and risk – 1991*, American Society of Mechanical Engineers, San Diego, 23–27 June 1991, 117–122.

34. S.J. Brett: 'The long-term creep rupture of nickel-based "cold" welds in Cr–Mo–V components' (Paper C288/87), Proc. Conf. on *Refurbishment and life extension of steam plant*, Institution of Mechanical Engineers, London, 14–15 October 1987, 253–260.

35. R.H. Leggatt and L.M. Friedman: 'Residual weldment stresses in controlled deposition repairs to $1\frac{1}{4}$Cr-$\frac{1}{2}$Mo and $2\frac{1}{4}$Cr-1Mo steels', Proc. Conf. on *Pressure vessels and piping*, ASME, Montreal, 21–26 July, 1996.

36. I. Hrivnak, J. Lancos, S. Vejvoda and P. Bernasovsky: 'Relaxation over-stressing of huge spherical storage vessels repaired by welding', Proc. 3rd International Conf., on *Joining of Metals* (JOM-3) O.A.K. Al-Erhayem ed., Helsingor, 19–22 December 1986, 336–343.

37. BS 1113: 1992, 'Design and manufacture of water-tube steam generating plant (including superheaters, reheaters and steel tube economisers)', the British Standards Institution.

38. BS 2633: 1987, 'Arc welding of ferritic steel pipe-work for carrying fluids', the British Standards Institution.

39. The American Society of Mechanical Engineers Boiler and pressure vessel code, Section VIII Division 1, '1995 ASME Boiler and Pressure Vessel Code Rules for construction of pressure vessels'

40. ASME Boiler and pressure vessel code, Section XI 'Rules for service inspection of nuclear power plant components', Paragraph IWA-4633.1 Alternative repair welding methods – Shielded metal arc welding, The American Society of Mechanical Engineers, New York, July 1995.

41. W.D. Doty: 'History and need behind the new NBIC rules on weld repair without PWHT', *Welding Research Council Bulletin 412*, June 1996, 3–8.

42. W. Lucas and D.J. Howse: 'Activating flux-increasing the performance and productivity of the TIG and plasma processes', *Welding and Metal Fab.*, 1996, **64**(1), 11–17.

43. D.J. Allen: 'Developments in materials and welding technology for power generation', *Welding and Metal Fab.*, 1996, **64**(2), 70–75.

44. W. Lucas: 'Shielding gasses for arc welding -part 1', *Welding and Metal Fab.*, 1992, **60**(5), 218–225.

45. A.J. Kinsey: 'Heat affected zone hydrogen cracking of C-Mn steels when welding with tubular cored wires', *TWI Members Report 578/1996*, November 1996.

46. J.M. Race and H.K.D.H. Bhadeshia: 'Precipitation sequences during carburisation of Cr-Mo steel', *Materials Science and Technology*, 1992, **8**(10), 875–882.

Material Data Requirements for Assessing Defect Integrity at High Temperatures

S. R. HOLDSWORTH

ALSTOM Energy Ltd, Steam Turbines, Newbold Road, Rugby CV21 2NH, UK

1 INTRODUCTION

The material properties required for the design and assessment of present and future high temperature power generation and process plant are wide ranging and no longer limited to traditional uniaxial deformation and failure data. For example, during the past thirty years, high temperature fracture mechanics property data have evolved from being essentially the product of forward-thinking industrial R&D activities to playing an integral part in the fitness-for-purpose evaluation of many critical components.[1-3]

Fundamental design calculations still largely rely on physical properties and uniaxial descriptions of mechanical monotonic, creep and fatigue properties. Nevertheless, the consideration of defect acceptability using fracture mechanics property data is now an essential design assessment activity. For plant operation, high temperature crack growth and fracture toughness properties are an integral part of cost effective in-service inspection management, and the assessment of remaining life and life extension feasibility.

Design and in-service defect assessments of high temperature components require a knowledge of material resistance to crack propagation as a consequence of steady and/or cyclic loading. Moreover, the adopted assessment methodology invariably depends upon the deformation and fracture characteristics of the material, the defect size with respect to ruling section dimensions and the stress state. The material data requirements for assessing high temperature integrity and the current means of representing crack growth properties for this purpose are reviewed.

High temperature fracture mechanics property data are available for a number of the low alloy ferritic and austenitic stainless steels commonly used in the power generation and process plant industries during the past 30 years. Some of this information is referred to for illustrative purposes. However, the present paper focuses mainly on the data available for the advanced high temperature alloys now being exploited for present and future plant applications in these industrial sectors.

2 DEFECT ASSESSMENT

The assessment of defect integrity at high temperatures requires a knowledge of the solutions for a number of parameters characterising local and global states of

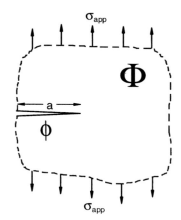

Fig. 1 Terminology used to define applied stress (σ_{app}), defect size and shape (a and ϕ), and component geometry (Φ).

stress (and strain) associated with the flaw in the geometry under investigation.[1-3] The stress state parameters referred to in this review are simply defined in the following section, primarily as a means of introducing the terminology used for assessing defect integrity.[†]

The stress intensity factor, K, is used to characterise the state of stress at the crack tip for linear elastic loading conditions. The parameter is a function of the remotely applied stress, the dimensions and shape of the defect, and the component geometry (Fig. 1), ie.

$$K = \sigma_{app} \cdot \sqrt{a} \cdot Y(a, \phi, \Phi) \tag{1}$$

where solutions for the geometry function, $Y(a, \phi, \Phi)$, are widely available.[4]

Under non-linear loading conditions, the states of stress and strain at the crack tip are defined in terms of a contour integral referred to as J. In addition to being a function of the applied stress, the dimensions and shape of the defect, and the component geometry, J requires a knowledge of the flow characteristics of the material,[5] ie.

$$J = \sigma_{app}^{\beta+1} \cdot a \cdot k' \cdot H(a, \phi, \Phi, \beta) \tag{2}$$

where

$$\varepsilon = \sigma/E + k' \cdot \sigma^{\beta} \tag{3}$$

[†]the terminology used is defined in Section 7

The contour integral most widely used to characterise crack tip states of stress and strain rate at elevated temperatures is the C^* parameter,[6] ie.

$$C^* = \sigma_{app}^{n+1} \cdot a \cdot B \cdot G(a, \phi, \Phi, n) \tag{4}$$

The C^* parameter requires a knowledge of the time dependent deformation characteristics which, for simplicity in eqn 4, are represented by the constants in a creep law such as

$$\varepsilon_C = B \cdot \sigma^n \cdot t^p \tag{5}$$

Plastic collapse and reference stresses are represented by the parameter defined in eqn 6.

$$\sigma_{ref} = \sigma_{app} \cdot a \cdot Q(a, \phi, \Phi) \tag{6}$$

where solutions for $Q(a,\phi,\Phi)$ are available in the literature.[7]

The common feature of the stress state parameters listed is that they are all functions of applied stress and defect size in conjunction with a factor characterising defect/component configuration. The C^* parameter is similar but is also a function of deformation rate.

Defect assessment is typically performed to evaluate the conditions under which flaw(s) in a component will not exceed the critical size responsible for unstable fracture (or a size which will propagate) during service (Fig. 2).

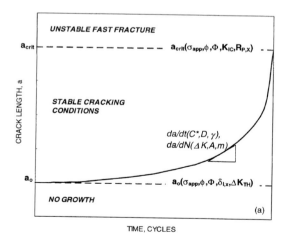

Fig. 2 Material data requirements for fracture assessment in terms of (a) crack size versus component life and (b) a failure assessment diagram.

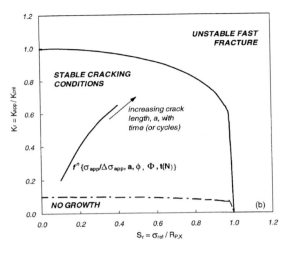

Fig. 2b

3   CRITICAL FLAW SIZE

3.1   LEFM Toughness Properties

Irrespective of whether defect assessment is part of a design or remaining life evaluation, the critical event to be avoided is component failure. At low temperatures, critical defect sizes are almost invariably established, at least in part, on the basis of a linear elastic fracture mechanics (LEFM) analysis. The key material property is the fracture toughness parameter, K_{IC}, and critical defect size is essentially determined by re-arranging eqn 1, ie.

$$a_{crit} = (K_{IC}/\sigma_{app})^2 / Y(a, \phi, \Phi) \tag{7}$$

Values of K_{IC} for a number of high temperature materials are summarised in Fig. 3.[8–12] Creep resistant low alloy ferritic and martensitic stainless steels typically exhibit a transition from low to high toughness involving a fracture mechanism change at temperatures close to ambient. Most of the data for these alloys are available in this low temperature regime, primarily because the critical loading conditions for many high temperature components are encountered during start-up and/or shut-down, rather than during operation. Another reason is that it becomes increasingly difficult to meet the K_{IC}/K_{ICJ} validity requirements[13] for these alloys at temperatures above 100°C. Nevertheless, there is data to demonstrate that the toughness of 1CrMoV rotor steel maintains its upper shelf level to ~400°C.[8] At temperatures above 450–500°C, toughness deteriorates in terms of the LEFM parameter (Fig. 3). Austenitic stainless steels and nickel base alloys do not exhibit a fracture mechanism transition and retain their toughness until the creep range is approached.

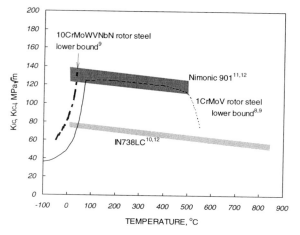

Fig. 3 Variation of fracture toughness with temperature for a range of high temperature alloys.

The toughness variations shown in Fig. 3 reflect the properties of the virgin material. In service, the ambient temperature toughness of certain alloys can deteriorate with time at certain elevated temperatures. For example, the maximum deterioration in toughness occurs after long time service at 480°C in 1CrMoV steels, whereas it occurs at 600°C in the newer advanced 10Cr steels and Type 316 stainless steels and at 650–750°C in Ni base superalloys (Fig. 4). In the case of the low alloy ferritic steel, the deterioration in toughness is due to a

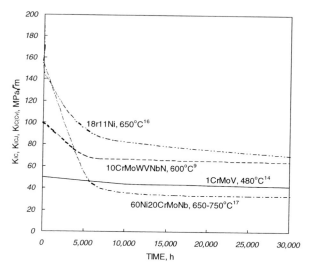

Fig. 4 Variation of ambient fracture toughness with time at temperature for a range of high temperature alloys.

temper embrittlement process resulting from the migration of phosphorous to grain boundaries.[14] The higher alloyed steels are not susceptible to temper embrittlement, but experience a loss in toughness with time at the higher temperature due to the formation of Laves phase.[9,15,16] Loss in toughness due to ageing in Ni base superalloys occurs due to the transformation of gamma prime precipitate to delta Ni_3Cb at grain boundaries.[17]

3.2   Plastic Collapse Properties

The strength/toughness characteristics of many strategic materials are such that plastic collapse is also an important overload condition to be assessed and, in the limit, critical flaw sizes are determined on the basis of yield strength or an appropriate flow strength, $R_{P,X}$,[1] ie.

$$a_{crit} = (R_{P,X}/\sigma_{app})/Q(a, \phi, \Phi) \tag{8a}$$

Modern defect assessment procedures establish critical flaw sizes on the basis of a two criteria approach combining the results of linear elastic and plastic collapse analyses[1,2] (eg. Fig. 2b). The ambient temperature failure envelopes for a number of high temperature materials are shown in Fig. 5. At an advanced level, these procedures require not simply a knowledge of $R_{P,X}$, but also the monotonic flow properties of the material, eg. eqn 3.

At higher temperatures, the applicability of K_{IC} and/or $R_{P,X}$ becomes limited by the increasing influence of time dependent deformation behaviour. In the limit, critical defect sizes are determined by the creep-rupture properties of the material in a reference stress calculation,[3] ie.

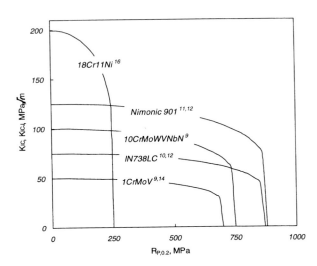

Fig. 5 Ambient temperature failure envelopes for a selection of high temperature materials.

$$a_{crit} = (R_{r/T/t}/\sigma_{app})/Q(a, \phi, \Phi) = R_{r/T/t}/\sigma_{d,ref} \tag{8b}$$

On this basis, two-criteria approaches are being developed for high temperature assessment.[18,19]

There are a number of sources of creep-rupture properties for engineering materials.[20,21]

Having determined the critical flaw size, defect acceptance standards for components entering high temperature service may be established on the basis of permissible-growth or no-growth during the design/remaining lifetime.

4   PERMISSIBLE CRACK GROWTH

Permissible-growth assessments are performed when sub-critical crack extension is tolerable during component lifetime. In this case, the objective is to ensure that the initial defect size, a_o, does not extend to the critical defect size, a_{crit}, during the design (or remaining) life of the component (Fig. 2). The properties required for permissible-growth assessments are those which characterise the sub-critical crack growth behaviour of the material for the appropriate loading conditions. Sub-critical crack growth from pre-existing defects in high temperature components may occur as a consequence of steady and/or cyclic loading.

4.1   Crack Extension due to Steady Loading

4.1.1   Creep Crack Growth Properties
Creep crack growth (CCG) properties are most commonly presented as $da/dt_C(C^*)$ data plots (eg. Fig. 6). These are similar in appearance to the more familiar and characteristic $da/dN_F(\Delta K)$ fatigue crack growth profile (see Section 4.2.1) and, like $da/dN_F(\Delta K)$ in the mid-ΔK regime, $da/dt_C(C^*)$ in the Stage-II CCG regime is relatively insensitive to component geometry, such that:

$$da/dt_C = D(\varepsilon_r) \cdot (C^*)^{\gamma} \tag{9}$$

and $D(\varepsilon_r)$ and γ are regarded as material characteristics, neither of which in practice are strongly dependent on temperature. Alternatively, $D(\varepsilon_r)$ may be replaced by $D'/\varepsilon_{r/T/t}$, where $\varepsilon_{r/T/t}$ is the creep rupture ductility for a given temperature and time.[23] The transition from Stage-I to Stage-II CCG typically occurs after $\Delta a \sim 0.5$–1 mm creep crack extension.[22] Stage-II CCG scatter bands for a spectrum of high temperature alloys are compared in Fig. 7 showing, for example, the superior creep crack growth resistance of the new martensitic stainless steels at 600°C[24] relative to that of 1CrMoV rotor steel at 550°C.[22]

An important difference from the characteristic $da/dN_F(\Delta K)$ fatigue crack growth profile is that the 'tail' or Stage-I regime of the $da/dt_C(C^*)$ profile for an individual test does not represent a true creep crack growth C^* threshold (Fig. 6), but the development at C_o^* of crack tip damage to the formation of an

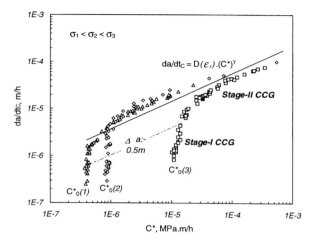

Fig. 6 Creep crack growth rate regimes illustrated with data from three tests on a 1CrMoV rotor steel at 550°C.[22]

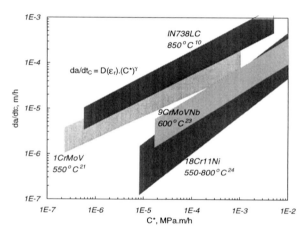

Fig. 7 Stage-II CCG scatter bands for a range of high temperature alloys.

engineering size crack and the start of Stage-II growth.[22] In this respect, the $da/dt_C(C^*)$ diagram does not indicate an essential piece of information for the design engineer, namely the time to the onset of crack extension (sometimes referred to as the incubation time, $t_{i,x}$).

4.1.2 Creep Crack Initiation Properties
Creep crack incubation periods are very dependent on creep ductility such that $t_{i,x}$ occupies a small fraction of life for creep brittle materials, eg.

0.5Cr0.5Mo0.25V CGHAZ, but a significant fraction of life for creep ductile materials, eg. 2.25CrMo parent steels.[26] Design assessments which do not acknowledge the crack incubation times exhibited by creep ductile materials can lead to excessively conservative life predictions.

The terminology adopted for incubation time purposely qualifies the parameter in terms of crack initiation criterion. This is because the time to generate a 50 μm deep creep crack ahead of a pre-existing defect can be significantly lower than the time for 0.5 mm crack extension, in particular in creep ductile materials.[22]

Incubation times may be predicted using a critical strain parameter such as the crack tip opening displacement at the onset of crack extension, $\delta_{i,x}$ as in eqn 10,[27] or by using an empirically determined $t_{i,x}(C_o^*)$ model[28,29] (eg. Fig. 8) which absorbs $\delta_{i,x}$ and the uniaxial creep property parameters into a simple power law constant, ie.

$$t_{i,x} = \left\{ \frac{1}{B} \cdot \left(\frac{B \cdot \delta_{i,x}}{C_o^*} \right)^{n/n+1} \right\}^{1/p} \tag{10}$$

Values of $\delta_{i,x}$ are becoming increasingly available for a number of steels.[22,27–30]

4.1.3 Defect Assessment

The full assessment of permissible sub-critical crack growth when the applied loading is steady includes a consideration of the time to initiate creep crack extension and the time for the crack to propagate between a_o and a_{crit},[3] ie.

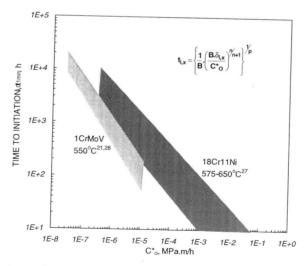

Fig. 8 Comparison of creep crack initiation times as a function of C_o^* for 1CrMoV and 18Cr11Ni steels.

$$t_r = t_{i,x} + t_{CCG} \tag{11}$$

where the incubation period is determined for a given initiation criterion, x. A pragmatic strategy is to adopt an initiation criterion of $x = \Delta a \sim 0.5$ mm enabling growth times to be simply based on the integration of a Stage-II CCG model (eg. eqn 9).[30] In these circumstances, crack propagation behaviour may be simply modelled on the basis of the $D(\varepsilon_r)$ and γ values determined from CCG data of the type given in Fig. 7.

4.2   Crack Extension due to Cyclic Loading

4.2.1   General

Fatigue crack growth behaviour is traditionally characterised by three regimes (Fig. 9). These are a low-ΔK regime close to the fatigue crack threshold, ΔK_{TH}, a mid-ΔK regime in which propagation rates are modelled by a power law (eqn 12),[31] and a high-ΔK regime in which K_{max} approaches K_C (or σ_{ref} approaches R_m).

$$da/dN_F = A(T,v) \cdot \Delta K^m \tag{12}$$

In eqn 12, $A(T,v)$ and m are material constants dependent on temperature, environment and frequency (below a limiting value).

At low ΔK levels close to ΔK_{TH}, the magnitude of da/dN_F is very sensitive to small increases in ΔK and dependent on the same factors which influence ΔK_{TH};

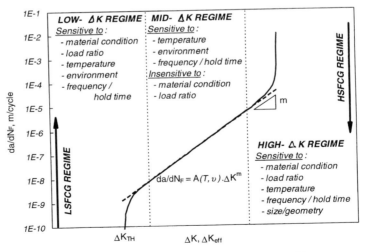

Fig. 9 Fatigue crack growth rate regimes.[34]

these being material, microstucture and yield strength, temperature, environment and load ratio ($R = K_{min}/K_{max}$) (see §4.2.3). Propagation rates in the mid-ΔK regime are less sensitive to microstructure and mean stress (R) effects.

In the high-ΔK regime, da/dN_F becomes increasingly sensitive to the level of ΔK and, in particular, to K_{max} (and/or $\sigma_{ref,max}$) as K_C (and/or plastic collapse) is approached (Fig. 2). Depending on the deformation and fracture characteristics of the material, crack growth rates can be strongly influenced by size and geometry. In these circumstances, a simple LEFM defined ΔK is not the most effective correlating parameter and alternative energy based cyclic loading parameters such as ΔJ or ΔK_{eq} are employed in eqn 12 (see §4.2.3).[32,33] In addition to the factors already mentioned, da/dN_F in the high-ΔK regime is strongly dependent on microstructure, load ratio (mean stress), temperature, environment and frequency (strain rate).

At elevated temperatures, the $da/dN_F(\Delta K)$ diagram is alternatively split into two crack growth rate regimes (Fig. 9).[34] In the low strain fatigue crack growth (LSFCG) regime, load/ displacement transients result in linear elastic loading cycles and low to mid-ΔK crack growth rates for which ΔK (or ΔK_{eff}[‡])[33,35] still provides the most appropriate correlating parameter. Load/displacement transients responsible for cyclic loading involving a degree of general yield (in particular in tension) are referred to as high strain fatigue crack growth (HSFCG) cycles. HSFCG rates are due to higher apparent ΔKs and are influenced by the load/displacement control mode, in particular in the material's creep regime. The overlap shown between the two regimes in Fig. 9 is due to the fact that higher ΔKs can be generated under linear elastic conditions with displacement controlled loading because of shakedown into compression.

Fatigue crack growth properties are increasingly sensitive to frequency with increasing temperature.

4.2.2 High Frequency

At loading frequencies above those for which time dependent effects are influential (ie. at elevated temperatures where oxidation is not responsible for significant crack tip oxide-wedging and/or creep damage enhancement), the factors affecting $da/dN_F(\Delta K)$ behaviour are as summarised in §4.2.1.

The effect of temperature may be quantified in the mid-ΔK regime by rewriting eqn 12 to give:

$$da/dN_F = A_{20} \cdot (E_T/E_{20})^m \cdot \Delta K^m \qquad (13)$$

---

‡ΔK_{eff} is the part of ΔK responsible for crack opening in transients involving a compressive component of loading in the cycle. Such transients are commonly associated with thermal (displacement controlled loading) cycles in power/process plant components.

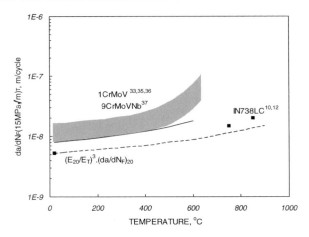

Fig. 10 Effect of temperature on high frequency fatigue crack growth rate for a range of engineering alloys.

where A_{20} is equal to $A(T,v)$ at ambient temperature in eqn 12. The effectiveness of this approximation is demonstrated in Fig. 10 with data for a selection of high temperature materials.[10,12,34,36-38] Clearly, at relatively high frequencies, there is a temperature (dependent on material) above which da/dN_F becomes increasingly influenced by thermal effects.

Unlike creep crack growth, there is not an incubation period preceding the onset of crack extension from a pre-existing defect by fatigue. However below ΔK_{TH}, cracks do not extend (Fig. 9). Values for high frequency ΔK_{TH} are available in the published literature.[10-12,39]

4.2.3 Low Frequency

At lower frequencies, oxidation and creep interaction effects become more influential (eg. Fig. 11).[40] For load controlled situations, below a limiting frequency, crack growth rates may be regarded as being dominated by the time dependent component of the damage mechanism (eg. creep crack growth, Fig. 12). The high temperature behaviour demonstrated by the new 9%Cr steels and the IN738LC cast nickel base super alloy in Fig. 12 is typical for many engineering materials in the mid-ΔK regime, for load controlled situations.[34]

In contrast, crack growth rates in the low-ΔK regime can reduce and ΔK_{TH} values increase with decreasing frequency due to oxide wedging and premature crack closure (Fig. 11). Data to illustrate this observation for 1CrMoV steels at 530/550°C is shown in Fig. 13.

At high temperatures, the high growth rates associated with the high-ΔK (or HSFCG) regime (Fig. 9) can be generated as a consequence of relatively high remote cyclic loading applied to a long crack, or (usually at an initial stage of thermal fatigue crack development) high local strain transients applied to a small

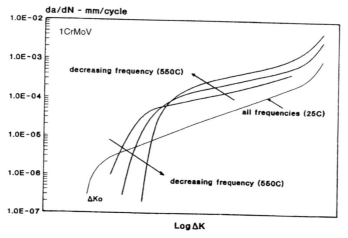

Fig. 11 Effect of frequency on high temperature LSF crack growth rates.[40]

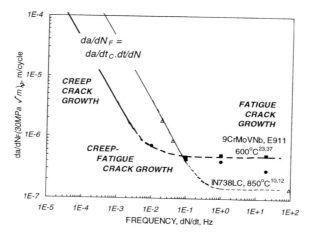

Fig. 12 Summary effect of frequency on fatigue crack growth rate for a range of engineering alloys.

crack contained in the cyclic plastic strain field.[41] In the former case, cyclic crack growth rate behaviour is modelled using a modified form of eqn 12 (see §4.2.1),[33] eg.

$$da/dN_F = A(T, v) \cdot (\Delta K_{eq})^m \qquad (14)$$

The high strain fatigue crack growth rates associated with small cracks contained in local cyclic plastic strain fields ($a \leq 5$ mm) are most effectively modelled as a function of $\Delta \varepsilon$[42–44] (eg. Fig. 14):

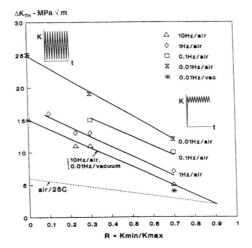

Fig. 13 Effect of frequency, temperature and mean stress on ΔK_{TH} for 1CrMoV rotor steel at 530/550°C.[34,40]

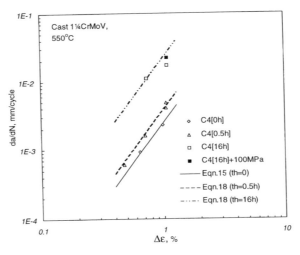

Fig. 14 Comparison of crack growth rates after 0.5mm crack extension in creep-fatigue tests on cast 1.25CrMoV steel at 550°C.[44]

$$da/dN_F = B' \cdot a^b \cdot \Delta\varepsilon^q \quad \text{(for } a \leq 5\,\text{mm)} \tag{15}$$

4.3   Crack Extension due to Cyclic/Hold Loading

The creep-fatigue crack growth properties required for the defect assessment of components subject to fatigue cycles involving hold (steady operating) periods may be derived from pure fatigue and pure creep crack growth rate data in a construction of the form given in Fig. 12, when the loading is directly applied (ie. in load controlled situations). In such circumstances, the effective frequency may be simply determined from a knowledge of the total cycle time (ie. transient + hold time). Alternatively, creep-fatigue crack growth behaviour is analytically modelled on the basis of fatigue and creep crack growth rate characteristics for the material, ie.

$$da/dN_{CF} = da/dN_F + da/dN_C \qquad (16)$$

In eqn 16, da/dN_F represents eqn 14 where $A(T,v)$ is a function of the oxidation and creep damage established during the prior hold period(s) through its frequency dependence,[33] and da/dN_C is:

$$da/dN_C = \int_{t_{i,x}}^{t_h} D(\varepsilon_r) \cdot (C^*)^\gamma \, dt \qquad (17)$$

An analytical approach is particularly appropriate when the cyclic loading experienced by critical regions in high temperature components is due to the consequences of relatively local, thermally induced strain transients; the high local strains generated during the transient leading to a residual stress state which relaxes during the hold (steady operating) period (eg. Fig. 15).[33]

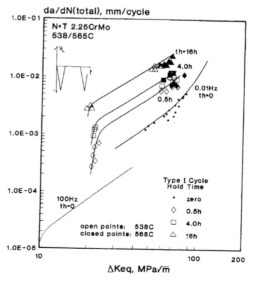

Fig. 15 Long crack cyclic/hold creep-fatigue crack growth test data for 2.25CrMo cast turbine steel at 538/565°C.[33]

For relatively short cracks contained in high cyclic plastic strain fields, da/dN_{CF} is effectively modelled using a refinement of eqn 15 (eg. Fig. 14).[43,44]

$$da/dN_{CF} = B' \cdot a^b \cdot \Delta\varepsilon^q/(1 - D_c)^2 \quad \text{(for } a \leq 5\,\text{mm)} \tag{18}$$

5   NO CRACK GROWTH

For certain components, any crack extension during the design/remaining life-time is unacceptable (Fig. 2). The material property data used to determine high temperature defect acceptability in such circumstances are reviewed in the following sub-sections.

5.1   Avoidance of Creep Crack Extension

Creep crack extension of pre-existing defects is avoided by procedures which effectively establish those conditions under which cracks will not extend in a given life time.[18,19,27–29] Existing knowledge does not support the concept of a creep crack growth threshold (see §4.1.1). Consequently, it is necessary to determine the magnitude of the employed loading parameter associated with no growth in $t_{i,x}$ hours, eg. $C^*_{0/t_{i,x}}$ from Fig. 8.

The same principle applies in the construction of a two-criteria no-growth diagram[18,19] (eg. Fig. 2b). These are prepared for a given time and temperature using the material properties appropriate for these conditions.

5.2   Avoidance of Fatigue Crack Extension

Fatigue crack extension is avoided by ensuring that the ΔK acting at the crack tip of the pre-existing defect does not exceed ΔK_{TH} for the appropriate load ratio (eg. Fig. 13), ie.

$$a_o \geq (\Delta K_{TH}/\Delta\sigma_{app})^2/Y(a, \phi, \Phi) \tag{19}$$

High frequency ΔK_{TH} values are available for a wide range of materials.[10–12,39] The available information for low frequency ΔK_{TH} is relatively sparse,[40] reflecting the position that low frequency applications are usually evaluated by a permissible crack growth assessment. For those circumstances where this is not the case, the material property data is typically generated for the specific application.

5.3   Avoidance of Creep-Fatigue Crack Extension

There is no published guidance for the no-growth defect assessment of components subject to creep-fatigue loading. Such assessments typically assume component loading to be creep or fatigue dominated and analyses performed accordingly (§5.1,5.2).

6   CONCLUDING REMARKS

The material data requirements for assessing defect integrity at high temperatures have been reviewed. Attention has been focused primarily on the properties required to assess the integrity of defects which may or may not propagate in a stable manner as a consequence of steady, cyclic or cyclic/hold loading, or in an unstable manner, during component lifetime. Where possible, the properties required for such defect assessments have been illustrated for a range of high temperature alloys.

TERMINOLOGY

a, a_o, a_{crit}	crack (flaw, defect) depth, initial crack depth, critical crack depth
Δa	crack extension
$A(T,v)$, A_{20}	material constants in mid-ΔK power laws (eqns 12,13,14), dependent on metallurgical condition, temperature, environment and frequency
b	crack length exponent in eqns 15,18
B	constant in Norton law (eqn. 5)
B'	constant in creep-fatigue crack growth rate equations (eqns 15,18)
CCG	creep crack growth
CGHAZ	coarse grain heat affected zone
C^*, C_o^*	parameter characterising stress and strain rate fields at tip of crack in material deforming due to creep, initial C^* at start of test (duty)
da/dN_C, da/dt_C	creep crack growth per cycle, creep crack growth per unit time
da/dN_F, da/dt_F	fatigue crack growth per cycle, fatigue crack growth per unit time
da/dN_{CF}	total crack growth per cycle
$D(\varepsilon_r)$, D'	material constant in creep crack growth law (eqn 9) – dependent on creep ductility, alternative when ε_r is specifically included in eqn 9
D_C	creep damage fraction
E_{20}, E_T	elastic modulus at ambient, elastic modulus at temperature, T
$G(a,\phi,\Phi,n)$	function characterising defect size/shape with respect to component geometry (in C^* expression)
$H(a,\phi,\Phi,\beta)$	function characterising defect size/shape with respect to component geometry (in J contour integral expression)
HSFCG	high strain fatigue crack growth

J, ΔJ	parameter characterising stress and strain fields at tip of crack in material deforming plastically, cyclic J
k'	hardening constant in Ramberg–Osgood model (Eqn. 3)
K, ΔK	stress intensity factor, cyclic stress intensity factor
K_{max}, K_{mean}, K_{min}	maximum, mean, minimum stress intensity factor
K_C, K_{IC}	critical stress intensity factor
ΔK_{eff}	part of ΔK responsible for crack opening in transients involving a compressive component of loading in the cycle
ΔK_{eq}	equivalent cyclic stress intensity factor[33]
ΔK_{TH}	threshold cyclic stress intensity factor below which cracks do not propagate
LEFM	linear elastic fracture mechanics
LSFCG	low strain fatigue crack growth
m	exponent in mid-ΔK power laws (eqns 12,13)
n	stress exponent in creep law (eqn 5)
N	number of cycles
p	time exponent in creep law (eqn 5)
q	strain range exponent in eqns 15,18
$Q(a,\phi,\Phi)$	function characterising defect size/shape with respect to component geometry (general yielding/creep deformation)
R	load ratio, $R = K_{min}/K_{max}$
R_Y, $R_{P,X}$, R_M	yield strength, X% proof strength, tensile strength
$R_{r/T/t}$	creep rupture strength for a given temperature and time
t, t_h, t_r	time, hold time, time to rupture
$t_{i,x}$	time to the onset of creep crack extension (incubation time), for a given crack initiation criterion, x
t_{CCG}	creep crack growth time
T	temperature
x	crack initiation criterion, eg $x = \Delta a = 0.5$ mm
$Y(a,\phi,\Phi)$	function characterising defect size/shape with respect to component geometry (linear elastic deformation)
β	hardening exponent in Ramberg–Osgood model (eqn 3)
δ, $\delta_{i,x}$	crack tip opening displacement, crack tip opening displacement at the onset of creep cracking (for crack initiation criterion, x_c)
ε, $\Delta\varepsilon$, ε_C	strain, cyclic strain, creep strain
ε_r, $\varepsilon_{r/T/t}$	creep-rupture ductility, creep-rupture ductility for a given temperature and time
γ	C^* exponent in creep crack growth law, eqn 9
ϕ, Φ	flaw shape parameter, component geometry parameter
v	frequency
σ, σ_{app}, σ_{ref}	stress, applied stress, reference stress (eqn 6)
$\Delta\sigma_{app}$	cyclic applied stress

REFERENCES

1. R6: Assessment of the Integrity of Structures Containing Defects, Rev. 3, R.A. Ainsworth *et al.* eds, Nuclear Electric, Barnwood.

2. BS 7910:1998: Guide on Methods for Assessing the Acceptability of Flaws in Structures (Replacing PD 6493 and PD 6539), British Standards Institution, London.

3. R5: Assessment Procedure for the High Temperature Response of Structures, Issue 2, R.A. Ainsworth *et al.* eds, Nuclear Electric, Barnwood, 1994.

4. Y. Murakami ed.: *Stress Intensity Factor Handbook*, (a) Vols 1 & 2, Pergamon Press, Oxford, 1987, (b) Vol. 3, Japan Soc. Mat. Sci, Kyoto, 1992.

5. A. Zahoor: *Ductile Fracture Handbook*, EPRI Report No. NP-6301-D, June 1989.

6. H. Riedel: *Fracture at High Temperatures*, MRE-Springer Verlag, Berlin, 1987.

7. A.G. Miller: 'Review of limit-loads of structures containing defects', *Pressure Vessels & Piping*, 1988, **32**, 197.

8. V.P. Swaninathan and J.D. Landes: 'Temperature dependence of fracture toughness of large steam turbine forgings produced by advanced steel melting processes', *Fracture Mechanics: Fifteenth Symposium*, ASTM STP 833, 1984, 315.

9. D.V. Thornton and K-H. Mayer: 'New materials for advanced steam turbines', *Proc. 4th Intern. Charles Parsons Turbine Conf. on Advances in Turbine Materials, Design and Manufacturing*, 1997, Newcastle, A.Strang *et al.* eds, The Institute of Materials, 1997, 203.

10. S.R. Holdsworth: 'The significance of defects in gas turbine alloys', *Proc. COST Conf. on High Temperature Alloys for Gas Turbines*, Liège, 1978, D.Coutsouradis *et al.* eds, App. Sci. Pub., 469.

11. S.R. Holdsworth: 'The significance of defects in gas turbine disc alloys', *Proc. ASM 4th Intern. Conf. on Superalloys*, Seven Springs, 1980, J.K. Tien *et al.* eds, 375.

12. S.R. Holdsworth and W. Hoffelner: 'Fracture mechanics and crack growth in fatigue', *COST Conf. Proc. on High Temperature Alloys for Gas Turbines*, Liège, 1982, R. Brunetaud *et al.* eds, Reidal, 345.

13. BS 7448:1991: *Fracture Mechanics Toughness Tests - Part I: Method for Determination of K_{IC}, Critical CTOD and Critical J Values of Metallic Materials*, British Standards Institution, London.

14. S.R. Holdsworth and D.V. Thornton: 'Prediction of isothermal embrittlement in modern 1CrMoV rotor steels', *Proc. Conf. on Microstructures and Mechanical Properties of Ageing Materials*, ASTM/TMS Materials Week, Chicago, November 1992.

15. B. Nath, E. Metcalfe and J. Hald: 'Microstructural development and stability in new high strength steels for thick section applications at up to 620°C', *Microstructural Development and Stability in High Chromium*

Ferritic Power Plant Steels, A. Strang and D.J. Gooch eds, The Institute of Materials, 1997, 123.

16. R. Hales: private communication.

17. J.F. Radavich and A. Fort: 'Effects of long-time exposure in Alloy 625 at 1200°F, 1400°F and 1600°F', *Proc. Conf. on Superalloys 718, 625, 706 and Various Derivatives*, E.A. Loria ed., The Minerals, Metals & Materials Soc., 1994, 635.

18. J. Ewald and K-H. Keienberg: 'A two-criteria-diagram for creep crack initiation', *Proc. Intern. Conf. on Creep*, Tokyo, April 1986, 14.

19. R.A. Ainsworth: 'The use of a failure assessment diagram for initiation and propagation of defects at high temperatures', *Fatigue Fract. Engng Mater. Struct.*, 1993, **16**(10), 1091.

20. NRIM Creep Data Sheets, Japanese National Research Institute for Metals.

21. PD6525:1990: *Elevated Temperature Properties for Pressure Purposes; Part 1 - Stress Rupture Properties*, British Standards Institution, London, Issue 2, February 1994.

22. S.R. Holdsworth: 'Initiation and early growth of creep cracks from pre-existing defects', *Materials at High Temperatures*, 1992, **10**(2), 127.

23. G.J. Neate: 'Creep crack growth behaviour in 0.5CrMoV steel', *Mater. Sci. Eng.*, 1986, **82**, 59.

24. N. Taylor, E. Lucon, V. Bicego and P. Bontempi: 'Crack stability assessment for advanced 9Cr steels in boiler components', *Proc. Intern. Conf. on Advanced Steam Plant*, I.Mech.E Trans, 1997, C52/026, 201.

25. H. Huthmann *et al*: 'Creep crack growth behaviour of Type 316L steel - Round-robin and post test investigations including metallurgical examinations', *Final Report for CEC Contract Nos. RAP-071-D and RA1-0111-D*, 1991.

26. S.R. Holdsworth: 'Creep crack growth properties for the defect assessment of weldments', *Proc. Intern. Conf. on Integrity of High Temperature Welds*, Nottingham, I.Mech. E., November 1998, 155.

27. R.A. Ainsworth: 'The initiation of creep crack growth', *Int. J. Solids Structures*, 1982, **18**(10), 873.

28. P. Bensussan, R. Piques and A. Pineau: 'A critical assessment of global mechanical approaches to creep crack initiation and creep crack growth in 316L steel', *Non-Linear Fracture Mechanics, Vol. 1 - Time Dependent Fracture*, ASTM STP 995, 1989, 27.

29. E. Molinie, R. Piques and A. Pineau: 'Long term behaviour of a 1CrMoV steel, Part II: Creep crack initiation and creep crack growth', *Fat. Fract. Engng. Mater. Struct.*, 1989, 10.

30. S.R. Holdsworth:, 'Creep crack growth in low alloy steel weldments', *Materials at High temperatures*, 1998, **15**(3/4), 203.

31. P.C. Paris and F. Erdogan: 'A critical analysis of crack propagation laws', *J. Basic Eng. (Trans ASME)*, 1963, **85**, 258.

32. N.E. Dowling: 'Crack growth during low cycle fatigue', *Cyclic Stress-Strain and Plastic Deformation Aspects of Fatigue Crack Growth*, ASTM STP 637, 1977, 97.

33. S.R. Holdsworth: 'Factors influencing the high temperature HSF crack growth rates in turbine casting steels', *Behaviour of Defects at High Temperatures*, ESIS 15, R.A. Ainsworth & R.P. Skelton eds, Mechanical Engineering Publications, London, 1993, 327.

34. S.R. Holdsworth: 'High temperature fatigue crack growth', *High Temperature Crack Growth in Steam Turbine Materials*, J.B. Marriott ed., Commission of the European Communities, COST Monograph EUR 14678EN, 1994.

35. R.P. Skelton, S.M. Beech, S.R. Holdsworth, G.J. Neate, D.A. Miller and R.H. Priest: 'Round robin tests on creep-fatigue crack growth in a ferritic steel at 550°C', *Behaviour of Defects at High Temperatures*, ESIS 15, R.A. Ainsworth and R.P. Skelton eds, Mechanical Engineering Publications, London, 1993, 299.

36. P.K. Liaw, A. Saxena and V.P. Swaminathan: 'Effects of load ratio and temperature on the near threshold fatigue crack propagation behaviour in a CrMoV steel', *Metall. Trans. A*, 1983, **14A**, August, 1631.

37. J. Ewald, C. Berger and H. Brachvogel: 'Investigation on crack initiation and propagation under static, cyclic and combined loading conditions of 1CrMoNiV steels at 530°C', *COST 505 D20/D21 Final Report, Siemens Report No. TW1187/89*, June 1989.

38. A. Vanderschaeghe, C. Gabiel-Cousaert and J. Lecoq: 'Influence of stress ratio, temperature and thermal ageing on the threshold values of steel 91' *Proc. Conf. on Materials for Advanced Power Engineering*, D. Coutsouradis ed., 1994, Part 1, 383.

39. D.A. Taylor: *A Compendium of Fatigue Crack Growth Thresholds and Growth Rates*, EMAS, Cradley Heath, 1985.

40. R.P. Skelton and J.R. Haigh: 'Fatigue crack growth rates and thresholds in steels under oxidising conditions', *Mat. Sci. Engng.*, 1978, **36**, 17.

41. S.R. Holdsworth: 'Factors influencing crack extension in high temperature turbine castings', *Proc. 2nd Parsons Intern. Conf. on Materials Development in Turbo-Machinery Design*, 1988, Churchill College, Cambridge.

42. C. Levaillant and A. Pineau: 'Assessment of high temperature low cycle fatigue life of austenitic stainless steels using intergranular damage as a correlating parameter', *Low Cycle Fatigue and Life Prediction*, ASTM STP 770, 1982, 169.

43. R.P. Skelton: 'Damage factors during high temperature fatigue crack growth', *Behaviour of Defects at High Temperatures*, ESIS 15, R.A. Ainsworth & R.P. Skelton eds, Mechanical Engineering Publications, London, 1993, 191.

44. S.R. Holdsworth: 'Creep-fatigue crack growth from a stress concentration', *Materials at High Temperatures*, 1998, **15**(2), 111.

A Review of Service Problems during High Temperature Operation

R. D. TOWNSEND

Consultant

ERA Technology Ltd., Cleeve Rd, Leatherhead, Surrey KT22 7SA, UK

ABSTRACT

This paper provides an historical perspective of the service problems experienced during operation of creep resistant steels at high temperatures. The broad theme of the paper is that overall, the operation of plant at temperatures within the creep range has been extremely successful and there are many examples world-wide of high temperature components lasting in service for well over 25 years. This is because the overall design concept is generally conservative. Where failures have occurred, it is invariably because the original designs did not take into account, or the designers were unaware of, the impact of fabricating methods, operating practices and material specification on the stresses imposed during service and the changes produced in metallurgical structure which can markedly change properties. The paper describes how metallurgical investigations of high temperature incidents have identified the essential factors contributing to the failure(s) and that from this understanding, new specifications, fabricating methods and operating procedures have been developed to reduce the incidence of similar failures. However, an essential requirement in achieving these improvements is to ensure that plant designers and operating engineers are made fully aware of the effects different fabrication techniques and operational practices can have on material properties. Success in this will improve future plant reliability but failure to do so will cause failures as indeed occurred recently on a header end cap in a UK coal-fired power station. In this case the replacement header was constructed in the newly developed P91 steel but apparently the essential message that this material is susceptible to Type IV cracking, was not conveyed to the design engineers.

1   INTRODUCTION

In this paper, giving an historical perspective of service experience, attention will be focussed on those situations where changes in metallurgical structure and ductility have given rise to component failure. At the outset, however, it must be emphasised that overall the incidence of cracking in service is low and that examples of catastrophic failures leading to major steam leaks and damage are extremely rare. Such events have attracted the wide-scale interest of the metallurgical fraternity and it is important to note that the investigations initiated after these incidents have greatly improved our understanding of high temperature materials and the impact of operating factors.

This knowledge will certainly reduce the incidence of similar failure in future.

Plant for power generation has now been designed and operated at temperatures in the creep range, for the materials concerned, for over 50 years. There are many examples in this country and overseas of boiler and turbine components giving satisfactory performance for times well in excess of 10^5 h.

The oldest high temperature pipe work for example in the United Kingdom was on Blyth Power Station and that has operated for times in excess of 200 000 h. It would therefore appear that the original design criteria for high temperature components has been effective and also reasonably conservative. In this paper a brief explanation will be given of these criteria and then an examination made of those situations where the design methods have not been successful in avoiding failure. In the main it will be demonstrated that the original designs in these cases were not fully able to account for variations in fabrication and operating parameters and for concomitant changes in metallurgical structure giving rise to marked changes in rupture ductility and susceptibility to cracking.

2 HIGH TEMPERATURE DESIGN CRITERIA

The design criteria for high temperature components are covered in a number of national and international standards such as BS 1113 for headers and boiler tubing. These take into account the material used, the design temperature, the design lifetime, the effects of metal wastage, ligament efficiency for complex pressure vessels such as headers, and allowances made for bending. For boiler components such as tubes, pipework and headers where close dimensional and clearance tolerances are not necessary, the design stress is obtained from the mean rupture stresses at 10^5 h such that;

$$\text{Design Stress} = \frac{\text{Stress to Rupture in } 10^5 \text{ h}}{\text{Safety Factor}}$$

plus a 10% increase in thickness as an allowance for bending.

For turbine components for which there is a requirement for high accuracy in dimension and tolerance the design stress is determined by the stress to give 0.1% strain in 10^5 h;

$$\text{Design Stress} = \frac{\text{Stress to give } 01\% \text{ Strain in } 10^5 \text{ h}}{\text{Safety Factor}}$$

It should be noted that for both these criteria the design is based on the strength of the material (either rupture strength or creep strength) and that there is no consideration given to the overall strain tolerance of the material ie. the rupture ductilily. It should also be noted that no explicit allowance is given in the codes for welded structures.

The conservatism in the design comes from the dependence on the 10^5 h strength criteria and clearly if these are known accurately, failures ought not to occur before such time. However, this depends on the accuracy by which the 10^5 h rupture creep strength was known and for early designs considerable extrapolation was necessary. To some extent this effect was allowed for by choosing a high safety factor of say 1.5 where considerable extrapolation was required in the rupture data to obtain the 10^5 h rupture stress compared to more recent factors of ~ 1.3 where the 10^5 h rupture stress is more clearly defined. The safety factors are also used to account for variation in materials properties for which there can be considerable scatter.

As indicated above, these simple criteria have in most cases allowed successful designs to be evolved, but in a few cases the designers were not able to fully allow for variations in operating parameters or metallurgical structure, and failures have occurred. Thus in these cases the designers were not able to estimate correctly nor allow for the effects of residual welding and fabrication stresses, thermal stresses due to temperature gradients, and system stresses due to differential expansions. In other cases no allowance was made for operation at excessive temperatures nor for variation in metallurgical structure due to welding, fabrication and operating mode.

3   OVERALL SERVICE EXPERIENCE

Table 1 gives a listing of some of the high temperature failures which have occurred over the last 40 years. These have occurred on all types of component and over the years have been thoroughly investigated such that the factors contributing to failure(s) are well understood and that in all cases metallurgical and engineering procedures are now available to prevent and minimise problems in future. In this paper it will not be possible to discuss all the failures in detail. It has therefore been decided to concentrate on two types, (i) those which have given rise to catastrophic steam leaks and (ii) the failures on pipework which give good examples of the impact of rupture ductility and how changes in this contributed to different types of failure.

It should be noted that major enquiries were conducted after the incidents of catastrophic failure and that in each of the countries concerned new operating procedures were adopted to prevent similar events in future. In nearly all cases the cracking/distortion listed in Table 1 can be attributed to a combination of both operating/fabrication condition and metallurgical condition which were not or could not be taken account of in the design. In nearly all cases also the combination of these factors resulted in a reduction in the creep rupture ductilities of the material and propagation of damage by creep crack growth. In one or two instances however, the reverse pertained and the metallurgical changes in service resulted in too much ductility or to a reduc-

Table 1 Incidents of high temperature cracking/distortion in service.

Component/Description	Country/Year	Operating/Fabricating condition not taken account of in design	Metallurgical condition not taken account of in design
1 Pipework			
(i) Weldments (CrMoV)		Residual welding stresses	Residual elements
Reheat cracking	UK 1965/85	Inadequate weld procedures	Course grains
Weld metal cracking	UK 1965/85	Improper heat treatment	
Type IV Cracking	Global 1980s	System stresses	Weak zone in weld HAZ
(ii) Cracking of seam welded pipes (1CrMo. 2CrMo)	USA 1985/90		Inclusions along fusion line
(iii) Bend failures (CrMoV, 12CrMo)	Germany Russia 1980/87	No allowance made for bend wall reduction	Over estimate of 10^5 h rupture strength
(iv) Failure of austenitic Pipework	USA 1985	Residual stresses due to thermal cycling	Sigma phase formation
(v) Distortion of austenitic pipework	UK 1975	Thermal cycling	Thermal stresses exceed yield
(vi) Failures of cold bent pipework CMn	UK 1965/86	Residual stresses and bend system stresses	Strain hardening due to bending creep in service
(vii) Dissimilar metal weld failures	UK1 1975/85	Stresses due to mismatch of parent and weld metals	Brittle interfaces, MnS particles, weak zones at interface
2 Bolting			
(i) Ferritic (1CrMoV (Nb)	UK Germany 1965/79	Superimposed bending stresses due to thermal expansion	Coarse grained structures, residual elements, high hardness
(ii) Nimonic 80A	UK Germany Mid 80's	Increased stresses due to lattice ordering and contraction	Embrittlement due to ordering and S/P segregation
3 Rotors			
(i) Distortion of CrMoV rotors	UK 1975	Incorrect heat treatment	Variations in circumferential creep
(ii) Cracking in heat release grooves	USA, UK Japan 1975/80	Increased stress due to groove	Structures associated with weld repairs
4 Chests/Casings	Japan UK 1980s	Residual stress associated with weld repair	Structures associated with weld repairs
5 Headers			
(i) Catastrophic failure	UK 1969	Excessive temperature	Low rupture ductility
(ii) Stub weld cracking	Global 1970/80	Excessive temperature	Weld structures/system stresses
(iii) Interligament cracking	Global 1988/90	Thermal stresses due to cycling	Normal structures
6 Boiler tubes Superheater tubing Reheater tubing Waterwall tubing	All countries 1950/98	Severe stressing, temperature and environmental conditions in fossil boilers	Incorrect materials, wrong heat treatments, tubing thinning

tion in creep strength. Macroscopic examples of this were the distortion of austenitic pipework at Drakelow C (Table 1, 1v) and bending of 1CrMoV rotors due to uneven heat treatment (Table 1, 3.i). Excess ductility/creep strain can also occur on a microscopic scale if this occurs in restricted zones as in welds discussed in the next section the result can appear macroscopically brittle.

Of the failures listed, it should be noted that most have been observed in several of the advanced industrial nations. Two or three however, seem to have assumed 'national characteristics'. Thus in the mid 70s UK experience of welds in 0.5CrMoV steel pipework and terminal joints was particularly unsatisfactory[1] as was German experience with pipe bends in 0.5CrMoV[2] and American experience with seam welded reheat pipework made in 1Cr0.5Mo and 2.25Cr1Mo steels.[3,4] In each case the root causes have been identified as due to particular design/fabricating criteria in the countries concerned. In each case also, as discussed below, the problem has been controlled by a vigorous inspection, repair/replace policy as determined by safety and economic factors.

The above statement is particularly true for boiler tube failures and it is the common experience world-wide that the main causes of forced outages on power plant are those due to tube failures on fossil units. The main reason for the large numbers of failures are the severe environmental conditions in fossil boilers as the effects of stress, temperature, temperature gradients, corrosion, erosion and vibration combine to produce degradation of the tubing steel. Corrosion by oxidation, by combustion products and by impure water can significantly reduce the tube wall thickness and result in failure of a tube many years before its designed service life. Errors can also occur in the design, manufacture, storage, operation and maintenance of boiler tubing, and the wrong type of material installed in a critical location can lead to premature failures. To offset the damaging effects of boiler tube failures many years of work, particularly by engineers and scientists in the UK and the USA from many different disciplines, have identified seven broad categories of boiler tube failure mechanisms and these are listed in Table 2.

In recent years EPRI in the United States has put considerable resources into the prevention of boiler tube failures. The overall approach was developed from procedures established by Ontario Hydro, Dooley and Westwood 1983.[5] These in turn had been developed from the methodology used by the UK Central Electricity Generating Board (CEGB) in the 1970s which was based on a systematic boiler tube failure reporting system and proper procedures to control fireside corrosion and water chemistry. The EPRI approach is incorporated in the 'Manual for Investigation and Corrosion of Boiler Tube Failures' 1985[6] which establishes a systematic approach based on Root Cause Analysis of each tube failure. More recently, National Power in the UK has produced its own Boiler Tube Failure Manual[7] incorporating the extensive CEGB experience in this field.[8]

Table 2 Boiler tube failure mechanisms.

1	Stress-rupture	5	Fatigue
	Short-term overheating		Vibration fatigue
	High temperature creep		Thermal shock
	Dissimilar metal welds		Thermo-mechanical cycling
2	Steam/Water-side corrosion		Corrosion fatigue
	Caustic corrosion	6	Mechanical damage
	Acidic corrosion		Falling slag
	Pitting		During overhaul work
	Stress corrosion cracking		Steam oxidation
3	Fire-side corrosion	7	Defects and lack of QA
	Dew point corrosion		Chemical excursion damage
	Furnace wall		Material defects
	Superheater/Reheater corrosion		Welding defects
4	Erosion		Design Inadequacies
	Dust erosion		
	Sootblower erosion		

4 CATASTROPHIC FAILURES

Most of the instances of cracking listed in Table 1 were detected as creep cracks during inspection prior to actual steam leakage. At least four however, were catastrophic failures with major steam leaks and extensive secondary damage. All were caused primarily by low rupture ductility of the material concerned. Figure 1 is a photograph of the Wakefield header failure in 1969.[9,10] This par-

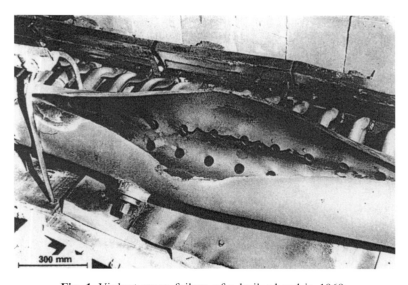

Fig. 1 Violent creep failure of a boiler head in 1969.

ticular header had been operating at excessive temperatures for two days immediately prior to the failure. The subsequent investigation of the incident highlighted the low rupture ductility of the steel concerned as a major contributing factor. The steel was Fe–0.5% Mo with a rupture ductility < 5% which is much inferior to the ductilities found in modern steels used for headers such as 1Cr0.5Mo and 2.25Cr1Mo which typically display ductilities >25%. A positive benefit of this failure was that it caused the then CEGB to initiate a monitoring programme for high temperature headers which in a modified form as detailed in the CEGB Operations memoranda GOM 101 Procedure For Boiler Header Creep Life Assessment,[11] is still used today by National Power, PowerGen and Eastern Electricity and has formed the basis of EPRI guidelines for life extension of fossil plant.[12]

Figure 2 shows a photograph of the damage caused due to the failure of 24 Durehete 1055 studs on a steam chest in 1979. The operating condition contributing to this failure, and not accounted for in design, was a severe temperature gradient through the chest cover. This had two effects – firstly to impose a bending stress on each of the studs and secondly to concentrate creep strain (during relaxation of the studs in service) in the hottest region of the stud, the first engaged thread located just below the flange cover. The situation was further exacerbated by the metallurgical structure of the studs which were predominantly coarse grained, of high hardness, with high residual element content and low ductility.[13] Again benefit was achieved by improvement to the CEGB operating procedure for bolts GOM 85B[1] and improvement to the metallurgical

Fig. 2 Catastrophic failure of 24 low alloy steel bolts on a steam valve body 1979.

specification which lowered the hardness and residual element content and by means of improved heat treatment resulted in a fine grained structure.

The other examples of catastrophic failures were those that occurred in seam welded pipes at Mahave Power Station in Southern California in June 1985 and at Monroe Power Station in January 1986 and the explosive failures which occurred in West Germany on CrMoV bends during the mid 80s. These are discussed in detail below.

5   PERFORMANCE OF PIPEWORK

All sorts of different steels are used for pipework in power stations. They include C–Mn steels for integral boiler water feedpipes, and for main and reheat pipework the low alloy steels 1Cr0.5Mo, 2.25CrMo, 0.5CrMoV, high alloy steels such as 9CrMoVNb and 12CrMoV and austenitic materials such as 316 and Esshete 1250. The main technical feature determining the selection of material is temperature. Thus the relatively weak C–Mn is adequate for boiler feed water pipework because it only operates at 300–350°C.

In most countries for example, in US, Japan, and Europe, the main and reheat temperature is 540°C (1000°F) and 1Cr0.5Mo and 2.25CrMo steels are used extensively, in the UK however, the main steam condition is 565°C (1050°F) and 0.5CrMoV has been selected. In Germany and Scandinavia they have a variety of steam temperatures between 540–575°C and use 0.5CrMoV and 12CrMoV steels. The newer 9Cr steel such as T91 has recently been used in the Kawagoe Ultra Supercritical Plant in Japan, which operates at 565°C, in UK coal-fired plant as replacement super-heater header material, and for Pipework and headers in the supercritical units under construction in Denmark. Austenitic materials are used either for very high temperature, 600–650°C, such as on the original Ultra Supercritical Plant, Eddystone in the United States and Drakelow C in the UK, as pipework operating at 565°C in the Dungeness B AGR plant, or at comparatively low temperatures of a 340°C on BWR plant.

5.1   Seam Welded Pipework

The nature of the service problems encountered with these different steels is also a reflection of the design and the operating parameters. Thus, in America a particular feature of their design codes in early plant was to permit the use of seam welded pipe made by the submerged are welding process for reheat pipework in 1.25Cr0.5Mo and 2.25Cr1Mo steels.

These pipes often entered service in a renormalised and tempered condition and thus the structure of the welds was substantially modified. In particular the HAZs were removed. The presence of a seam weld on pipework is, to say the least, a doubtful proposition. This is because the maximum hoop stress acts perpendicularly across the weld. Nevertheless, the Americans had good experience with this pipework, until about 12 years ago a catastrophic and fatal accident occurred at Mahave PS in Southern California.[3,4] On examination it was found that failure occurred along the original weld interface. The suspected mechanism was creep crack growth from original welding defects. Subsequently, further failures have occurred and a massive programme of pipe inspections, largely initiated by EPRI,[14] has been conducted. As a result of these, many examples of damaged pipes have been found containing *original fabrication* flaws such as porosity, slag inclusions and lack of fusion along the weld interface and also *service-induced* flaws comprising creep cavitation and cracking and thermal fatigue cracking. In some cases the original fabrication flaws were only a few inches in extent but in some instances the flaws were intermittent over several feet. A consequence of the inspection programme is that a number of pipe runs have already been replaced and others monitored. In the damaged pipes it is clear that there is a marked reduction in rupture ductility along the original weld interface. What is remarkable, however, is that these potentially dangerous situations did not become evident until almost 100 000 h service and this is possibly attributable to the relatively low hoop stress in reheat pipework.

5.2   Cracking Of Weldments

In the UK, the codes fortunately do not permit seam welds in pipes for high pressure/high temperature pipework and indeed detailed assessment and measurement of creep strain on CEGB pipework[15] indicates that service lives of well over 200 000 h will be attainable in straight lengths of wrought 0.5CrMoV pipes and several stations have achieved operating lives in excess of 150 000 h. Where problems have been experienced in the UK has been with the performance of weldments. These were made with 2.25Cr1Mo weld metal and gave rise to a variety of problems during fabrication and early service with cracking in the weld metal or HAZ. As described below the problems were successfully contained by research into the causes of cracking and the introduction and systematic monitoring scheme of welds in service.

5.2.1   Weld Metal Cracking

The instances of weld metal cracking were primarily caused by solidification cracking or cracking early in service under the action of the operational stresses plus inadequately relieved residual welding stresses. A significant fact was that cracking resulted from the formation of impurity films at grain boundaries due to low ratios of Mn/S and Mn/Si. This was largely overcome by lowering the S content and increasing Mn to achieve a ratio of ~40:1.[1]

5.2.2   Reheat Cracking

Reheat Cracking occurs in the coarse grained region of the heat affected zone, associated features include a fine matrix distribution of V_4C_3 giving very high matrix strength, the presence of MnS on grain boundaries, the presence of residual elements, and inadequate stress relief after welding.

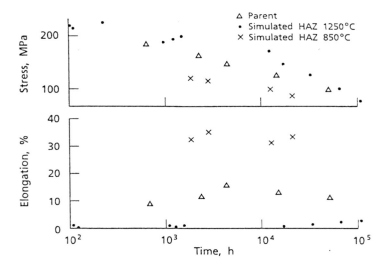

Fig. 3 Creep rupture strength and ductility of parent and HAZ simulated 0.5Cr0.5Mo0.25V 565°C.

This type of cracking is initiated by the formation of creep cavitation on grain boundaries and their subsequent coalescence into microcracks and growth into macrocracks. Grain size is clearly an important factor, (Fig. 3)[15,16] shows creep tests performed on 0.5CrMoV steel at 565°C. Although the simulated coarse grained tests HAZ structure is stronger in creep than the fine grained normalised and tempered structure its creep ductility is less than 1% cf to 20% in the N&T material. An evident aspect of HAZ cracking in service would therefore seem to be that the coarse grained structure simply exhausts its rupture ductility as the residual welding stresses relax in service. A breakthrough in our understanding of the mechanism of cavitation was made by Middleton.[17,18] In a study of reheat cracking in coarse grained simulated HAZ structures in 0.5CrMoV, he observed that MnS inclusions on prior austenite grain boundaries acted as nuclei for cavities. He was subsequently able to show that three microstructural features were present in materials with high susceptibility to cavitation. These were:

(i) A fine dispersion of alloy carbides in the grain interiors
(ii) A weak (precipitation-free) zone immediately adjacent to prior austenite grain boundaries
(iii) A dispersion of sub-micron incoherent particles (MnS) on the prior austenitic grain boundaries.

The way in which these features promote cavitation is illustrated in Fig. 4. Under such circumstances any particles on the boundaries will tend to act as cavity nuclei and indeed carbide particles have been observed to act in this

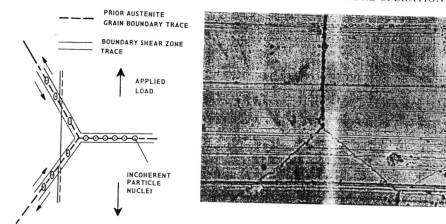

Fig. 4a Schematic of deformation processes operating during hot tensile deformtion of 0.5CrMoV bainitic steel.

Fig. 4b Optical micrograph showing surface deformation fiducial mark displacement and cavitation.

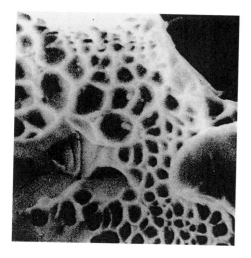
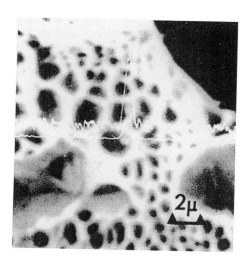

Fig. 5 Cavitites nucleating on MnS particles.

capacity. However, MnS particles are much more potent as cavity nuclei since they are unstable in ferrite and hence incoherent with the matrix (Fig. 5).

Middleton found that the elimination/modification of any of these features drastically reduced the susceptibility to cavitation. Thus, the relatively coarse carbide distribution within grain interiors of tempered ferritic/bainite structures of N&T material, as compared to the fine distribution found in completely bainitic structures, permits creep strain within the grain interiors and prevents

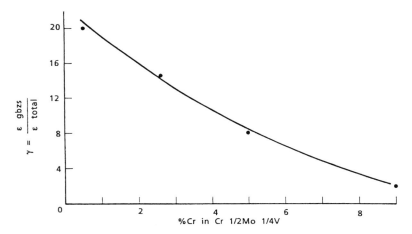

Fig. 6 Effect of chromium content Fe0.5Mo0.25V in reducing grain boundary shear.

the onset of cavitation until late in creep life. The most dramatic effect however, was found with reducing the S content to less than 0.005. This prevented the formation of MnS on prior austenitic grain boundaries, eliminated cavitational damage and significantly improved high temperature ductility. It should be noted that steels deliberately doped with residual embrittling element such as As, Sb and P retained good ductility provided no MnS particles were present to act as cavity nuclei. The role of these elements is apparently associated with the way in which they modify the size and distribution of the MnS particles which in turn controls the size and distribution of cavities. Thus increasing the Sn content from 0.01 to 0.06 wt% reduced the cavity diameter from (1–5) μ to (0.5–1) μ and significantly impaired ductility.

The final way in which significant changes can be made to cavitational susceptibility is to reduce the amount of grain boundary zone shear. Middleton found that Cr additions are particularly effective in this respect (Fig. 6).

The explanations given above for the factors controlling cavitation in ferritic steels underwrites modern development in secondary refining techniques for steel making and the benefits to accrue by using steel of low S and P content and controlled residual element content Jaffee.[19]

It also underwrites the greater use of high Cr alloys such as 9CrMoNV(N) (T91) and high strength 12Cr steels such as TR1200 12CrMoV(W) for improved rupture strength and good ductility.

5.2.3 Type IV Cracking
A disadvantage with ferritic steels, certainly in comparison to austenitic steels, is the complex variety and distribution of phases/structures formed in weldments as a direct consequence of the ferrite to austenite transformation and the reverse

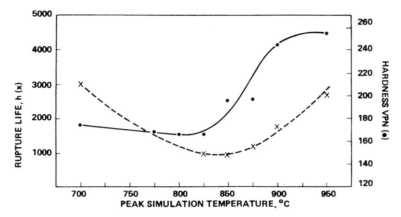

Fig. 7 Rapture lives of simulated HAZ structures in 0.5CrMoV tested at 620°C and hardness variation in as-simulated specimens.

reaction during welding.[20] A particularly important structure is the softened zone which occurs at the very edge of the HAZ immediately adjacent to the unaffected parent metal. This zone is the region of the HAZ which does not achieve full austenisation during the welding cycle but may have been duplex (austenite plus carbide) at temperature if heated above Ac_1, triplex (austenite, ferrite and carbide) if heated into the intercritical zone or simply overtempered if the temperature had not even achieved Ac_1 (ferrite plus carbide). The effect is that, in terms of hardness, this region is very much softer than adjacent regions of the HAZ and is fine grained (Fig. 9).[15,16]

Creep tests on this structure also indicate that it is much weaker (Figs 3 and 7), and also more ductile than other weld structures, (Fig. 3).[15,16] Given this variation in properties it is perhaps somewhat surprising that more service failures have not been associated with this region. The probable explanation is that in respect to the hoop stress (the maximum stress for a pipe butt weld) redistribution during creep will reduce the stress on the intercritical zone which will be supported by the adjacent creep stronger regions. For the axial stress the pressure component is only half the hoop stress so a significant system stress would be required to initiate failure.

As it happens there is extensive world wide experience with ferritic steel pipe lines in excess of 75 000–100 000 h without cracking. Over the last few years, however, there have been reports of type IV cracking on 0.5CrMoV pipework system after 60 000 h at 568°C,[20] on 12CrMoV pipework and on 1Cr0.5Mo pipework[21] again after extended operation. It now appears however, that this form of cracking may be the ultimate end-of-life failure mode for ferritic welds as service lives approach times in excess of 200 000 h.

When observed the mode of type IV cracking seems similar for most ferritic steels – following a long incubation period, creep cavities nucleate in the inter-

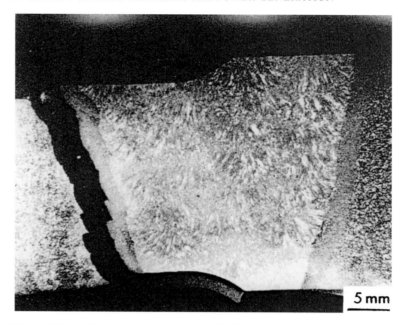

Fig. 8 Type IV cracking on the parent metal side of the HAZ in a 0.5CrMoV steel welded with 2.25Cr1Mo weld metal.

critical zone, coalesce and form micro cracks and ultimately large cracks which in one or two cases have penetrated to the bore and caused steam leaks. On a macroscopic scale it would seem that this type cracking is a low ductilily situation. The same pertains for tests on cross weld specimens – at high stresses failure occurs in the parent metal with significant ductility but at low stresses the failures occurs in the type IV region with very little ductility in the parent metal, (Fig. 8).[15,16] On a microscopic scale however, extensive cavitational damage occurs within the intercritical zone, (Fig. 9),[15,16] indicating considerable strain within this region – so that macroscopic appearance of low ductility is almost certainly a consequence of the very restricted width of these zones which typically are less 3 mm wide.

5.3   Austenitic Pipework

Because of its relative cost the use of austenitic steels in pipework systems has been restricted to specialised applications such as fast reactors, AGRs and BWRs and in fossil plant, the ultra-supercritical units operating at temperatures in excess of 600°C. Potential disadvantages with austenitic material at high temperatures is their high coefficient of thermal expansion and low conductivity which gives rise to much higher thermal stresses compared to ferritic materials.

A further disadvantage is that austenitics enter service in the solution treated condition and hence have a propensity for intermetallic and carbide precipitation

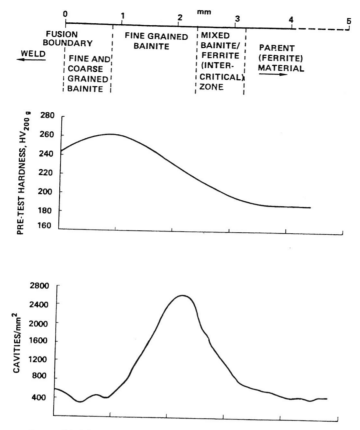

Fig. 9 Comparison of initial microhardness and post creep test cavity density across weld HAZ.

during high temperature exposure which frequently causes marked reductions in rupture ductility. The actual rate of carbide and intermetallic formation is of course a strong function of composition, time, temperature and applied stress and has been extensively studied by Lai and Wickens.[22]

The effect on rupture strength and ductility is shown in Fig. 10. The striking point to note is the marked variation in rupture ductility which occurs over the complete range of time from a few hundred hours to tens of years. This variation in ductility is caused by small variations in the composition and thermo-mechanical treatment experienced by the material which in particular controls the amount of sigma phase precipitated during testing. In most cases the design criteria based on rupture strength has ensured that the operating stress is too low for the material to be subjected to high creep strain and consequently a loss of rupture ductility has not caused problems. However, on the ultra supercritical plant Eddystone in the US the operating conditions on the 316 pipework were

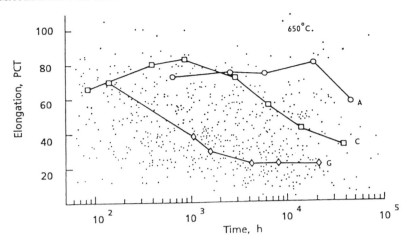

Fig. 10 Rupture ductility in Type 316 steel 650°C.

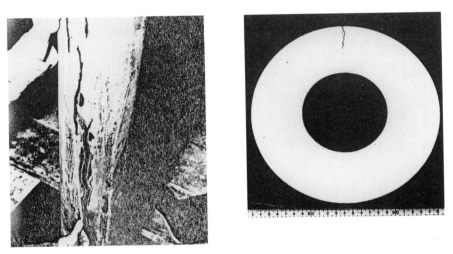

Fig. 11 Cracking of main steam 316 pipework on Eddyston No. 1 unit after 130 000h at 600°/650°C.

such as to cause extensive cracking (Fig. 11) which required large sections of pipework to be replaced.[23] The circumstances were as follows:

Eddystone had been operating for some time in excess of 130 000 h in the temperature range 620°C–650°C. With the particular casts of steel used these service conditions gave extensive precipitation of sigma phase on grain boundaries (Fig. 12) and in subsequent post service testing the material was

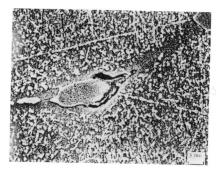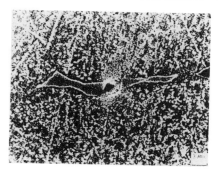

Fig. 12 SEM micrographs showing micro formation on sigma phase particles in failed 316 pipework from Eddyston P.S.

found to have very low rupture ductility. Superimposed on this phenomenon was the cyclic nature of the actual service. This meant that the failed sections of the pipework experienced thermal shocks during operation and because of the thermal properties of the material these gave rise to high residual tensile stresses and ultimately the combination of high stresses and low ductility gave rise to extensive cracking on the OD of the pipes.

There are many ways forward to avoid a repetition of this type of incident:

1. It should be recognised that the thermal properties of austenitic material can give rise to high thermal stresses and therefore austenitic steels should not be subjected, particularly in thick section, to severe thermal cycling. If austenitic steels have to be used, methods of preheating/insulation need to be considered to reduce the thermal transients. In the UK this pragmatic approach has been usefully applied by using heating coils to prevent damage to thick section components such as boiler stop valves and heat exchanger tubeplates.[24]
2. It should also be recognised that austenitic steels do form embrittling precipitates – this effect can be offset by modifying the composition of the steel or use a different steel less prone to such embrittlement. The former technique can be used with 316 by reducing the C content to less than 0.5% as demonstrated by Lai.[25] Alternative materials can be used such as Esshete 1250 which displays excellent rupture ductility for times in excess of 10^5 h and is not prone to sigma precipitation.
3. The final solution is in fact to avoid the use of austenitics for high temperature applications. For a long time this has not been a valid approach simply because ferritic steels have not had sufficient strength at high temperatures. However, over the last 10–15 years there has been a marked improvement in the development of strong 9–12% Cr steels which have very attractive properties for high temperature use. Already with steels such as T91 and HCM12M we have materials to replace austenitics at temperatures up to

600°C and newer steels are currently being developed for use at even higher temperatures.

In complete contrast to the behaviour of the 316 pipework at Eddystone was the response of this same material to thermal cycling on the CEGB Ultra-supercritical unit C11 at Drakelow C. In this case thermal ratchetting under cyclic conditions caused movement of the pipework by several feet which could only be compensated for by cutting sections out. Experience on the Drakelow C, Unit 12 with Esshete 1250 pipework has been much better with very little distortion in 102 000 h at 600°C probably because its higher tensile strength reduces the susceptibility to thermal ratchetting.[25]

5.4 Bends

Bends in both reheat and main steam pipework represent a potential source of concern. A number of reports have been made which give details of failures on bends in West German Power Stations and also Russian Stations. The materials affected have been 0.5CrMoV and 1Cr0.5Mo and also 12CrMoV.[2,26,27] In a number of cases there have been major steam leaks and also one or two instances of catastrophic failures (Fig. 13). The sequence of failures appears to be firstly the initiation of intergranular creep cavitation, microcrack formation and ultimately axial cracks along the extrados of the bend. The damage pattern has, however, been extremely variable and significant cavitation has also been observed at the intrados and tangential bend positions. Failure times have also been very variable and quoted as 46 000 h at 540°C for one set of reheat pipework to 120 000 h at 525°C for 1Cr0.5Mo bends in a boiler dead space. Because of these problems a systematic scheme of inspection is now applied on West German stations. Replication has been used extensively and a system of characterising

Fig. 13 Burst pipe bend caused by creep damage 540°C 46 000h.

this, at least qualitatively, developed by Neubauer.[28] Typical results from the 176 MW unit B a Westphalia power station after 145 000 h operation were of 32 hot reheat bends:

2 bends had orientated cavitation (Category 4)
3 bends had many unorientated cavities (Category 3)
6 bends with isolated cavities (Category 2)
17 bends with no cavitation (Category 1)
4 bends were not investigated as they had been recently installed.

Two bends were particularly damaged – one over a length of 1400 mm and the other over 1600 mm–grinding these bends indicated substantial damage 3.7 mm below the surface ie. 18% of the wall thickness. The immediate action on finding damage of this form in 1984 was to reduce the operation temperature from 535°C to 510°C. However, because the station had recently been fitted with f.g.d. equipment it was decided in 1985 to replace all 4 pipe runs on the A and B units.

Despite a fairly extensive inspection programme, damage of the type described above has not been detected on UK plant and there seems to be three possible reasons why this is:

1. It is believed that a full renormalising heat treatment after hot bending was not always applied in Germany as in the UK although recent evidence by Jesper and Meyer[29] indicates that failures can occur in properly heat treated pipework.
2. Secondly in the UK a 10% increase in wall thickness is allowed in the design to accommodate increases in stress due to bending and this was not applied in Germany.
3. For 0.5CrMoV the German design stresses prior to 1960 were higher than the corresponding British Standard BS806.

It would appear therefore that as a result of these differences, for the same duty, German pipework would be operating with thinner section and at higher stresses than in the UK. This would also mean that the German pipework would also be more susceptible to misalignment in hanger supports which would increase the system loading stress.

Of interest to this conference is whether these pipe bend failures on the continent were primarily due to a loss of rupture ductility. My own belief is that at least initially this was not the case but a loss of creep rupture ductility does become important as the creep damage builds up.

The reasons for saying this are that from many different investigations it appears that the prior cold and hot work has very little effect on the rupture strength and ductility of the low alloy and high alloy ferritic steels and that in this respect there is a marked contrast to the behaviour of the austenitic steels for which prior strain increases rupture strength and reduces ductility.

Thus it would seem that the damage build up on bends is due primarily to increase in stress and that this gives rise to a more rapid accumulation of creep cavities than on the lower stressed straight sections of the pipe (Fig. 13).

As the cavitation damage accumulates and eventually gives rise to micro-cracking then clearly the material sustained a major loss of creep ductility. This is indicated by post exposure tests on specimens taken from a 12CrMoV bend after 60–80 000 h service when the rupture ductility was less than 2% compared to 20–30% from straight pipe sections.[29]

The main concern with pipe bend failures is the safety issue since it is impossible to sustain a leak-before-break argument. This is why a major monitoring programme on creep cavitation has been initiated in Germany. In the main this has been performed by use of plastic replicas and more recently research has been initiated on technologies such as density changes and ultrasonic velocity changes. Any sign of cavitation on a bend is sufficient grounds to consider replacement at the earliest opportunity and operation at reduced temperature until then.

5.5   Transition Joints

An important type of component which frequently occurs in the superheater section of boilers and on pipe runs are the welded transition joints or dissimilar metal joints between low alloy (ferritic) components and austenitic steel components. These joints must accommodate significant differences in chemical,

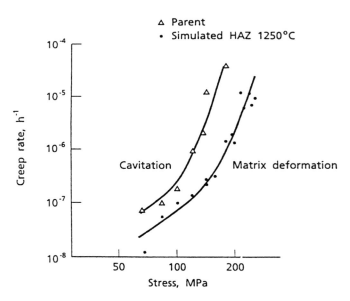

Fig. 14 Creep strength of parent and simulated HAZ 0.5Cr0.5Mo0.25V 565°C.

Fig. 15 Dissimilar metal weld showing interface failure following weld bead contours.

mechanical and physical properties and often operate at temperatures well within the creep range of the ferritic steel.

Experience indicates that the service lives of transition joints can be significantly shorter than those achieved on joints between similar metals.[30] The most common type of dissimilar metal joints are those between a low alloy ferritic steel such as 2.25Cr1Mo and an austenitic steel such as AISI 316. Typically, failures occur due to low ductility cracking in the heat affected zone of the ferritic steel or close to the fusion line. Welds made using an austenitic weld metal as filler are invariably weaker than similar welds made with a nickel base weld metal. Macroscopically, the failures appear to be entirely brittle as shown in Fig. 15 where the fracture follows the original fusion line and even outlines the weld bead contours.

However, it must be emphasised that with transition joints marked changes in the mechanism and location of failure can occur depending on the precise details of weld chemistry, heat treatment and microstructure and the imposed conditions of stress and temperature. The point illustrated in Fig. 16. This shows the decrease in rupture strength and concomitant decrease in rupture ductility as the failure point moves from the parent ferritic steel at high stresses (short rupture times) to the interface at low stresses (long rupture times).

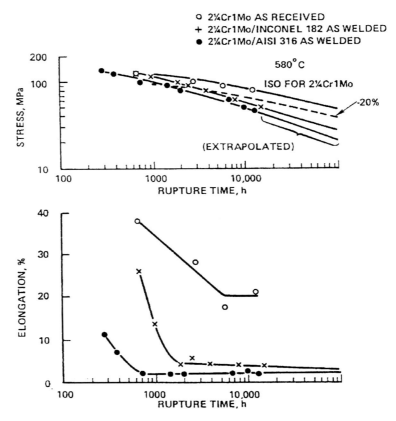

Fig. 16 (a) Variation of rupture strength and ductility on 2.25Cr 1Mo/316, (b) dissimilar metal welds and 2.25Cr 1Mo parent metal tested at 580°C.

Indeed, depending on the particular combination of the above variables, the location of failure can be, along the weld fusion line, in the coarse grained HAZ of the ferritic, in the fine grained (intercritically annealed zone) of the ferritic HAZ, within the weld metal or in some rare instances within the HAZ of the austenitic.[31]

It is extremely important therefore, with transition joints to regard each combination of parent and filler metal as a unique entity requiring detailed investigation over the complete spectrum of service conditions. The fitness for purpose of the particular combination can then be judged by the R5 assessment route developed by the CEGB[32] or by the PODIS route developed by EPRI.[33]

6 CONCLUSIONS

(i) Power Plant has operated successfully at temperatures in the creep range for over 50 years.

(ii) In the main the design criteria used for high temperature components are conservative.

(iii) The incidence of cracking of high temperature components is low and catastrophic failures have been extremely rare.

(iv) The reasons for these cracking incidents were that the original designs did not allow for variations in fabrication and operating parameters and for the concomitant changes in metallurgical structure which caused marked changes (reductions) in rupture ductility.

(v) Investigations of the different types of high temperature cracking incidents have identified the fabrication, operational and metallurgical parameters contributing to the failure(s).

(vi) From this understanding new specifications, fabricating methods and operational procedures have been developed to reduce the incidence of similar failures.

(vii) The use of these measures will increase future plant reliability. To implement these, it is essential that designers and plant operators are made fully aware of these developments.

ACKNOWLEDGEMENT

The author is indebted to the staff and management of ERA Technology in the preparation of this paper.

REFERENCES

1. I.D. Hall and G.H. Whitley: 'A review of the developments in material properties and quality which have led to improvements in availability in the CEGBs large fossil-fired generating plant', *Proceedings of an International Conference on Advances in Material Technology for Fossil Power Plants*. Chicago, Illinois, USA, ASM International, 1–3 Sep 1987.

2. W. Bendick, K. Haarmann and M. Teuke: '20 years of mannesmann experience with the assessment of service exposed main steam components', *Proceedings of International Conference on Residual Life of Power Plant Equipment - Prediction and Extension*, Hyderabad India, 23–25 Jan 1989, paper 4B4, 1989.

3. The National Board of Boiler and Pressure Vessel Inspectors: 'Boiler Reheat Line Explosion', *National Board Bulletin*, Columbus, Ohio, Oct 1985, **43**(2).

4. *Detroit Edison Today*, Feb 1986, **14**(4).

5. R.B. Dooley and H.J. Westwood: *Analysis and prevention of boiler tube failure*, Canadian Electrical Association CEA 83/237/931, 1983.

6. G.A. Lamping, R. Arrowood and R.B. Dooley: 'Manual for the investigation of boiler tube failures', *EPRI Report CS 3945*, 1985.

7. R.D. Townsend (Editor): *The boiler tube failure manual*, A National Power Publication, 1991.

8. R.D. Townsend: Keynote Paper. Boiler tube failure prevention in fossil fired boilers, 1st International Symposium Heat Exchanger and Boilers (HEB93), AEA Egypt, Alexandria, Jun 1993.

9. B.T. Bates and R.D. Townsend: *Creep life assessment technical and managerial aspects*, VGB Conference, The Hague 1981.

10. D.J. Gooch: 'How much longer? Remanent life assessment of high temperature components', *CEGB Research Special Issue No. 21 on Structural Integrity*, published by CEGB, May 1988.

11. R.D. Townsend: Procedures used by Central Electricity Generating Board for Components Life Assessment, I Mech E paper C312/87, 1987, pp 223.

12. C.S. Pillow, L.W. Perry and R.B. Dooley: 'Generic Guidelines for the Life Extension of Fossil Fuel Power Plant', *EPRI Report CS4778*, 1986.

13. R.D. Townsend: 'Performance of high temperature bolting in power plant', *Performance of bolting materials in high temperature Plant Applications*, Institute of Materials, 1995, 15.

14. F.L. Becker, S.M. Walker, B. Dooley and J.G. Byron: 'Guidelines for the Evaluation of Seam-Welded Pipes', *EPRI Report CS-4774*, Feb 1987.

15. D.J. Gooch and R.D. Townsend: 'CEGB Remanent Life Assessment Procedures', *EPRI International Conference Life Extension and Assessment of Fossil Power Plants*, 2–4 Jun, Washington 1986.

16. S.T. Kimmins and D.J. Gooch: 'Type W cracking in 1/2CrMoV steam pipe steels', *CEGB Report RD/L/3383/R88*.

17. C.J. Middleton: *CERL Reports Part 1 RD/L/N37/79, Part 2 RD/L/N38/79 and Part 3 RD/L/N47/79*, CEGB, Aug 1987.

18. C.J. Middleton: *Metal Science*, 1981, **15**, 154.

19. R.I. Jaffee: 'Materials and Electricity', *Metallurgical Transactions*, May 1986, **17A**, 755.

20. C.J. Middleton: 'An Assessment of the Relative Susceptibilities to Type IV Cracking of High and Low Ferritic Steel Pipework Systems', *I Mech E Steam Plant for the 1990's paper C386/CJM*, 1990.

21. W. Answald, R. Blum, B. Newbauer and K.E. Paulson: 'Remaining Life Affected by Repair Welds', *Proceedings, International Conference Creep*, Tokyo, 1986.

22. J.K. Lai and A. Wickens: 'Microstructural changes and variations in creep ductility of 3 casts of stainless steel', *Acta Met*, 1979, **27**.

23. J.E. Delong, W.F. Siddell, F.V. Ellis, H. Haneda, T. Tsuchiya, T. Daikoku, F. Masuyama and K. Setoguchi: 'Operational experiences and reliability evaluation of main steam line pressure parts of Philadelphia Electric Co. Eddystone No 1', Translated from November 1984 issue of *The Thermal and Nuclear Power*, 1984, **35**(11), 1225.

24. B. Plastow, B.I. Bagnall and D.E. Yeldham: 'Welding and fabrication in the nuclear industry', *BNES Conference*, London 1979.

25. J.K. Lai: 'Optimising the stress rupture properties of AISI Type 316 stainless steel', *CERL Report RD/L/N210/80*, CEGB, 1981.

26. H. Jesper: 'Creep damage to pipework following long periods in operation', *ECSC Steel Conference on Residual Life*, Brussels, Nov 1984.

27. N.S. Sirotenko, V.M. Ezhov and V.M. Kobrin: 'The effect of hot bending on the rupture strength of thick walled 12CrMoV steel pipelines', *Teploenergetika*, 1979, **26**(41), 53.55.

28. B. Neubauer: '1984 Creep damage evolution in power plants', *Proc 2nd Int. Conf. Creep and Fracture of Engineering Materials and Structures Swansea*, Pineridge Press, April 1984, 1271–1226.

29. H. Jesper, H. Meyer and H. Remment: 'Location, appearance and detection of creep damage in pipelines after long periods of service', *Paper No 11 of the VGB 1985 Conference*, 1985.

30. R.D. Nicholson, A.T. Price and J.A. Williams: *Conference on Dissimilar Welds in Fossil Fired Boilers*, EPRI CS 3623 1985, New Orleans Louisiana, USA. 1985.

31. J.N. Soo: 'Creep deformation and fracture of a dissimilar metal weld between 9CrMo ferritic steel and alloy, 600', *Institute of Metals Conference, Rupture Ductility of Creep Resistant Steels*, York 1990, Papers 25 and 26.

32. B. Nath, J. Phillips and J.A. Williams: 'Dissimilar metal welds, a study of failure incidence, failure characterisation and potential improved design methods', *Seminar on Design of Welds for High Temperature Applications*, I Mech E, Jun 1990.

33. D.I. Roberts, R. Ryder *et al.*: 'Dissimilar weld failure analysis development program', *Report CS 4252 Vol 1-9 EPRI*, Palo Alto Ca, 1985.

Ferritic Power Plant Steels: Remanent Life Assessment and the Approach to Equilibrium

H. K. D. H. BHADESHIA, A. STRANG[†] and D. J. GOOCH[‡]

University of Cambridge, Department of Materials Science and Metallurgy, Pembroke Street, Cambridge CB2 3QZ, UK

† ALSTOM Energy Ltd., Steam Turbines, Newbold Road, Rugby, Warwickshire, CV21 2NH, UK

‡ National Power PLC, Windmill Hill Business Park, Whitehill Way, Swindon, Wiltshire, SN5 6PB, UK

ABSTRACT

The steels used in the power generation industry are almost always given a severe tempering heat–treatment before they enter service. This might be expected to give them a highly stable microstructure which is close to equilibrium. In fact, they undergo many changes over long periods of time. This paper is a review of some of the methods which exploit the changes in order to estimate the life that remains in alloys which are only partly exhausted.

INTRODUCTION

Many of the safety-critical components in power plant are made of steels developed to resist deformation when used in the range 450–600°C and 15–100 MPa.[1,2] Many of these components are expected to serve reliably for a period of about 30 years,[3] and it is these which are the subject of this review.

Given the long life span, the typical tolerable creep strain rate may be about 3×10^{-11} s^{-1} (approximately 2% elongation over the 30 years). The design stress must be set to be small enough to prevent creep rupture over the intended life of the plant. Other damage mechanisms which must also be considered include fatigue, thermal fatigue, creep-fatigue, progressive embrittlement, corrosion/oxidation and, where relevant, deterioration caused by hydrogen.

The steels are able to survive for such long periods because the operating temperature is only about half of the absolute melting temperature, making the migration of atoms very slow indeed. Figure 1 illustrates the typical distance ($\simeq$10 nm) diffused by iron or substitutional solute atoms during one hour at 500°C. The longevity of iron alloys in power plant relies on the fact that the diffusivities are incredibly small, Fig. 1. Notwithstanding this, slow but significant changes are expected over the long service life.

Another reason why the changes that are expected over the years must occur at a slow rate is that the steels are generally severely tempered before they enter

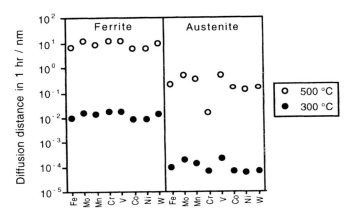

Fig. 1 A $2\sqrt{Dt}$ estimate of the distance diffused in one hour, for iron and some substitutional solutes in iron, as a function of temperature.[4] D is the diffusion coefficient and t the time.

service. This gives them a very stable microstructure. The stability can be discussed in terms of the free energy stored in the material, as illustrated in Table 1. The stored energy of the plain carbon steel (Fe–Mn–C) is described relative to the equilibrium state which is a mixture of ferrite, graphite and cementite without any defects. The phases in cases 1 and 2 involve the equilibrium partitioning of all elements so as to minimise free energy. In cases 3–5 the substitutional solutes are configurationally frozen.

Table 1 The stored energy as a function of microstructure, defined relative to the microstructure for which the stored energy is arbitrarily set to zero. Bainite and martensite have higher stored energies relative to ferrite because of strain energy contributions and solute trapping. The term 'paraequilibrium' means that the phase has the same ratio of iron to substitutional solute atoms as the parent phase from which it formed.

Case	Phase mixture in Fe–0.2C–1.5Mn wt.% at 300 K	Stored energy / J mol^{-1}
1.	Ferrite, graphite & cementite	0
2.	Ferrite & cementite	70
3.	Paraequilibrium ferrite & paraequilibrium cementite	385
4.	Bainite & paraequilibrium cementite	785
5.	Martensite	1214
Case	Phase mixture in Fe–0.1C–2.2Cr–1Mo wt.% at 873 K	Stored energy / J mol^{-1}
6.	Ferrite & paraequilibrium cementite → ferrite & $M_{23}C_6$ (e.g. after extensive service)	0
7.	Ferrite & paraequilibrium cementite → ferrite & M_2X (e.g. when steel enters service)	63

Cases 6 and 7 are for a classical power plant alloy. It is seen that the reduction in stored energy when the microstructure changes from the M_2X to $M_{23}C_6$ microstructure is very small. This means that it will occur at a correspondingly slow rate.

Another consequence of the small diffusion coefficients is that the dominant creep mechanism is the climb of dislocations over obstacles with the help of thermal energy. The obstacles are mainly carbide particles which are dispersed throughout the microstructure.

Suppose that the microstructure and the operating conditions do not change during service. The accuracy with which component life might then be predicted would depend only on the quality of the experimental data. Uncertain data lead to more conservative design with larger safety factors. There are many different kinds of design criteria, but as an example, the design stress σ for boiler components where close dimensional control is not vital is given by:[5,6]

$$\sigma = \frac{\sigma_r\{t_r\}}{1.1 \times \text{safety factor}} \tag{1}$$

where $\sigma_r\{t_r\}$ is the stress required to rupture a specimen over a time period t_r, and the factor of 1.1 allows for material thinning during manufacture. It was the practice 40 years ago to design for $t_r = 10^5$ h but power plant design lives are more typically at 2.5×10^5 h.

For turbine components where dimensional tolerance is crucial, the design stress is defined in terms of the creep strain, for example,

$$\sigma = \frac{\text{stress to give specified strain in } t_r \text{ h}}{\text{safety factor}} \tag{2}$$

The specified strain may be in the range 0.1–0.5%. The safety factor is typically $\simeq 1.5$ but can be reduced to $\simeq 1.3$ if reliable long term data are available. Safety factors for assessment, as opposed to for design, may be reduced if additional long term data become available.

The steels used are always heterogeneous and the service conditions vary over a range of scales and locations. The design life is therefore set conservatively to account for the fact that measured creep data follow a Gaussian distribution with a significant width. In spite of this, experience has shown to date that plant reaching its original design life generally has many further years of safe operating life remaining. To take advantage of this observation requires methods for the reliable estimation of the *remaining life*. The techniques used for this purpose are summarised in Table 2. Of the properties listed, no single measurement is sufficiently comprehensive to describe the steel with all requisite completeness. The implementation of a life–extension procedure must consequently be based on broad considerations backed by more frequent inspections. Unfortunately, many

Table 2 Some of the methods used in the estimation of remaining life.

Category	Property/Test method
Mechanical	Hardness
	Tensile test
	Small specimen tests
	Impact toughness
	Component creep strain monitoring
Microstructural	Interparticle spacing
	Cavitation parameter
	Number density of cavities
	Volume fraction of cavities
	Area fraction of cavities
	Misorientation measurements
Other	Component temperature monitoring
	Ultrasonics
	Resistivity
	Density

inspection procedures can only be carried out on equipment which has been shut down.

Remanent life assessment is a very large subject. The focus of this paper is creep–limited life, with only minor excursions into the important subjects of fatigue and embrittlement during service.[3] The components considered are those which might be expected to survive for many decades rather than components such as bolts which can be replaced relatively easily. The relative costs of replacement and life extension are also not addressed.

CRITICAL REGIONS FOR ASSESSMENT

The first step in any assessment procedure is to check for the absence of any gross defects or embrittlement effects during service, which might limit safe operation before creep damage becomes important. Assuming that such defects are absent or can be repaired, regions are selected which are likely to be at the most advanced stages of deterioration.

Components may be neglected in the assessment procedure if they are not life limiting. For example, failures of stub welds on steam headers are quite common but can be easily repaired.[5,6]

High and Intermediate Pressure Rotors

The critical regions for the accumulation of damage in high temperature rotors are the bore (if one is present), the blade root fixings, balance holes and stress concentrating features such as relief grooves, on the outer surface.[7]

In many rotors the position of maximum steady state strain accumulation is the rotor bore at the inlet stages. The potential failure mode which has to be guarded against is the initiation and growth of intergranular creep cracks to a size which could result in brittle fracture, either at cold start-up due to centrifugal and thermal stresses or during overspeed. Ultrasonic and eddy current techniques are used to detect macroscopic defects but the direct examination of the bore surface for microscopic damage provides a significant challenge. It is however now possible to prepare and examine surfaces and also to remove samples for examination using remotely controlled devices.[8]

A potential complication is that the region of maximum triaxial stress, which may be the area most susceptible to creep cavitation, is actually sub-surface from the rotor bore.

Some rotors are provided with bands or pips for the measurement of creep strain during service using datum measurements taken prior to service. Bore strains and remaining safe life can then be estimated through the use of critical strain data determined from experimental programmes.

Potential failure modes of blade root fixings[7] are either creep crack initiation and growth or shear failure due to excessive steady state creep strain accumulation. Axial entry root fixings can be examined directly or by the taking of plastic replicas without removing the blades but circumferential entry fixings are less accessible. Fortunately the latter are generally less highly stressed.

On some rotor designs the critical region is at the balance holes in the discs or at the transition between disc and body. These regions are more accessible for examination.

It is not often practicable to remove sufficient material from critical areas to produce specimens for direct determination of remaining life by accelerated creep testing. In this case samples may be prepared from 'cold' regions with the aim of positioning the original creep properties of the specific forging within the scatter band assumed by the design procedure.

High Temperature Headers

Steam headers (*i.e.* nodes where many large diameter, steam-carrying pipes converge) are regarded as critical components because they are subject to large variations in temperature and pressure. They are also extremely expensive and difficult to replace.[5,6]

Boiler headers are built to many different designs. However in general they involve butt and end cap welds between thick section pipes as well as branch and stub welds. Experience shows that creep damage generally occurs first in the vicinity of these weldments. Cracks may also be formed in the parent material ligaments between stub penetrations as a result of creep fatigue interactions generated by thermal stresses. In addition to these observed damage modes headers are accumulating microscopic or sub-microscopic creep damage throughout operation which has to be ssessed if the component is to be used beyond its original design life.

In general, steady state relaxed stresses are at a maximum at outer surfaces of cylindrical pressure vessels so that access is not a major problem. Direct observation, replication and material sampling are then options for damage assessment by microstructural techniques. There are also techniques for removing samples for accelerated creep testing without compromising the integrity of the component.[9] However, the critical region is not always accessible. For example (1) damage in welds generally initiates sub-surface due to microstructural variations and (2) the position of maximum stress at stub penetrations is at the bore. In these cases stress analysis, service experience and ultrasonic examination become the primary assessment tools.

Pipework

The areas most susceptible to creep damage in high pressure pipework systems are, as in headers, the welds. However, whereas in headers the stresses are generally determined by the pressure and thermal stresses alone, which are known or can be calculated, pipework systems are also subject to highly variable and largely unquantified external forces. This means that life management procedures for welds have remained largely experience based despite a great deal of effort expended in trying to devise quantitative prediction procedures.[10] Qualitative assessment of the severity of creep cavitation through examination of surface replicas is frequently used for pipe weld life management.

For operation beyond design life the condition of the parent material also has to be determined. Obtaining material for this is not a problem but the huge number of heats and individual forgings and castings in a typical system presents difficulties. Attention may then be concentrated on critical components within the system such as T-pieces and, in particular, bends. The latter are more highly stressed than straight pipes because of thinning and ovality produced during the bending process as well as the greater potential for stresses generated by system forces. Experience in continental Europe has confirmed the analytical prediction that outer surfaces are the most susceptible to creep damage but a key question is the profile of damage blow the surface. However, this does not appear to always be the case; Viswanathan[11] has found sub-surface cavitation to be the most severe. Methods which have been examined for quantification of the damage include the usual metallographic procedures as well as ultrasonic, magnetic and resistivity based techniques. Surface replication techniques are obviously not appropriate when the damage is most severe within the bulk of the pipe.

Chests and Casings

Valve chests and turbine casings are large complex components which are usually castings but sometimes closed die forgings. Internal cracking due to thermal fatigue is the most common mode of degradation encountered; procedures have been developed for this but are not within the scope of this review. There have been instances where poor design has led to creep damage in internal ligaments of valve chests but, in general, remanent life assessment of chests and casings has not been as great an issue as with the foregoing components (The greater concern has been with the welds joining valve chests to pipework systems.)

DAMAGE SUMMATION

A satisfactory way of representing creep damage (C) is to use a parameter (ω) which is normalised by its value at failure (ω_r). The magnitude of ω_r will depend on the precise values of stress (σ), temperature (T) and any other variable which influences the creep process. Since these variables are not necessarily constant, the extent of damage is often written[12,13]

$$C = \sum_i \frac{\omega_i}{(\omega_r)_i} \qquad (3)$$

where ω is typically the time or the creep strain. Failure occurs when the sum achieves a value of unity. In practice, failure by creep or fatigue occurs in three stages: initiation of a crack, growth of the crack to a critical size and finally, much more rapid failure. The ultimate failure may be ductile if it occurs at high temperatures or brittle when it occurs at low temperature, for example during the start–up or shut–down of plant.[11] The life fraction rule clearly does not distinguish between these stages, i.e. it does not embody *mechanism*. It should not be used for predictive purposes through a mechanism change.

Evans[12] argues that it is more appropriate to use the strain rather than the time, since the latter is not considered as a 'state variable'. In the context of thermodynamics, the state of a system can in principal be specified completely by a number of state variables (such as temperature, pressure) such that its properties do not depend at all on the path by which those variables were achieved. This clearly cannot be the case even for the creep strain. This is because the extent of damage *is* expected to depend on the path by which a given value of strain is achieved, for example, whether the strain is localised at grain boundaries or uniformly distributed. This necessarily means that eqn 3 is an approximation; as Evans states, it should be a reasonable approximation if the mechanism of creep does not change between the components of the summation. Thus, Cane and Townsend[14] conclude that the use of the life fraction rule in taking account of temperature variations is more justified than for variations in stress. This is because for the latter case, the dislocation networks become finer (relative to the carbide spacings) at large stresses. The network nodes then do not coincide with carbide particles, thus changing the mechanism of deformation. This is not the case with variations in temperature because the dislocation network then scales with the particle spacing. The failure of the life fraction rule is sometimes accommodated by empirically setting the limiting value of C to some positive value which is not unity.

HARDNESS

Figure 2 shows a hardness scan taken on the bore surface of a 1CrMoV HP–IP[†] rotor which had been in service for 174 000 h with an inlet steam temperature of

---

[†]HP and IP stand for high–pressure and intermediate–pressure.

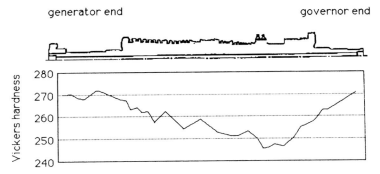

Fig. 2 Hardness distribution on the bore of a 1CrMoV steel rotor after long term service. Note that the steam entry point corresponds to the location of the minimum hardness. The corresponding hardness distribution in an unused rotor is more uniform with a variation of about ±10 HV about the mean.[15] Specific steel compositions are listed in Table 3 and discussed later.

538°C. The hardness in the cold region near the generator is close to the value expected when the rotor entered service whereas there is a minimum in the region where the service conditions are expected to be most severe.

The hardness can in principle be used as an indicator for the state of the steel in its life cycle. Changes in hardness occur due to recovery, coarsening of carbide particles, and recrystallisation. All creep-resistant power plant steels are severely tempered before they enter service. They are therefore beyond the state where secondary hardening is expected and the hardness can, during service, be expected to decrease monotonically. In these circumstances, an Avrami equation adequately represents the changes in hardness,

$$\xi = 1 - \exp\{-k_A t^n\} \tag{4}$$

where t is the time, k_A and n rate constants and ξ is given by

$$\xi\{t\} = \frac{H_0 - H\{t\}}{H_0 - H_\infty} \tag{5}$$

where H_0 is the initial hardness, H_∞ is its hardness at the end of useful life and $H\{t\}$ the hardness at time t. The initial hardness may not be available when making life assessments, in which case an approximate value might be determined from the cold end of a rotor. Simpler linear relationships are used in practice. Thus, for 1CrMoV rotor steels, Gooch et al.[16] find

$$\frac{H\{t, T\}}{H_0} \simeq 1.605 \times 10^{-3} P - 4.962 \times 10^{-8} P^2 - 11.99$$

$$\text{where} \quad P = T(16.2 + \log\{t\})$$

with the temperature and time expressed in units of Kelvin and hours respectively. P is the Larson–Miller parameter which helps rationalise time and temperature effect. The scatter and range of the data on which this equation is based is illustrated in Fig. 3.

Judging from data in the published literature for 1CrMoV steels, the hardness at the point where the microstructure is extensively annealed is likely to be around $H_\infty = 150$–190 HV for most power plant steels. Its main components include the intrinsic strength of iron and solid solution strengthening. The starting hardness is likely to be in the range $H_0 = 220$–280 HV. Under power plant operating conditions, all that can be expected is a change in hardness of about 30–70 HV over a period of some 30 years. Thus, Roberts and Strang[18] have shown that the hardness can decrease by about 20% in the stressed regions of long–term creep test specimens when compared with the unstressed parts. This is consistent with an approximately 25% reduction found by Maguire and Gooch[19]. Figure 4 shows the nature of the changes in hardness to be expected typically, as reported by Maguire and Gooch[19] for a 1CrMoV steel which was tempered at 700°C for 18 h prior to the ageing at temperatures in the range 600–640°C.

Gooch and co-workers[16,17,19] have studied the hardness of a large number of 1CrMoV rotors. They find again that the reduction in hardness in the gauge length is always greater than that in the head of creep test samples:

$$H\{t\}_{gauge} - H_0 = 1.234(H\{t\}_{head} - H_0) - 14.55 \qquad (6)$$

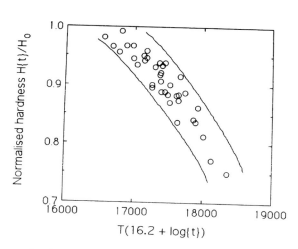

Fig. 3 The change in the normalised hardness as a function of the Larson–Miller parameter for a 1CrMoV rotor steel. The units of time and temperature are hours and Kelvin respectively. After Strang *et al.* Ref. 17.

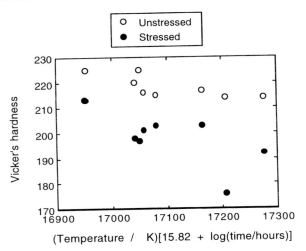

Fig. 4 Changes in the hardness of a 1CrMoV steel during ageing in the temperature range 600–640°C . The samples were tempered at 700°C for 18 h prior to testing; the effect of the 700°C treatment is not included in the calculation of the time–temperature parameter plotted on the horizontal axis. The open circles are data from the grips of a creep test specimen, and the filled circles represent measurements from the gauge length. The large filled circles correspond to samples whose hardness values were measured at failure, *i.e.* after the complete exhaustion of life. The stress applied during each of the tests was 93 MPa. Data from Maguire and Gooch.

where the subscripts gauge and head refer to the positions where the hardness tests were conducted. The observations indicate that gauge length softening is proportional to that in the head, apparently irrespective of the stress, temperature or test duration. This is inconsistent with an empirical relationship used by Tack *et al.*[20]:

$$\frac{H\{t\} - H\{t,\epsilon\}}{H\{t\} - H_\infty} = b_1 + b_2 \ln\{\epsilon\}$$

$$\equiv b_2 \ln\{\frac{\epsilon}{\epsilon_0}\}$$

(7)

where b_1, b_2 and ϵ_0 are empirical constants. Unfortunately, values of these constants were not presented in the original paper. The equivalence that we have expressed in eqn 7 illustrates the fact that the equation can only be applied for strains greater than ϵ_0. For $\epsilon < \epsilon_0$ the equation predicts an increase in hardness which was not intended in the original representation. ϵ_0 can therefore be regarded as a threshold strain below which there is no effect of strain softening. If it is now assumed that ϵ_0 is large then the proportionality implied by Gooch *et al.* is recovered. Furthermore, the idea of a threshold strain seems to correlate with a threshold stress reported by Goto[15] and by Strang *et al.*,[17] below which they did not observe any difference between the gauge and head regions.

Consistent with all the work described above, Goto[15] has also observed a greater rate of softening in strained regions, but with the difference with the unstrained regions being a strong function of the stress (Fig. 5). He therefore described the hardness in terms of an empirical relationship of the following form:

$$\frac{H\{t, T, \sigma\}}{H_0} = 1 - b_{10}\sigma^{b_{11}} \frac{t}{t_r} \exp\{\frac{-b_9}{T}\} \tag{8}$$

where as usual, b_9, b_{10} and b_{11} are empirical constants and t_r is the time to rupture. On fitting these results to experimental data it was found that $b_{10} = 5500$, $b_{11} \simeq 0$ and $b_9 = 8511$ K. This is a surprising result which suggests that the stress has no effect on the change in hardness, contradicting the data presented in Fig. 5. This anomaly arises because stress is implicitly included twice in eqn 8. The rupture time t_r is a strong function of stress.

Given the variety of relationships proposed, it would be useful to reanalyse the entire collection of data either empirically or with a model that has some fundamental basis.

Goto has emphasised a quantitative relationship between the hardness and the rupture life, and hence the remaining life via the parameter t/t_r. Hardness is expected to be a crude indicator of remaining life. It does not, for example, relate easily to creep cavitation *damage*. Hardness tests are nevertheless useful in the regime of steady–state creep before the onset of gross damage. This is evident

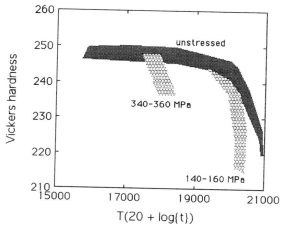

Fig. 5 Hardness as a function of the Larson–Miller parameter and as a function of the stress, for a 1CrMoV rotor steel. The time and temperature are in units of Kelvin and hours respectively. Simplified from a version by Goto.[15]

from Fig. 5 where it is seen that specimens with the same hardness are at different life-fractions.

There is a further complication, that the hardness of welded regions is likely to be inhomogeneous even when the welds are made with matching compositions. The potential location of failure is then difficult to identify since creep ductility, creep strength and creep strain may vary with position. The weld metal is likely to have a lower ductility than the parent material because of (a) the larger inclusion content, (b) the presence of regions of large grain size such as un-transformed or partially transformed columnar grains or, (c) trace levels of creep embrittling species such as As, Sn, Sb and P. It cannot therefore be assumed that failure will always occur in the softest part of the joint. A harder weld may be needed to ensure the same rupture life as the parent steel (Fig. 6).

The heat affected zone of the parent material also has a highly inhomogeneous microstructure with the potential for narrow bands of either very large or very fine grains due to the different time–temperature history during welding. In general a fine grain size leads to high ambient temperature hardness but low creep strength, although the latter may be compensated by high creep ductility. However, local variations in time–temperature history also result in extreme differences in carbide distribution. These may be the principle determinators of local hardness and creep strength although not necessarily ductility or rupture life.

Finally, the presence of (a) narrow bands of weak material in the heat affected zone and (b) the interface between weld metal and heat affected zone produces local regions of multi-axial stress. Increasing tri-axiality reduces creep de-

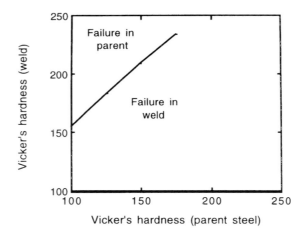

Fig. 6 Data for 2.25Cr1Mo steel and matching weld metal. The curve represents the locus of all points along which the weld metal and parent metal have equal rupture lives. Note that for a given life, the weld must be harder than the parent steel (after Tack et al.[20]).

formation rates but increases creep cavitation. The conflict between these two effects may result in either reduced or increased rupture life. Reduced ductility because of this effect is almost certainly a contributory factor in the so-called 'Type IV' cracking in fine grained fegions of low alloy ferritic weld heat affected zones. Conversely, the supporting effect of a strong parent material is the reason why it is often possible to specify a weld metal of lower creep strength if there is a concern over the ductility of a matching weld metal.

In short, the relationship between hardness and life of a weld is amazingly complex. Hardness may therefore only be used as a general indication of the state of the microstructure and only then if there are ample specific data for the materials and detailed welding processes concerned. This is well illustrated by the classic case of creep failures of low alloy ferritic welds, which may be in the weld metal or coarse or fine heat affected zones depending on chemical composition, weld process details, weld geometry or stress or residual welding stress level.

Mechanism of Strain Softening

The observation that the hardness in the gauge length of the creep test specimen is smaller than that in the unstressed part is consistent with much basic work on the tempering of ferrite, bainite and martensite. In this section we review the basic mechanisms by which strain softening might occur. There are two kinds of studies, one in which the steel is deformed first and then tempered (strain tempering) and the other in which tempering occurs during creep deformation (creep tempering).

Strain tempering experiments involve the heat treatment of plastically deformed bainite or martensite. The process leads to large increases in strength with some loss in ductility and toughness. The strengthening that is obtained increases with the level of prior deformation but is not just a reflection of the effects of deformation on the microstructure. The subsequent tempering also causes a significant rise in strength by the precipitation of very fine alloy-carbides on deformation induced defects.[21] This work emphasises that an increase in the dislocation density accelerates precipitation, probably by providing a larger number density of heterogeneous nucleation sites. The latter is well-established for the control-rolling of microalloyed steels, where the strain-induced precipitation of carbo-nitrides on the dislocations introduced during the deformation of the steel is common.[22] Precipitation rarely occurs in the matrix in the absence of the dislocations.[23] There are many studies reviewed by Gladman which show that the acceleration of precipitation varies in proportion with the degree of deformation.[24–26]

It is established that the recovery and recrystallisation that occurs when bainite is tempered is accelerated during creep deformation.[27–28] Ridal and Quarrell have also shown that in Mo containing bainitic steels, Mo_2C gives way more rapidly to the more stable M_6C carbide when the tempering occurs during creep deformation. The acceleration is greater for tests at lower temperatures, for higher creep strain rates and for steels where the kinetics of the $Mo_2C \rightarrow M_{23}C_6$

reaction are slow in the absence of creep. The results are consistent with an enhancement of diffusion and nucleation rates, both of which can be attributed to the dislocations. Such effects are expected to be noticeable at low temperatures, high strain rates and when the reaction rates are intrinsically slow. In the same work, Ridal and Quarrell also showed that the $M_7C_3 \rightarrow M_{23}C_6$ reaction during the tempering of Fe–Cr–C martensitic alloys is, if anything, slightly retarded by creep deformation. If this observation is certain then it implies that dislocations stabilise the M_7C_3. This effect is known for other carbides in the context of martensite[29] and bainite,[30] where a large dislocation density retards the precipitation of ϵ-carbide by trapping carbon atoms at the dislocations. Whether this is applicable to the precipitation of alloy carbides remains to be demonstrated.

INTERPARTICLE SPACING

Precipitates impede the motion of dislocations and any strength in excess of H_∞ is often related to the spacing (λ) between the particles (Cane *et al.*, 1986, 1987):

$$H - H_\infty \propto \frac{1}{\lambda} \propto t^{-\frac{1}{3}} \tag{9}$$

where it is assumed that $\lambda \propto t^{\frac{1}{3}}$ in order to be consistent with coarsening theory. There can be complications in this simple interpretation because there are generally two populations of particles, those on lath boundaries and other smaller precipitates in the matrix as well. The latter naturally tend to dissolve during any coarsening process.

Some results are illustrated in Fig. 7 for a 1Cr0.5Mo steel.[31] The spacings are typically measured using transmission electron microscopy at a magnification of about ×20 000 with approximately 100 fields of view covering 30 μm^2 taken at random. The actual measurement involves counting the number of particles per unit area (N_A) and it is assumed that $\lambda = N_A^{-\frac{1}{2}}$. The amount of material examined in any transmission microscope experiment is incredibly small, so care has to exercised in choosing representative samples of steel. In some cases, the microstructure may be inherently inhomogeneous. One example is the 12Cr and 9Cr type steels where there is a possibility of regions of δ-ferrite where the precipitation is quite different from the majority tempered martensite microstructure.

Obviously, hardness tests are much simpler to conduct when compared with the effort required to measure properly the particle spacings. A further complication is that there is frequently a mixture of many kinds of particles present, some of which continue precipitation during service whereas others dissolve. Thus, Battaini *et al.*[32] found that in a 12CrMoV steel, precipitation continues to such an extent during service that there is a monotonic *decrease* in λ. In fact, the distribution of particles was bimodal with peaks at 30 nm and 300 nm diameters.

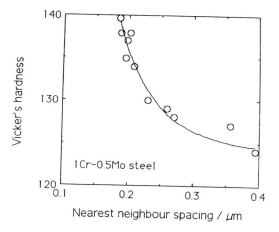

Fig. 7 The hardness as a function of the near neighbour spacing of carbides following creep tests at 630°C for a variety of time periods. After Askins and Menzies, 1985.[33]

It is strange that they were only able to correlate the hardness against the changes in the coarser particles. For another steel (12CrMoVW), Battaini *et al.* found an even more complex variation in the interparticle spacing with a maximum value in λ for the coarse particles.

It should be emphasised that the standard error in λ measurements is quite large, frequently larger than the variations observed. Given the difficulties of interpretation, and the experimental error, it is unlikely that λ measurements can be used as a satisfactory general measure of remaining life. The data can nevertheless be of use in the design of physically based creep models.

EQUILIBRIUM PHASES

The simplest assumption in kinetic theory is to take a 'flux' to be proportional to the 'force', where the magnitude of the latter depends on the deviation from equilibrium. It follows that to understand the microstructural changes that happen in a steel during service, it is necessary first to consider the equilibrium state. Table 3 lists some of the common power plant alloys together with quotes from material specifications. It seems that the specifications are compiled by agreement between many vested interests and therefore tend to be unrealistically broad. Thus, many of the steels have only a maximum carbon concentration specified! The actual alloys are purchased by users to much tighter ranges than those given in Table 3.

The results of thermodynamic calculations which give the phase fractions and compositions as a function of the overall alloy composition and temperature, are given in Table 4 for the common power plant steels. The calculations have been

Table 3 Typical compositions (wt.%) of creep-resistant steels used in the power generation and petrochemical industries. The range of compositions stated lie within the technical specification, but are really intended to reflect the variations observed in practice. The sulphur concentration is usually within the range 0.005–0.02 wt.%, and that of phosphorus within the range 0.005–0.025 wt.%.

Designation	C	Si	Mn	Ni	Mo	Cr	V	Others
1Cr0.5Mo	0.15	0.25	0.50	—	0.6	0.95	—	
range	0.08–0.18	0.10–0.35	0.40–1.00	—	0.40–0.60	0.70–1.10	—	
0.5Cr0.5Mo0.25V	0.12	0.25	0.50	—	0.6	0.45	0.25	
range	0.10–0.15	0.10–0.35	0.40–0.70	<0.30	0.50–0.70	0.30–0.60	0.22–0.28	
1CrMoV	0.25	0.25	0.75	0.70	1.00	1.10	0.35	
range	0.24–0.31	0.17–0.27	0.74–0.81	0.60–0.76	0.65–1.08	0.98–1.15	0.27–0.36	
2.25Cr1Mo	0.15	0.25	0.50	0.10	1.00	2.30	0.00	
range	< 0.16	< 0.5	0.3–0.6	—	0.9–1.1	2.0–2.5	—	Ti=0.03
Mod. 2.25Cr1Mo	0.1	0.05	0.5	0.16	1.00	2.30	0.25	B=0.0024
3.0Cr1.5Mo	0.1	0.2	1.0	0.1	1.5	3.0	0.1	
range	< 0.16	< 0.5	0.30–0.60	—	0.45–0.65	4.0–6.0	—	
3.5NiCrMoV	0.24	0.01	0.20	3.50	0.45	1.70	0.10	
range	< 0.29	< 0.11	0.20–0.60	3.25–4.00	0.25–0.60	1.25–2.00	0.05–0.15	
9Cr1Mo	0.10	0.60	0.40	—	1.00	9.00	—	
range	< 0.15	0.25–1.00	0.30–0.60	—	0.90–1.10	8.00–10.00	—	Nb=0.08 N=0.05
Mod. 9Cr1Mo	0.1	0.35	0.40	0.05	0.95	8.75	0.22	Nb 0.06–0.10
range	0.08–0.12	0.20–0.50	0.30–0.60	< 0.2	0.85–1.05	8.00–9.50	0.18–0.25	N 0.03–0.07 Al < 0.04
9Cr0.5MoWV	0.11	0.04	0.45	0.05	0.50	9.00	0.20	W=1.84 Nb=0.07 N=0.05
range	0.06–0.13	< 0.50	0.30–0.60	< 0.40	0.30–0.60	8.00–9.50	0.15–0.25	Nb 0.03–0.10 N 0.03–0.09 Al < 0.04
12CrMoV	0.20	0.25	0.50	0.50	1.00	11.25	0.30	
range	0.17–0.23	< 0.5	< 1.00	0.30–0.80	0.80–1.20	10.00–12.50	0.25–0.35	W=0.35
12CrMoVW	0.20	0.25	0.50	0.50	1.00	11.25	0.30	W< 0.70
range	0.17–0.23	< 0.5	< 1.00	0.30–0.80	0.80–1.20	10.00–12.50	0.25–0.35	Nb 0.30
12CrMoVNb	0.15	0.20	0.80	0.75	0.55	11.50	0.28	N 0.06

Table 4 The mole fractions of precipitate phases in the steels listed in Table 2, when at equilibrium at 565°C (838 K).

Designation	M$_2$X	M$_3$C	M$_7$C$_3$	M$_{23}$C$_6$	M$_6$C	Laves	NbC	NbN	VN
0.25CrMoV	0.0053			0.0247					
1CrMoV	0.0089			0.0412					
2.25Cr1Mo				0.0335					
Mod. 2.25Cr1Mo				0.0211	0.0019				
3.0Cr1.5Mo				0.0185	0.0057				
3.5NiCrMoV	0.0009		0.0161	0.0285					
9Cr1Mo				0.0222					
Mod. 9Cr1Mo				0.0222				0.0009	0.0030
9Cr0.5MoWV				0.0248		0.0135		0.0008	0.0032
12CrMoV				0.0443					
12CrMoVW				0.0444	0.0007				
12CrMoVNb				0.0318			0.0006	0.0029	0.0018

done using the *MTDATA* computer program and *SGTE* (Scientific Group Thermodata Europe) database, taking into account the carbide phases and Laves phase listed[†]. The chemical elements considered are carbon, silicon, manganese, chromium, nickel, molybdenum, vanadium and niobium; nitrogen is not included but is known to be important in forming MX particles. Note also that some very careful work has recently identified an M$_5$C$_2$ in 1Cr–0.5Mo steels;[34] this phase has not been included in the analysis.

CHANGES IN CARBIDE CHEMICAL COMPOSITIONS

The cementite that forms in association with bainite or martensite has at first, a chemical composition which is far from its equilibrium composition. It therefore enriches with elements such as chromium during service in power plant, and might be used as a built-in recorder for the thermal energy felt by the steel.[35–44] The method has been used to estimate the effective service temperature of 0.5Cr0.5Mo0.25V steel to an accuracy of ±12°C.[5,6]

Virgin Cementite
There is strong evidence to show that cementite and other iron-based transition carbides grow by a paraequilibrium, displacive transformation mechanism when the precipitation is from martensite or bainite. This subject has been reviewed recently.[45] In this mechanism, the substitutional lattice is displaced into the new

---

[†]The *SGTE* database contains assessed thermodynamic data which are interpreted by *MTDATA* to estimate phase diagrams *etc*. *MTDATA* has been developed by the Thermochemistry Group of the National Physical Laboratory, Teddington.

structure at a rate controlled by the diffusion of carbon to the transformation front. This is not terribly surprising given the very small distances over which the susbtitutional atoms can diffuse during the precipitation event (Fig. 1). Cementite, for example, precipitates in a matter of seconds during the tempering of martensite.

Of course, not all cementite in power plant steels comes from the bainite or martensite reactions. Many low-alloy steels contain small quantities of pearlite. The cementite which is a constituent of pearlite grows by a reconstructive transformation mechanism in which all of the atoms diffuse in a way which minimises strain energy. But the diffusion at the same time allows species of atoms to partition into phases where they are most comfortable. Consequently, the cementite within pearlite always has a chemical composition which is closer to equilibrium. This is illustrated in Fig. 8a, where the cementite when it first forms, is seen to be richer in chromium for pearlite than for bainite. This is the case even when the pearlite and bainite grow at the same temperature.[46]

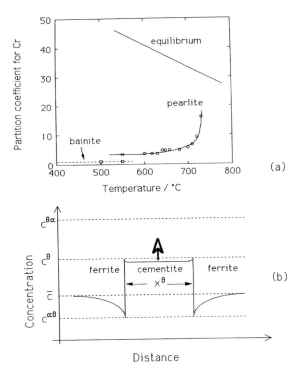

Fig. 8 (a) The partition coefficient for chromium in cementite, when the cementite is a part of bainite or pearlite, together with equilibrium data.[46] The partition coefficient is the ratio of the concentration in cementite to that in the ferrite. (b) The solute concentration profile that develops during the enrichment of a cementite particle.

Naturally, the cementite will tend towards its equilibrium chemical composition when exposed to elevated temperatures for prolonged periods of time. This generally means that its substitutional solute content increases with time. It can be shown[39] that this change is given by:

$$t = \frac{\pi[x^\theta(c^\theta - \bar{c})]^2}{16D_\alpha(\bar{c} - c^{\alpha\theta})^2} \qquad (10)$$

where it is assumed that the particle is in the form of a thin plate of thickness x^θ, t is the time at the annealing temperature, c^θ is the average solute concentration in cementite (θ) at time t, D_α is the diffusion coefficient of the substitutional solute in ferrite, $c^{\alpha\theta}$ is the solute concentration in ferrite which is in equilibrium with that in cementite, and $\bar{c}$ is the initial concentration of solute in the carbide (Fig. 8b). Note that c^θ is not the solute concentration $c^{\theta\alpha}$ in cementite which is in *equilibrium* with ferrite. The diffusion flux in the ferrite is so small that the solute concentration in the cementite at the boundary cannot build up to the equilibrium level since any enrichment at the interface is relieved by diffusion within the cementite. This is because $(\bar{c} - c^{\alpha\theta})$ which drives the diffusion within the ferrite is much smaller than $(c^{\theta\alpha} - \bar{c})$, *i.e.* the solubility of elements such as chromium is much larger in cementite than in ferrite. It follows that in spite of the diffusion flows, the concentration distribution in the cementite is always likely to be quite uniform. This is a useful result since it makes it reasonable to determine the particle composition using microanalysis techniques which gather information from the entire particle. Thus the solute concentration in the cementite rises uniformly from $\bar{c}$ to $c^{\theta\alpha}$.

Equation 10 indicates that the change in composition $(c^\theta - \bar{c})$ should vary with $t^{\frac{1}{2}}$ during isothermal annealing; this is described as a *parabolic* law. This contrasts with the common assumption made even today, that the change should follow $t^{\frac{1}{3}}$ kinetics. It is often not possible in practice to distinguish between the two exponents; the experimental data give reasonable linear fits with either relationship. However, values of the diffusion coefficient extracted from such plots are only correct in the case for parabolic kinetics.[47] This confirms the physical significance of eqn 10, which can extrapolated with more confidence than the $t^{1/3}$ relationship. An additional advantage is that the effect of the steel chemistry is explicitly included in eqn 10 via the equilibrium and average compositions, so that data from many different alloys can be interpreted together.

Figure 9 shows experimental measurements of the composition of cementite as a function of tempering time, for a microstructure which was initially fully bainitic.[48,49] The steel composition is Fe–0.1C–0.24Si–0.48Mn–0.84Cr–0.48Mo wt.%; the reason why the cementite composition is richer than the average is because a stress-relief heat treatment was given prior to ageing. There are a number of important features in this graph other than the fact that the parabolic law is verified. It is obvious that the largest changes are observed for

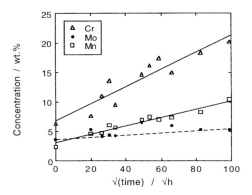

Fig. 9 Measured changes in the chemical composition of cementite particles as a function of the square root of time, during ageing at 550°C. The steel composition is Fe–0.1C–0.24Si–0.48Mn–0.84Cr–0.48Mo wt.%. It was heat treated to give a fully bainitic microstructure, stress-relieved at 730°C for one hour and then tempered at 550°C for the periods illustrated. Data from Afrouz, Collins and Pilkington.[38]

chromium, which would therefore be a good element to monitor. This is because the solubility of Cr in the cementite is much larger than that of Mn or Mo (Table 5). Notice that the solubility varies with temperature, becoming smaller with increasing temperature.

It should also be emphasized that the equilibria listed in Table 5 are steel specific and will vary greatly within the range of power plant alloys used typically in industry. A simple method for estimating the values is to write the partition coefficient of alloying element Z as follows:

$$k_Z = \frac{c^{\theta\alpha}}{c^{\alpha\theta}} \simeq \exp\left\{\frac{A + BT}{RT}\right\} \tag{11}$$

where the parameters A and B are listed in Table 6.[50] The partition coefficient only gives a ratio – to calculate the actual concentrations requires the following further relationship which derives from mass balance:

Table 5 Calculated equilibrium concentrations for Fe–0.1C–0.24Si–0.48Mn–0.84Cr–0.48Mo wt.% alloy used by Afrouz *et al.* These calculations are specific to this steel.

| | Equilibrium wt.%, 703°C | | Equilibrium wt.%, 550°C | |
	$c^{\alpha\theta}$	$c^{\theta\alpha}$	$c^{\alpha\theta}$	$c^{\theta\alpha}$
Cr	0.52	21.3	0.30	36.0
Mo	0.40	5.8	0.34	9.3
Mn	0.38	6.7	0.29	13.2

Table 6 Compilation of parameters used for the calculation of partition coefficients.[50]

Element Z	A / J mol^{-1}	B / J mol^{-1}K^{-1}
Cr	47028	−17.45
Mn	42844	−20.21
Mo	27363	−5.86
Ni	−2619	−2.80
Si	0	−25.10

$$\bar{c} = c^{\theta\alpha} V_\theta + \frac{c^{\theta\alpha}}{k_Z} - \frac{c^{\theta\alpha} V_\theta}{k_Z} \tag{12}$$

V_θ, the volume fraction of cementite in the alloy, can be estimated using the lever rule with carbon, so that the equilibrium compositions of the two phases can easily be calculated as a function of heat treatment and average composition.

Another feature highlighted by the parabolic law is the dependence of enrichment on the particle size x^θ. Smaller particles enrich faster and saturate earlier because they are smaller reservoirs for solute (Fig. 10).

The model described above predicts parabolic kinetics, which can be considered to have been verified by published experimental data. Although the data refer to time periods as long as 10 000 h, the time scales of interest are much larger, of the order of 250 000 h. Data can of course be obtained from steels which have been in service but they can be quite unreliable because the history is then uncertain. It would not be safe to extrapolate the parabolic law to very long times, because the analytical expression does not account for a phenomenon known as *soft-impingement*. In this, the diffusion fields of adjacent particles begin

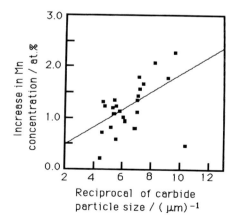

Fig. 10 Size dependence of cementite enrichment (annealed at 565°C for 4 weeks).[48,49]

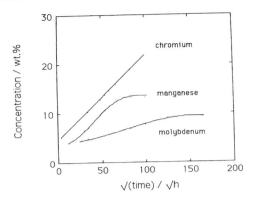

Fig. 11 Calculated changes in the chemical composition of cementite particles as a function of the square root of time, during ageing at 550°C. The steel composition is Fe–0.1C–0.24Si–0.48Mn–0.84Cr–0.48Mo wt.%. The calculations are done using a finite-difference method.[39]

to overlap and significantly interfere, slowing down the enrichment process until it stops completely when equilibrium is achieved. Such a process cannot easily be treated analytically. Bhadeshia has used a finite difference method instead; some results are presented in Fig. 11, for precisely the experiments reported by Afrouz *et al.* The parabolic law is a good approximation during the 'early' stages but the effect of soft-impingement is to cause deviations with a complete halt when the carbide saturates with the solute concerned. Manganese and molybdenum do so quite rapidly for this particular example. Thus, the numerical method (finite difference) can be used to extrapolate to very long times.

Pearlite

It has been emphasised that the substitutional solute to iron atom ratio is the same everywhere in a bainitic or martensitic microstructure as in the alloy as a whole. This is not the case for pearlite where the reconstructive mechanism of transformation ensures that the cementite is enriched with substitutional solutes, though often not to the extent expected from equilibrium. This makes it impossible to predict the initial composition of the cementite in pearlite. It turns out that for calculation purposes, this is not a serious problem for long ageing times, since the degree of enrichment then tends to be much larger than the starting solute level (Fig. 12).

Effect of Carbon

There are two effects which depend on the carbon concentration of the steel. The ternary Fe–Cr–C phase diagram on the $(Fe,X)_3C/\alpha$ field shows that an increase in the carbon concentration is accompanied by a decrease in the equilibrium concentration of chromium in the carbide. Thus, the carbide enrichment rate is

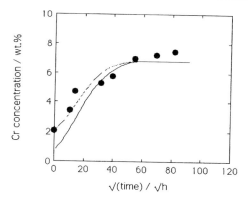

Fig. 12 Data for the enrichment of cementite associated with pearlite, and calculated curves assuming zero enrichment at $t=0$ (continuous curve), and a finite enrichment at $t=0$ (dotted line).[46]

expected to decrease. A further effect is that the volume fraction of cementite increases, in general leading to an increase in particle thickness and volume fraction. The thickness increase retards the rate of enrichment (eqn 10). If the carbide particles are closer to each other then soft-impingement occurs at an earlier stage, giving a slower enrichment rate at the later stages of annealing.

Local variations in carbon concentration may have a similar effect to changes in average concentration. Such variations can be present through solidification induced segregation, or because of microstructure variations caused by differences in cooling rates in thick sections. It is well known that the microstructure near the component surface can be fully bainitic with the core containing a large amount of allotriomorphic ferrite[†] in addition to bainite. In the latter case, the bainite which grows after the allotriomorphic ferrite, transforms from high carbon austenite. The associated carbides are then found to enrich at a slower rate (Fig. 13a). This discussion emphasises the role of carbon.

High-Alloy Steels, Alloy Carbides

The vast majority of steels used in the power generation industry have a total alloy concentration less than 5 wt.%, but there are richer martensitic alloys such as the 9Cr1Mo, 12Cr1MoV destined for more stringent operating conditions. For low alloy steels, the initial carbide phase found immediately after the stress-relief heat treatment (typically several hours at 700°C) is cementite. The cementite is very slow to enrich and to convert to alloy carbides, thus allowing the process to be followed over a period of many years during service. As the con-

---

[†]Ferrite which grows by a diffusional mechanism is termed allotriomorphic when its outward form does not reflect the symmetry of its crystal structure. It commonly nucleates and grows preferentially along the austenite grain surfaces.

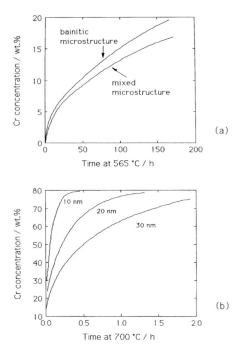

Fig. 13 (a) 2.25Cr1Mo steel, cementite enrichment in a fully bainitic microstructure and one which is a mixture of allotriomorphic ferrite and bainite.[48,49] (b) Rapid saturation of cementite in a 12Cr1MoV steel aged at 700°C, as calculated for a variety of particle sizes.[51,52]

centration of carbide forming elements is raised, the kinetics of cementite enrichment and alloy carbide precipitation become more rapid (Fig. 13b). Thus, relatively stable alloy carbides tend to dominate the microstructure after short periods in service or even after the stress relief heat treatment.[52] When compared with cementite, the formation of alloy carbides involves considerable diffusion so that they form with a chemical composition which is not far from equilibrium. Changes may nevertheless occur if, for example, if the heat treatment prior to service is at a different temperature when compared with the service temperature, since the carbide must then adjust to an equilibrium consistent with the service conditions.

Figure 14 shows microanalysis data from vanadium-rich carbides for two samples with average composition

$$Fe - 0.11C - 0.3Mn - 0.28Mo - 0.23V - 0.023N \text{ wt.\%}$$

The 'virgin' sample was produced by tempering at 700°C for an unspecified time period, following normalisation at 950°C for an unspecified period. The exposed

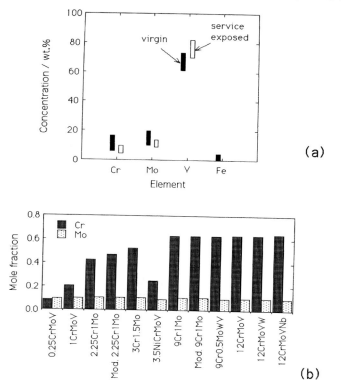

Fig. 14 (a) The chemical compositions of vanadium-rich carbides in virgin material (1Cr–0.3Mo–0.25V) and in samples which had been in service for ten years.[41] (b) The concentrations of Cr and Mo in $M_{23}C_6$ in the power plant steels listed in Table 2, for 565°C. The data are calculated using the *MTDATA* program.

sample had in addition spent 10^5 hours at 540°C and 50 MPa. There is no significant difference between the carbide compositions from the two samples. There is therefore little prospect of estimating the remaining life of high alloy steels, or of steels in general, by following changes in the chemical compositions of alloy carbides. As a matter of interest, Fig. 14b shows the chromium and molybdenum concentrations of the most prominent alloy carbide found in most power plant steels (see also Tables 2, 3).

One possibility is to follow the detailed microstructural changes. Figure 15 is a simplified presentation of the evolution of carbide and Laves' phases during the ageing of a 9CrMoVW steel (*NF616*) which is tempered from its martensitic state at a temperature of 600°C over six orders of magnitude of time. Obviously, an identification of the fractions of the phases could help to fix the state of the steel. Some of the precipitation reactions are sensitive to temperature. For example, the driving force for the precipitation of Laves phases decreases greatly as the temperature increases.

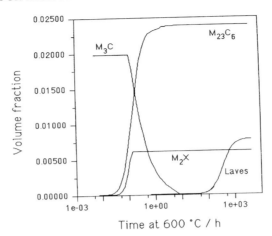

Fig. 15 Calculated evolution of the phase fractions in a 9CrMoVW steel (*NF616*) during tempering at 600°C.[53]

OVERALL CHANGES IN MICROSTRUCTURE

The main factor responsible for the good creep resistance of the steels used in the power generation industry is the formation of fine and highly stable dispersions of alloy carbides.[54] A significant contribution also comes from solid solution strengthening by substitutional solutes. The solid solution strengthening component becomes more important after prolonged service at elevated temperatures, as the microstructural contribution to strengthening diminishes due to annealing effects.

The heat treatments given to the steels prior to service are usually so severe that the precipitates are in an overaged condition.[55] The microstructures are therefore relatively stable prior to service, although further microstructural changes are inevitable given that service involves many decades at temperatures where substitutional atoms have significant mobility. We shall see that these changes can be used to monitor life.

1Cr0.5Mo Steel

The hardenability of this alloy is marginal so that it can be found in three microstructural states when ready for service: (i) ferrite and pearlite; (ii) ferrite and bainite; (iii) a mixture of ferrite, pearlite and bainite. Cementite and M_2C carbides are also present in the initial microstructure. The changes that occur in the microstructure during service are summarised in Table 7, in terms of the fraction of life expended (t/t_r). Dobrazanski and co-workers[56] have established atlases of microstructure which enable the remaining life to be characterised to an even higher precision than indicated in Table 7.

Table 7 Broad microstructural changes as a function of life fraction consumed, for a 1Cr0.5Mo steel. The observations of ferrite, bainite, pearlite and cavitation are on an optical microscope scale. The arrows indicate whether the phase concerned is increasing or decreasing with time. Adapted from Dobrzanski *et al.*, 1996.[56]

t/t_r	Microstructure
0	Ferrite plus bainite and/or pearlite
	M_3C and M_2C
0.2	Ferrite and partially disintegrated bainite/pearlite
	M_3C, $M_2C\uparrow$ and $M_{23}C_6\uparrow$
0.3–0.4	Ferrite and completely disintegrated bainite/pearlite
	$M_3C\downarrow$, $M_2C\downarrow$, $M_{23}C_6$ and $M_6C\uparrow$
0.4–0.6	Ferrite and coarse carbides
	$M_2C\downarrow$, $M_{23}C_6$ and M_6C
0.6–0.8	small number of irregularly spaced isolated cavities
	As above but more cavities
0.8–0.9	many cavities coalesced or about to do so
0.9–1.0	As above but crack-like coalescence of cavities
	Macroscopic cracking (removed from service)

2.25Cr1Mo

This steel usually has an upper bainitic microstructure but can contain substantial quantities of allotriomorphic ferrite, especially in large components which have been cooled slowly from the austenitisation temperature. The mixed ferrite/bainite microstructure actually is better in creep because, for reasons which are not clear, M_2C carbide persists longer in ferrite than in bainite or martensite, presumably because the higher defect densities associated with the last two microstructures enhance diffusion rates.

Heat treatment prior to service involves austenitisation at about 950°C for a time period dependent on the component size (typically several hours), followed by air cooling to ambient temperature. The component is then tempered at 710°C for 16 h to produce a 'stable' microstructure of overaged upper bainite with a prior austenite grain size of around 80 μm. A coarsened lath structure (width $\simeq 0.8$ μm) is seen using transmission electron microscopy, with $M_{23}C_6$ alloy carbide particles located mainly at the lath boundaries. These particles do not seem to coarsen much during service although there must be an overall coarsening of pinning particles since the lath size is found to coarsen (Fig. 16), although even in this case, the data are restricted to very small strains typical of the early stages of life.

CREEP STRAIN & RUPTURE DATA

Strain Monitoring

Creep strain is a direct measure of remaining life (Fig. 17) and is a quantity which can be monitored during service using a variety of techniques. It is estimated that

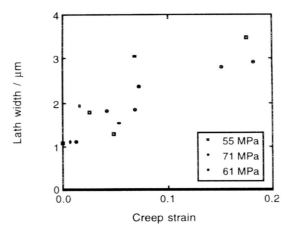

Fig. 16 The lath size in 2.25Cr1Mo steel as a function of the secondary-creep strain during tests at the variety of stresses indicated. The strain measurements have an error of about ±0.01, and the lath size data an error of about ±0.25 μm. Adapted from Lonsdale and Flewitt.[57]

to obtain an accuracy of 10% in the predictions of remanent life requires a strain measurement capability of 0.01%. The methods for measuring strain include continuous measurements using strain gauges, discontinuous measurements during maintenance shutdowns, the use of measurement grids plated using noble metal to avoid oxidation, special extensometers and studs welded to the surface as reference marks.[5,6,58] As pointed out by Parker,[58] care has to be exercised to

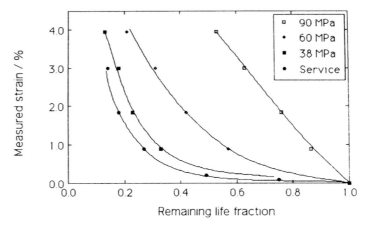

Fig. 17 The creep strain as a function of remaining life. The data are for creep samples loaded at three different stresses; the data marked 'service' refer to the conditions for a thick-walled pressure vessel. After Parker, 1996.[58]

ensure that the distance over which the measurement is carried out is appropriate for the component concerned. For example, a welded joint has an in-homogeneous microstructure; local strains in the heat affected zone may be much larger than the value measured across the whole joint. Notwithstanding these difficulties, it is possible to relate the strain directly to the remaining life using an appropriate model.[59]

In circumstances where the creep data are from short-term experiments and hence of small strains, the Evans and Wilshire[60] equation may be used to extrapolate to larger strains (ϵ):

$$\epsilon = \underbrace{\theta_1[1 - \exp\{-\theta_2 t\}]}_{\text{decaying rate}} + \underbrace{\theta_3[\exp\{\theta_4 t\} - 1]}_{\text{accelerating rate}} \tag{13}$$

where θ_i are obtained by fitting to experimental data, and t is the time at temperature. The first two of these parameters describe the primary or decaying strain rate component, whereas the remaining terms concern the accelerating regime. Parker has used this equation to predict successfully the remaining life of a CrMoV pressure vessel steel. He also suggests that the number of inspections should be increased once the creep strain has exceeded 1%.

In practice most power plant components are subjected to more complicated stresses than uniaxial tension. Plastic deformation occurs when the shear yield stress τ_y is exceeded. For uniaxial tension, yielding occurs when the principal stress σ_1 reaches the normal yield strength σ_y, which is identical in magnitude to the shear yield strength τ_y. For torsion, the maximum principal stress σ_1 must reach a magnitude $\sqrt{3}\tau_y$ before plastic deformation occurs; this comes from the application of the von Mises yield criterion which can be interpreted to mean that plastic flow occurs when the shear-strain energy reaches a critical value [e.g., Ref. 61]. It is convenient to use normal rather than shear stresses so plastic deformation is said to occur at an equivalent normal stress σ^*, which for tension is $\sigma^* = \sigma_y$ whereas for torsion $\sigma^* = \sqrt{3}\sigma_y$. This equivalent stress can be calculated for a variety of multiaxial stress systems using the von Mises criterion. The importance of multiaxial stresses in the development of cavitation damage is well established.[62,63] Cane[63] has pointed out that strain monitoring will lead to an underestimation of damage when $\sigma_1/\sigma^* > 1$ and vice versa when $\sigma_1/\sigma^* < 1$. Thus, for the same maximum principal stress σ_1, cavitation damage will be more severe in uniaxial tension than in torsion.[64]

Creep Rupture

Naturally, the creep rupture testing of service exposed material is considered to be one of the most reliable way of estimating the remaining life of a component.[17,18,65] This assumes that test samples can be removed 'non-destructively' from the component and that the samples are reliable. Several samples are required even when reliability is assured, because the tests have to be done at accelerated rates in order to be of use. The data are then extrapolated to the operating conditions in order to estimate the remaining life. Given the material limitations, the samples are usually much smaller in size than is usual in creep testing. The problems associated with the small size are discussed in a later section on small specimen testing; attention is focused here on the analysis of the creep data.

The steady-state creep rate $\dot{\epsilon}$ is in theory given by [e.g. Ref. 3]:

$$\dot{\epsilon} = b_3 \sigma^n \exp\left\{-\frac{Q}{RT}\right\} \tag{14}$$

where b_3 is an empirical constant, Q an activation energy, σ the applied stress, T the absolute temperature and R the universal gas constant. This can be integrated to find the creep rupture time t_r as

$$\ln\{t_r\} = \ln\left\{\frac{\epsilon_r}{b_3}\right\} - n\ln\{\sigma\} + \frac{Q}{RT} \tag{15}$$

Nevertheless, it is found in practice that slightly better fits are obtained by writing[65,66]

$$\ln\{t_r\} = b_4 + b_5\sigma + b_6T \tag{16}$$

where, b_4, b_5 and b_6 are fitting constants.

The small sample tests used to derive the constants typically extend over a temperature range 540–675°C and for a time period of 10 000 h. Any relationship for the creep rupture time must therefore be extrapolated to much longer times and service conditions. It has yet to be demonstrated whether the empirical relations extrapolate better than the physical relationship implied in eqn 15.

It has been argued that the use of very high temperatures during accelerated testing is unwise since microstructural changes might occur which are not representative of service at lower temperatures.[65] This is based on the fact that the creep rupture data from very high temperature tests exhibit a different behaviour from those at other temperatures in the range studied. Power plant steels are usually heavily tempered before they enter service, for many hours at temperatures in the range 680–750°C, in order to stabilise the microstructure and to relieve stresses. 'High-temperature' thus refers to temperatures in the vicinity of the original tempering temperature. It is recommended, therefore, that accelerated creep tests are confined to temperatures some 50°C lower than the original tempering temperature.

Scale Formation

When using isostress rupture tests for the prediction of remaining life in boiler components such as reheater tubes, the formation of scale on the steam side of the tube, and the fire-corrosion on the other side must be taken into account.[11] Fire-corrosion has the effect of reducing the thickness and hence of increasing the stress on the remaining section. The build up of scale on the other hand leads to a continuous increase in the temperature of the tube by insulating the inner surface. Any calculations therefore require a knowledge of the rate of oxide formation (and hence of tube temperature increase) and of fire-corrosion (and hence of stress increase), both of which can be incorporated into an assessment using an appropriate life fraction rule. In the absence of oxide-spalling, the current thickness of the oxide can be used to assess the effective temperature using experimental data of the kind illustrated in Fig. 18 for 2.25Cr1Mo steel.

CREEP DUCTILITY

The ability of a material to tolerate strain is important because it determines whether the failure occurs gradually with some warning, or whether it is sudden. The latter scenario is less desirable and would require a greater safety factor in design or in any life assessment. It is strange that many design criteria do not include a consideration of the rupture ductility.[5,6]

The reasons why creep ductility varies so much in practice are not well established. The variation can be from 2% to 30% in conditions which are nominally identical.[65] Contributing factors include residual elements, grain size, inclusions, solution treatment and tempering treatments, *etc*. With small ductilities, cavitation can be detected at correspondingly smaller strains (as low as

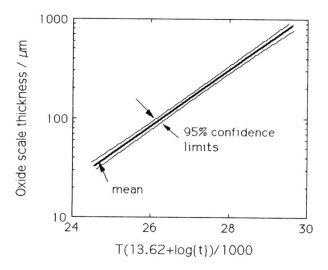

Fig. 18 Oxide thickness for 2.25Cr1Mo steel in steam at 538–593°C as a function of the Larson Miller parameter in which the temperature is in degrees centigrade and the time in hours. After Wells *et al.*[67]

0.5%) whereas higher ductility components do not reveal signs of cavitation with strains as large as 5%.[68]

It appears that many alloys show a creep ductility minimum as a function of the test temperature and service period. The life assessment should therefore be made bearing in mind the relationship between the service temperature and that at which the ductility minimum occurs. For 1CrMoV steel, the minimum occurs between 5000 and 30 000 h at 550°C.[65]

CAVITATION DAMAGE

We have already noted that cavitation occurs in steels that are relatively brittle in creep and in regions where there is a concentration of stress. In fact, deformation by creep is far from homogeneous even on a microstructural scale. One manifestation of this is the formation of cavities at the grain boundaries between differently oriented crystals. The boundaries contain incoherent particles, primarily sulphides, which can decohere and provide the sites where cavities generally form.[69] Cavitation plays a critical role in the events leading to ultimate failure and hence is a direct measure of remaining life.

The cavities tend to form mostly on boundaries with normals parallel to the principal tensile stress by a process of nucleation and growth, although damage also occurs on inclined boundaries if the applied stress is large.[63] Thus, at any instant of life, only a fraction of the transverse boundaries have cavitation damage. There are constraints to cavity formation when only some of the transverse boundaries are cavitated, and this has important consequences on the creep rupture life.[70] In Fig. 19a, all the transverse boundaries have identical cavities. The regularity of the structure allows the cavities to grow unconstrained, the extension along the stress axis being accommodated by grain boundary sliding. This is not the case when, as illustrated in Fig. 19b, only some of the boundaries have cavities. The extension due to the cavities at wx results in sliding which causes the adjacent grains to overlap at yz. This matter in the shaded region at yz must be carried away by diffusion before the cavities at wx can continue to grow. This in effect imposes a constraint on the growth of the cavities when compared with the situation in Fig. 19a. The physical basis of the retardation is that cavity growth now becomes controlled by the rate of matrix creep. The retardation is very significant, as seen in Fig. 19c.

The A-Parameter

One method of characterising cavitation is the 'A-parameter' determined as the number fraction of cavitated grain boundaries traversed by a scan parallel to the stress axis of a creep test specimen, whilst avoiding the necked region. The method presented below is due to Shammas[71] and relies on the assumption that any cavitated boundary is one which can be considered to have been completely damaged and incapable of supporting load. The creep rate as given by eqn 14 therefore becomes

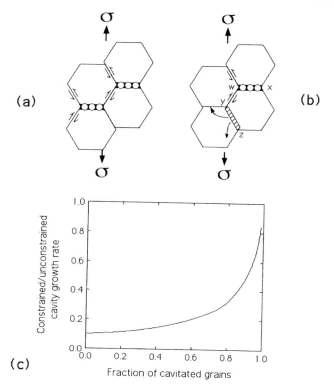

Fig. 19 The effect of constraint on the growth rate of cavities. (a) All transverse boundaries cavitated. (b) Non-uniform cavitation. (c) The reduced cavity growth rate when cavitation is not uniform.[70]

$$\dot{\epsilon} = \frac{b_7 \sigma^n}{(1-A)^n} \qquad (17)$$

where the term in the denominator accounts for the increase in stress due to a reduction in the cross-sectional area as a fraction A of the boundary normal to the stress becomes unavailable to support the load; b_7 is an empirical constant. This might be a reasonable assumption because it is well known that a boundary is severely damaged on a fine scale once cavities become optically visible.

If it is now assumed that

$$\dot{A} \propto \frac{1}{(1-A)^{b_8}} \qquad (18)$$

where b_8 is an empirical constant. Since $A=0$ at $t=0$ and $A=1$ at $t=t_r$, the integration of eqn 18 gives

$$(1 - A) = \left(1 - \frac{t}{t_r}\right)^{\frac{1}{b_8 - 1}} \qquad (19)$$

so that

$$\dot\epsilon = b_7 \sigma^n \times \left(1 - \frac{t}{t_r}\right)^{-\frac{n}{b_8 + 1}} \qquad (20)$$

This can be integrated subject to the conditions that $\epsilon = 0$ at $t = 0$ and $\epsilon = \epsilon_r$ at $t = t_r$, giving

$$\frac{\epsilon_r}{\epsilon} = \left(1 - \frac{n}{b_8 + 1}\right)^{-1} \qquad (21)$$

and $\left(1 - \dfrac{t}{t_r}\right) = \left(1 - \dfrac{\epsilon}{\epsilon_r}\right)^{\lambda}$

where $\lambda = \epsilon_r / \epsilon$. Comparison with eqn 19 gives

$$A = 1 - \left(1 - \frac{t}{t_r}\right)^{(1 - \frac{\epsilon}{\epsilon_r})/n} \qquad (22)$$

where as usual, t_r the time to rupture, ϵ the creep strain, ϵ_r the creep rupture strain and n a fitted constant. Any strain during primary creep is neglected. Figure 20 shows both experimental data and a curve calculated using eqn 22 and assumed values of $\epsilon_r / \epsilon = 2.5$ and $n = 3$.[71] The data seem to follow the form in-

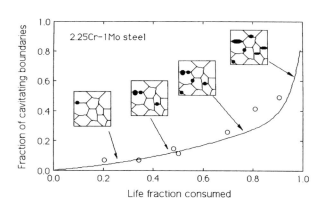

Fig. 20 The fraction of cavitating grain boundaries as a function of the life fraction consumed.[71]

dicated by the equation but there are other results[19,72] which suggest that the relationship remains linear up to the point of life-exhaustion, although the degree of scatter in the *A*-parameter increases significantly for a life fractions greater than 0.96.

The *A*-parameter is defined in terms of the number fraction of transverse boundaries which are cavitated. Many models for creep associate fracture with a complete loss of area available to support the load as the cavities coalesce along the appropriate plane.[73-75] A strict measure of the damage would then be the *area* fraction of transverse boundaries that is decorated by cavities. The *A*-parameter on the other hand could reached unity before the transverse boundaries are completely detached. However, this may not be a problem in practice if the rate of damage increase in the final stages is very high.

The magnitude of the *A* parameter at failure, *i.e.* A_r, is found to vary considerably with the test temperature, applied stress, creep ductility and austenite grain size although it is emphasised that these variables may not all be independent. For example, A_r increases as the test temperature decreases towards the ductility minimum for the steel concerned. For 1CrMoV type steels, A_r decreases from about 0.20 ± 0.1 to 0.02 ± 0.02 as the elongation to failure increases from 2 to 33%.[76] The scatter in A_r is reduced, but only slightly, when the experiments are confined to just one steel. Similarly, A_r varies from about 0.07 ± 0.05 to 0.35 ± 0.05 as the prior austenite grain size changes from 22 μm to

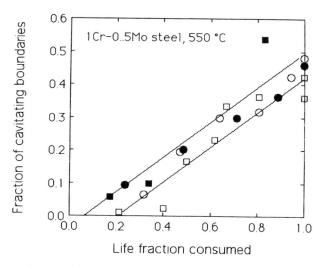

Fig. 21 Changes in the *A*-parameter as a function of the expended life fraction in 1Cr0.5Mo steel tested at 550°C. The filled points represent commercial purity steels and the others high-impurity steels. The squares are data acquired from interrupted tests on single specimens, whereas each circle is from a different sample. After Viswanathan Ref. 11.

75 μm. This latter effect may also be related to variations in creep ductility, which is dependent on the austenite grain size.

Cavity formation and creep deformation are sensitive to impurities in the steel and this might be reflected in the measured values of the A parameter. Viswanathan[11] has compared high purity and commercial purity 1Cr0.5Mo steels but there are no systematic trends within the scatter inherent in the measurements. Similarly, it does not seem to matter whether the cavitation measurements are done by interrupting a single test or by using a series of specimens.

The A parameter is considered to be of little use in life assessment when the creep ductility is greater than about 5% (or austenite grain size less than about 25 μm).[76] This is because such materials are unlikely to exhibit creep damage during the course of safe-service, the damage setting in at a stage close to failure. For steels with low creep ductility and large austenite grain size, the A parameter can be used to assess life fraction since changes in A can be observed as a function of life fraction. Measurements like these for 1CrMoV rotors indicate that the absence of any observable creep cavitation indicates a minimum remaining life fraction of approximately 0.45.[76] This obviously is steel- and technique-specific because work on 1Cr0.5Mo steels tested at 550°C shows that about 60% of the life remains at a stage when isolated cavities are discovered in the (optical) microstructure using a replica technique.[11]

A general procedure used successfully in the assessment of the remaining life of 1CrMoV rotors is shown in Fig. 22

Another method of assessing the remaining life using cavitation damage involves the measurement of the number density of cavities, which seems to increase in proportion to the creep strain during uniaxial loading.[57,62,77,78] This may not in fact amount to a different method (compared with the A-parameter) in the context of steels where the important cavities tend to form at the grain boundaries. The proportionality between the cavity population and strain is limited to small strains (up to 0.2). Deviations occur at large strains because some contribution to the creep strain comes from general deformation and also because the cavity nucleation sites become exhausted at the late stages of creep life.

The volume fraction of cavities increases approximately with the strain and can be used as a measure of remaining life,[75] although differences in the distribution of cavities (due to differences in microstructure) could introduce variations in creep life at constant cavity fraction.

Qualitative Methods

There are several general classifications of creep cavitation damage. Thus, Neubauer and co-workers[79,80] have categorised damage into the following categories based on optical microscope observations: no detectable damage, isolated cavities, oriented cavities[†], linked cavities (microcracks) and macrocracks

---

[†]meaning that there is an alignment of damage normal to the maximum principal stress

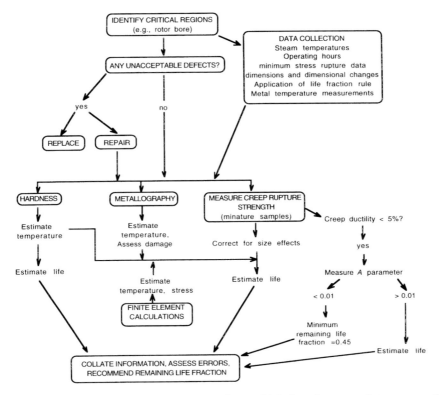

Fig. 22 Illustration of the general procedure which has been used to assess the remaining life of rotors.

about the size of a grain. Significant action is considered necessary, in the form of inspections, when oriented cavities (strings of cavities on boundaries normal to the stress axis) form and repair begins when the cavities begin to link. Macro-cracks are of course, unacceptable and must be repaired immediately.

As pointed out by Strang *et al.*,[17] these and more sophisticated qualitative classifications are useful in defining the safe service inspection intervals but not necessarily in the estimation of the remaining life for a given component. The prediction of life requires quantitative rules which give conservative estimates when extrapolated. On the other hand, the quantitative rules must take into account the heterogeneous nature of cavitation damage so that field parameters such as the cavity fraction are not on their own sufficient. Viswanathan[11] has proposed a compromise by linking the observations to direct measurements of the creep rupture life. His data for a 1Cr0.5Mo type steels are illustrated in Fig. 23, where the Neubauer categories are plotted against the expended life fraction measured using creep rupture tests. The rate of cavitation damage in this case seems more severe than indicated in Table 7.

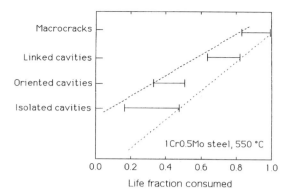

Fig. 23 Correlation between the cavity classifications and the life fraction consumed, for 1Cr0.5Mo steels tested at 550°C [After Viswanathan, Ref. 11]. The data are from high-purity and commercial purity steels.

One important use of purely qualitative classifications is that they help in establishing a methodology which allows a comparison between different research groups. Wu *et al.*[81] have developed a detailed classification of cavitation damage, which can readily be reduced to the Neubauer categories (Table 8). A variety of similar classifications have been incorporated in European guidelines for in-service damage assessment, for example, NORDTEST NT TR170 and VGB-TW507.

Techniques for Measurement of Cavities

Optical microscopy is frequently used to examine longitudinal sections from creep test specimens and the main difficulties during such examinations is the possibility of confusing cavities and localised pitting or boundary carbides unintentionally extracted from the sample surface. Therefore, one procedure is to use standard metallographic polishing to 6 μm diamond finishing. This is followed by repeated polishing with γ-alumina[57] or 1 μm diamond polish[76] and light etching in 2% nitric acid in methyl alcohol. This procedure minimises surface deformation effects which might obscure cavities. Reproducible results are obtained as long as these preparation conditions are standardised. However, the results from any technique must depend on resolution. It is well known that scanning electron microscopy of intergranularly fractured specimens exposes cavitation at a much smaller overall strain than is apparent using optical microscopy.[57†] High voltage transmission electron microscopy can reveal even finer dispersions of cavities; observations like these have confirmed the tendency for

---

†Intergranular fracture can be promoted by fracturing the samples at liquid nitrogen temperature.

Table 8 Damage ratings developed by Wu *et al.*[81] with experiments on rolled seamless pipes of outer diameter 220 mm and wall thickness 20 mm. 'Reheat cracking' occurs by cavitation on grain boundaries in the coarse region of the heat affected zone of a weld. It requires a matrix which is strong (strengthened, for example, by fine vanadium carbides) due to an inadequate heat treatment after welding, a precipitate-free zone at the boundaries and impurity particles (MnS) at the boundaries which help nucleate cavities.[5,6,82]

Rating	Appearance of cavitation
1	No cavities
Isolated cavities	
2B	1–100 cavities mm^{-2}
2C	100–1000 cavities mm^{-2}
2D	$\geq$ 1000 cavities mm^{-2}
Strings of cavities	
3B	Few strings and isolated cavities 2B
3C	Few strings and isolated cavities 2C or 2D
3D	Numerous strings and isolated cavities up to 2D
Microcracks $\leq$ 0.1 mm	
4A	Reheat cracks
4B	Few microcracks and damage 2B or 3B
4C	Few microcracks and damage 2C, 2D or 3C, 3D
4D	Numerous microcracks and damage 2C, 2D or 3C, 3D
Macrocracks $\geq$ 0.1 mm	
5A	Reheat cracks
5B	One macrocrack ($\leq$ 1 mm) and damage 2B, 3B or 4B
5C	Macrocrack ($\leq$ 5 mm) and damage 2C,3C or 4C
5D	Macrocracks ($\leq$ 5 mm) and damage 2D, 3D or 4D, or creep crack (> 5 mm)

cavities to form preferentially at carbide particles. The cavities also appear to be crystallographically faceted.

Many measurements are made on samples which are creep tested in air. It is then important to avoid heavily oxidised surface regions and regions where necking (intense localised deformation) has occurred. Strang *et al.*[76] therefore conducted measurements in the central $\frac{2}{3}$ rds region of the sample gauge length, keeping one gauge diameter away from the fractured surface. The samples are examined on their central longitudinal sections containing the tensile axis.

Cavitation damage methods also depend on a clear identification of the prior austenite grain boundaries, which can be difficult in heavily tempered ferritic microstructures.[76]

FRACTURE TOUGHNESS

It is well established that the fracture toughness of many power plant steels deteriorates during service for two reasons. Firstly, the carbide particles, particularly those located at boundaries, coarsen and hence provide easier sites for crack or void nucleation. Secondly, the segregation of impurities to interfaces has

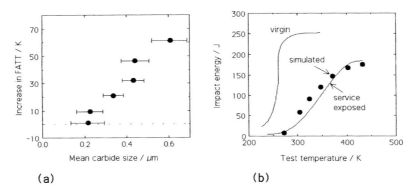

(a)   (b)

Fig. 24 Impact test data on 2.25Cr1Mo steel. (a) Change in the fracture appearance impact transition temperature with the mean carbide size as a function of ageing. (b) Comparison of the impact energies of the steel in its virgin condition, after service exposure at 813 K for 88 000 h and for a service-simulated sample (873 K for 10 000 h). After Wignarajah *et al.* Ref. 84.

an opportunity to proceed to its equilibrium extent during service. The rate of this embrittlement can sometimes be modelled.[83]

The tendency to embrittle during service is sometimes simulated by an accelerated step cooling experiment in which the sample is held at a variety of decreasing temperatures over the embrittlement range, and then allowed to cool to ambient temperature at a very slow rate. Unfortunately, although the procedure does allow the sensitivity of the sample to embrittlement phenomena to be tested, the data cannot be used to discover the state of the steel in its ageing process (a remaining life assessment).

Figure 24a shows how the fracture appearance transition temperature (FATT) assessed from impact tests varies with the mean carbide particle size in a 2.25Cr1Mo steel.[84] Figure 24b shows the complete transition curves for both the virgin steel (*i.e.* at implementation into service) and for an ex-service sample which had experienced 88 000 h at 813 K. The points represent samples of virgin steel subjected to 873 K for 10 000 h in an attempt to simulate prolonged service at 813 K. These conditions were chosen in order to achieve the same Larson–Miller parameters for the ex–service and simulated samples. It is clear that the fagreement between the simulation and service exposed steel is excellent, allowing the method to be added to the panoply of tools available for remaining life assessment.

MISORIENTATION MEASUREMENTS

It is well known from cold-deformation experiments that once a dislocation cell structure is established, further strain leads to a decrease in the cell size and a corresponding increase in the crystallographic misorientation between adjacent cells.[85] Dislocations move and their density and distribution also change during the course of creep deformation. These changes are more complex than those associated with cold-deformation since there are interactions with precipitates which are deliberately there to interfere with dislocation motion. Nevertheless, it has been known for some time that the orientation between adjacent cells in bainite at first changes linearly with the creep strain as shown in Fig. 25,[57] these measurements were for a 2.25 Cr1Mo steel. The data are limited to rather small strains and hence to the very early stages of service life. Similar results have been reported for creep in aluminium.[86]

Measurements have also been reported for a ferrite-pearlite microstructure in 2.25 Cr1Mo steel.[87] When compared with bainite, the scale of this microstructure is much larger. The measurements therefore refer to the minute misorientations that develop between dislocation cells within individual ferrite grains over a scale of about 10 μm. This contrasts with the orientation differences between adjacent laths of bainite, which are over distances less than 1 μm. The data for the ferrite/pearlite microstructure are strictly *intragranular*. Care has to be taken to avoid distortions due to thin-foil preparation for transmission electron microscopy. For the same reason, it is recommended that the orientations are measured in the thick regions of the sample using Kikuchi lines. Figure 26a shows the method used, in which the misorientation is defined as the mean of six measurements taken around a reference point. The misorientation is then divided by the distance between the points, and the mean from four repetitions is taken for comparison against overall creep strain.

The accelerated test data illustrated in Fig. 26b indicate that the misorientation per unit length increases with increasing creep strain, although there is a strong

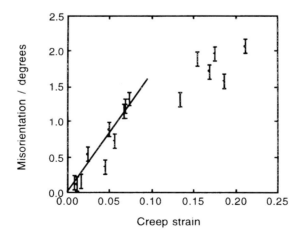

Fig. 25 The crystallographic 'misorientation' between adjacent laths of bainite in 2.25Cr1Mo steel as a function of the secondary-creep strain. The samples were tested at a variety of stresses and temperatures.[57]

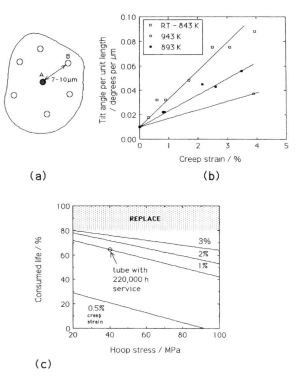

(a)   (b)

(c)

Fig. 26 (a) Measurement of misorientation. *A* is a reference point surrounded by six other points (*B*) where the relative misorientations are measured. (b) The 'misorientation per unit length' for a 2.25Cr1Mo steel tested at a variety of stresses and temperatures, as a function of the creep strain. The data are from accelerated tests in which either the stress or the temperature was increased. (c) Relation between creep strain and consumed life for a heat exchanger tube.[87]

dependence on the temperature. Thus variations in stress led to the same trend whereas variations in temperature led to different slopes. This is because recovery effects are more prominent at higher temperatures.[87] Accelerated tests are, in the context of misorientation measurements, best conducted with exaggerated stresses. Figure 26c shows how misorientation measurements were used to fix the level of creep strain in an ex-service heat exchanger tube, enabling its remaining life to be estimated using design curves.

A rigorous expression of the orientation relationship between identical crystals involves three degrees of freedom, the right-handed angle of rotation and the three direction cosines of the axis of rotation (only two of the direction cosines are independent since their sum of squares is always unity). A measurement of the angle on its own is not enough. The body-centered cubic structure has a high symmetry which means that each axis-angle pair can be expressed in twenty four crystallographically equivalent ways. Thus, a 5° rotation about [1 0 0] is approximately equivalent to a 3.5° rotation about [1 0 1]. It is surprising therefore that the data in Fig. 26b exhibit quite rational patterns.

HYDROGEN AND REMAINING LIFE

2.25 Cr1Mo steel is used routinely in chemical plant for pipes carrying hydrogen at about 4 MPa and 530°C. The hydrogen is there as a molecular gas H_2 and hence is not to be confused with nascent hydrogen which is detrimental to steel.

Experiments[88] indicate that this kind of exposure, for a period as long as 135 000 h, appears to have no effect on the creep rupture strength of the steel when compared with virgin samples which have not seen service (Table 9). This is surprising, because even if any potential effect of hydrogen can be neglected, the ex-service material does not appear to have significantly degraded. Measurements of creep strain (Fig. 27) seem to confirm these observations. One possibility is that degradation at 530°C is too slow when compared with the higher temperatures ($\simeq 565$°C) that 2.25 Cr1Mo steel is normally used at.

Bild *et al.* also found that the creep strain of samples which were annealed at 600°C for 1000 h in very high pressure hydrogen (5–25 MPa) was significantly greater than those which were similarly annealed in air (Fig. 27b). Scanning electron microscopy indicated that the effect of the hydrogen was to eliminate some unidentified fine precipitates from the microstructure, which would be consistent with a deterioration of creep properties. However, both the mechanism of the change in precipitation, and the reason why ex-service material compares well with virgin material (Table 9) remains to be clarified.

ULTRASONICS

Ultrasound refers to frequencies up to about 20 MHz. Its penetration into metals leads to a number of effects which can be utilised to detect creep cavitation or microcrack damage. The advantage of ultrasound methods is that they are non-destructive, rapid and can be used on site.

Table 9 The creep rupture strength of 2.25Cr1Mo steel in ordinary tests (A) and for samples which had been exposed to molecular hydrogen at 3.5 MPa for 135 000 h at 530° (B).[89]

Temperature / °C	Time / h	Creep strength A / MPa	Creep strength B / MPa
525	30 000	123	129
	100 000	98	92
550	30 000	90	89
	100 000	69	64
575	30 000	65	65
	100 000	48	48

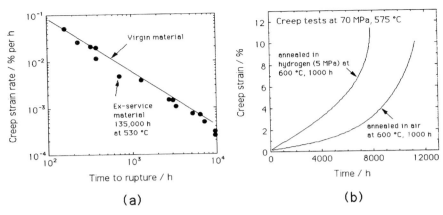

Fig. 27 (a) The minimum creep strain rates of 2.25Cr1Mo steel in ordinary tests (line) and for samples which had been exposed to molecular hydrogen at 3.5 MPa for 135 000 h at 530°C (points). (b) Creep curves of samples which were first annealed at 600°C for 1000 h in hydrogen (5 MPa) or in air. After.[89]

The introduction of cracks or cavities changes the compliance of the sample and hence must modify the velocity of the sound. However, the effect is so small (about 1% change) that it cannot be used with confidence especially when the scatter in experimental data is taken into account.

In a typical ultrasonics test, a spectrum of waves from a probe at the surface propagate through the sample. They interact with the metal and are eventually reflected from the opposite side and detected as echoes at the receiver. There are two major echoes, the first from high-frequency components and the second from low-frequency components. The former is attenuated by defects to a much greater extent than the latter and hence can be used to detect damage (Fig. 28a). Nakashiro et al.[90] point out that in spite of the acceptable sensitivity of the method, the attenuation tends to be sensitive to the quality of contact between the probe and the sample, and indeed to sample size and geometry, making it difficult to apply in practice.

Nakashio et al. propose instead a method based on the analysis of high frequency noise which precedes the arrival of the first back-wall echo. The noise in the absence of defects is first defined from a spectrum (decibels dB versus frequency) obtained using a defect free sample. The noise is the area of the spectrum in units of MHz.dB. The area measured from a damaged sample is then reduced by the noise associated with the defect-free sample to define a noise parameter N, which appears very sensitive to damage (Fig. 28b). The method has the additional advantages that it is free from the effects of the inner surface since it does not rely on back-wall echoes, and it can be optimised over the high-frequency range. It has been used, for example, to characterise the damage in 2.25 Cr1Mo welded samples where there is intense damage in the heat affected zone, leading to the famous Type IV cracking.

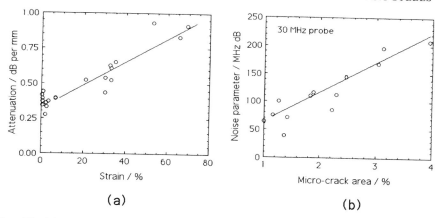

Fig. 28 (a) Relationship between the attenuation and the creep strain for a 2.25Cr1Mo steel. (b) Relation between microcrack area and noise N for a Type 321 stainless steel creep specimen, using a 30 MHz probe. After Nakashiro *et al.*, 1996.[90]

SMALL SPECIMEN TESTING

An excellent review has been published on small specimen tests in the context of irradiation effects, by Lucas.[91] Specific details relevant to the power plant industry are discussed below, but it is worth emphasising some general features highlighted by Lucas. Small samples require a higher standard of machining because they may be more sensitive to surface defects from punching or machining operations. Some creep tests are done on tubular samples which are pressurised, in which case variations in wall-thickness can lead to substantial differences in hoop stress. The machining tolerances have to be much better than for large samples. The chances of finding inclusions or other similar defects are reduced with small size specimens and this must be taken into account if scatter is an important issue, or if the property measured is sensitive to inhomogeneities. An obvious example in the latter context is the Griffith crack. Small specimens are also more susceptible to changes in temperature during the course of an experiment.

Creep Tension Tests

Miniature, cylindrical creep specimens can be as small as 2 mm in diameter and about 4 cm in total length (about 2 cm gauge length). It is also possible to use shorter samples with electron beam welded threaded grips. This compares with normal samples, which are typically 15 mm in diameter and 100 mm long. Normal creep tests are conducted in air but the small samples have to be tested in an inert environment (*e.g.* high-purity, dry argon) in order to avoid overwhelming oxidation. The results from small samples tested in this way are typically found to be about 25% better than those from large samples tested in air. This is because of the absence of oxidation, which not only reduces the effective cross-section but the surface oxide can penetrate along grain boundaries leading to notch effects.[65] These effects have to be taken into account when applying small sample data to life assessment procedures.

Many miniature creep rupture tests with 2 mm diameter specimens are conducted in air. Not surprisingly, it has been demonstrated in these tests that the full scale samples last some two to three times longer than the miniature samples.[11]

It is worth noting here that the ISO stress rupture data are based on tests conducted in air. The data are known to underestimate the actual life of thick components where surface oxidation effects matter less.

Bulge Tests

Bulge tests rely on parallel-sided sheet samples which are carefully machined from the surface of a component.[91,92] A typical sample would be 8 mm in diameter and 0.5 mm thick. Naturally, any high temperature tests must be conducted in inert gas given the rather small thickness. There are two kinds of test, one which is rather like the Ericsson cupping test in which a constrained or unconstrained blank is deformed by a punch with a hemispherical end. The punch moves into a die whose hole diameter is greater than the combined thickness of the punch and 2×sheet thickness.

The second method involves a shearing of the sheet along the punch/die interface; the punch in this case is flat-ended and has a cylindrical geometry. Because the deformation is localised when compared with the bulge test, the punch load versus displacement curve has characteristics akin to a tensile test. For example, the maximum in the load correlates with the ultimate tensile strength.

Purmenský and Wozniak[92] have applied the method to an unspecified 'CrMoV' power plant steel to determine the yield and tensile strengths, and some creep properties. It is not completely clear from their work whether the measurements are directly related to bulk properties or whether they have to be converted using calibration experiments. There has been a lot of work on modelling the deformation using finite element and other methods,[91] but inputs such as the punch/sample friction make the extraction of fundamental data quite difficult.

Tensile Tests

It is generally agreed that the average yield strength can be accurately assessed using miniature specimens as long as the smallest sample dimension is still much greater than the grain size of the microstructure. The size effect relative to the grain structure is expected since single-crystal deformation has different constraints when compared with polycrystalline materials where the grains must deform in a manner which maintains continuity. The required number of grains to produce representative deformation is believed to be about 10–25.[91] There may be differences in the level of scatter observed for large and small samples.

It has been known for a long time the elongation and other measures of ductility are sensitive to specimen geometry and absolute size. Elongation, which is quoted as the sum of uniform and non-uniform deformation, is measured over a gauge length and hence must be geometry dependent. Barba's law[93] expresses the non-uniform component of elongation as

$$\text{nonuniform elongation}\% = 100 \times \beta \frac{A_O^{\frac{1}{2}}}{L_O} \tag{23}$$

where L_O is the gauge length, A_O is the cross-sectional area and β is an empirical constant with a value of about 1.239. Many testing standards therefore state that the gauge length should be $5.65\sqrt{A_O}$. The ultimate tensile strength (UTS) is measured from the maximum load supported by the sample, which in turn depends on the onset of necking. The UTS is therefore more dependent on size effects than the yield strength.

Impact Testing

The Charpy test itself is recognised to be an empirical measure of toughness, of use in quality control but not as an aid to design. The use of miniature specimens is nevertheless fraught with difficulties because the mechanism of fracture is sensitive to the sample dimensions. The effects of plastic constraint are well documented in standard texts on the subject. There can be no general correlations between measurements on large and small specimens. Miniature specimens should therefore only be used for making comparisons, but with additional caution since brittle fracture may not occur at all.

The critical value K_{IC} of the stress intensity which must be exceeded to induce rapid crack propagation is the product of two terms:[94]

$$K_{IC} = \text{stress} \times \text{distance}^{\frac{1}{2}} \tag{24}$$

where the stress is a fracture stress σ_F which can measured independently using notched tensile specimens. It can be related to the microstructure via:[94,95,96]

$$\sigma_F \propto \left[\frac{E\gamma_p}{\pi(1 - v^2)c}\right]^{\frac{1}{2}} \tag{25}$$

where E is the Young's modulus and v is the Poisson's ratio. γ_p is the effective work done in creating a unit area of crack plane, estimated to be about 14 Jm^{-2} for many iron-base microstructures; it is much larger than a surface energy (typically 1 Jm^{-2}) because of the plastic zone which moves with the crack tip. This value of 14 Jm^{-2} seems to apply to a wide variety of steel microstructures, which is surprising given that they often have quite different deformation characteristics.

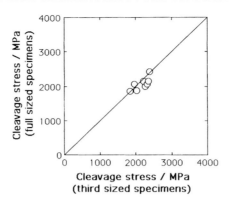

Fig. 29 The cleavage fracture stress determined from instrumented Charpy specimens. The plot compares the results from full-sized and third-sized samples. After Lucas, Odette, Sheckherd and Krishnadev Ref. 97.

In any event, there is no obvious way of relating γ_p to details of the microstructure. By contrast, the dimension c is usually attributed to the size of a sharp crack created by the fracture of a brittle microstructural constituent such as a cementite particle in wrought steels, or a non-metallic inclusion in a weld deposit.

Lucas and co-workers have found that cleavage fracture stress measured from miniature instrumented impact tests is in excellent agreement with similar measurements from full-sized samples (Fig. 29). Notice that the fracture stress is not a complete description of fracture toughness. The other parameter in eqn 24, distance$^{1/2}$, refers to a distance ahead of the crack tip, within which the stress is large enough to cause the fracture of brittle crack-initiators. The latter are unlikely to be distributed homogeneously in the microstructure and so that the critical distance may not be identical for small and large samples.

REGENERATIVE HEAT TREATMENTS

Cavitation and other similar irreversible creep damage does not occur until a late stage in the service ife of the creep resistant steels described above. During that period, any loss in properties is due largely to microstructural changes such as carbide coarsening, changes in the configuration of any dislocations, and in the general approach of the microstructure towards equilibrium. Some of these changes can in principle be reversed by regenerating the original microstructure by heating the component back into the austenite phase field, a process which is not often desirable due to the high temperatures involved.

A possible alternative is to regenerate just the carbides, by annealing the steel at a temperature above the service temperature ($\simeq 700°C$), but below that at which austenite can form.[98] The aim of such a heat treatment would be to dissolve some of the carbides, and reprecipitate them by ageing at a lower temperature, thereby regenerating a fine carbide dispersion. The work reported by Senior on a 1CrMoV steel indicates, however, that 700°C is not high enough to allow a substantial amount of carbide to dissolve, the annealing in fact leading to a further deterioration in the microstructure by accelerating its approach towards equilibrium. For example, in the 1CrMoV steel, the beneficial V_4C_3 carbide increases in volume fraction but also coarsens rapidly.

Embrittlement may also occur as a consequence of the segregation of impurities to the prior austenite grain surfaces during the course of service at elevated temperatures. The rate of segregation is slow at low temperatures and reduced at high temperatures because entropy effects then become dominant. There is, therefore, an intermediate temperature at which impurity embrittlement is most prominent. Thus, Yamashita *et al.* (1997) found the worst toughness for regions of a CrMoNiV rotor (which had experienced 16 years of service) to occur in regions which had experienced 714 K, the toughness improving for lower and higher temperatures. A heat treatment of the embrittled regions at 811 K for 24 h led to an improvement of toughness even though the hardness did not change. This is presumably because of the evaporation of segregated impurities but also because the carbides at the prior austenite grain boundaries were refined by the 811 K heat treatment.

CONCLUSIONS

It is now well understood that the management of ageing turbine plant is not simply a means of extending the service life, but is a method of optimising assets without compromising productivity and safety.[99] A large number of methods available for the assessment of remaining life have been reviewed in this paper. These methods can, to varying degrees, be used in combination with experience to design an integrated approach towards life assessment and monitoring.

ACKNOWLEDGMENTS

This paper is published with the permission of National Power and ALSTOM Energy Ltd, which make no warranty or representation whatsoever that the information is accurate or can be used for any particular purpose. Any person intending to use the information should satisfy himself of the accuracy thereof and the suitability for the purpose for which he intends to use it. The authors are also grateful to Dr R. Thomson, Dr J. Robson, Mr M. Lord and Mr N. Fujita for interesting discussions, and to Professor Alan Windle for the provision of laboratory facilities at the University of Cambridge.

This paper has already been published in Int. Mater. Rev., 1998, **43**(2), 45–69.

REFERENCES

1. R.W. Evans and B. Wilshire: *Creep of Metals and Alloys*, Institute of Materials, London, 1985.

2. F.B. Pickering: *Microstructural Development and Stability in High Chromium Ferritic Power Plant Steels*, Institute of Materials, London, 1997, 1–30.

3. R. Viswanathan: *Damage Mechanisms and Life Assessment of High-Temperature Components*, ASM International, Metals Park, Ohio, 1989.

4. H.K.D.H. Bhadeshia: *Bainite in Steels*, Institute of Materials, London, 1982.

5. R.D. Townsend: *Rupture Ductility of Creep Resistant Steels*, A. Strang ed., Institute of Metals, London, 1991, 1–16.

6. R. Townsend: *Electricity Beyond 2000*, Electric Power Research Institute, Palo Alto, California, U.S.A., 1991, 251–281.

7. R.G. Carlton, D.J. Gooch, E.M. Hawkes and H.D. Williams: *Proc. Conf. Refurbishment and Life Extension of Steam Plants*, Institution of Mechanical Engineering, London, 1987, 45–52.

8. J.D. Parker, A. McMinn and J. Foulds: *Proc. Conf. Life Assessment and Life Extension of Power Plant Components*, T.V. Narayanan ed., ASME, New York, book number H00486, 1989, 223.

9. D.J. Gooch and R.D. Townsend: *Proc. EPRI Conf. Life Assessment and Extension of Fossil Fuel Plants*, Pergamon Press, Oxford, 1987, 927–943.

10. D.J. Gooch, M.S. Shammas, M.C. Coleman, S.J. Brett and R.A. Stevens: *Proc. Conf. Fossil Power Plant Rehabilitation, Cincinnati, Ohio*, ASM International, Ohio, 1989, Paper 8901–004.

11. R. Viswanathan: *Materials Science and Engineering A*, 1988, **A103,** 131–139.

12. H.E. Evans: *Mechanisms of Creep Fracture*, Elsevier Applied Science Publishers, Essex, England, 1984.

13. E.L. Robinson: *Trans. Am. Soc. Mech. Engrs.* 1952, **74**, 777.

14. B.J. Cane and R.D. Townsend: *CEGB report TPRD/L/2674/N84*, Leatherhead, Surrey, 1984.

15. T. Goto: *Microstructure and Mechanical Properties of Ageing Materials*, Fossil Power Plants session at ASM Fall Meeting, Chicago, (unpublished), 1992.

16. D.J. Gooch, A. Strang and S.M. Beech: *5th Int. Conf. on Creep and Fracture of Engineering Materials*, Swansea, 1993, 1–17.

17. A. Strang, S.M. Beech and D.J. Gooch: *Materials for Advanced Power Engineering*, D. Coutsouradis *et al.* eds, Liege Part I, 1994, 549–560.

18. B.W. Roberts and A. Strang: *Refurbishment and Life Extension of Steam Plant*, Institution of Mechanical Engineers, London, 1987, 205–213.

19. J. Maguire and D.J. Gooch: *Proc. Int. Conf. Life Assessment and Extension*, The Hague, 1988.

20. A.J. Tack, J.M. Brear and F.J. Seco: *Creep: Characterisation, Damage and Life Assessment*, D.A. Woodford, C.H.A. Townley and M. Ohnami eds, ASM International, Ohio, USA, 1992, 609–616.

21. D. Kalish, S.A. Kulin and M. Cohen: *Journal of Metals*, February 1965, 157–164.

22. T. Gladman: *The Physical Metallurgy of Microalloyed Steels*, Institute of Materials, London, 1997, 189.

23. R.W.K. Honeycombe and H.K.D.H. Bhadeshia: *Steel, Microstructure and Properties*, Edward Arnold, London, 2nd edition, 1995.

24. T. Hoogendorn and M.J. Spanraft: *Microalloying '75*, ASM International, Ohio, USA, 1975, 75.

25. J.J. Jonas and I. Weiss: *Metal Science*, 1979, **13**, 238.

26. T. Chandra, M.G. Akben and J.J. Jonas: *Proc. 6th Conference on Strength of Metals and Alloys*, Melbourne, Australia, R.C. Gifkins eds, 1982, 499.

27. K.A. Ridal and A.G. Quarrell: *Journal of the Iron and Steel Institute*, 1962, **200**, 366–373.

28. M.C. Murphy and G.D. Branch: *Journal of the Iron and Steel Institute*, 1969, **206**, 1347–1364.

29. D. Kalish and M. Cohen: *Materials Science and Engineering*, 1970, **6**, 156–166.

30. H.K.D.H. Bhadeshia: *Acta Metallurgica*, 1980, **28**, 1103–1114.

31. M.C. Askins and K. Menzies: *Unpublished work*, referred to in Gooch and Townsend, (1985), 1987.

32. P. Battaini, D. D'Angelo, G. Marino and J. Hald: *Creep and Fracture of Engineering Materials and Structures*, Institute of Metals, London, 1990, 1039–1054.

33. M. Askins and K. Menzies: *Unpublished work*, Central Electricity Generating Board, Leatherhead, referred to in reference 9, 1985.

34. S.D. Mann, D.G. McCulloch and B.C. Muddle: *Metallurgical and Materials Transactions A*, 1995, **26A**, 509–520.

35. R.B. Carruthers and M.J. Collins: *CEGB Report NER/SSD/M/80/327*, 1980.

36. R.B. Carruthers and M.J. Collins: *Quantitative microanalysis with high spatial resolution*, Metals Society, London, 1981, 108–111.

37. R.B. Carruthers and M.J. Collins: *Metal Science*, 1983, **17**, 107.

38. A. Afrouz, M.J. Collins and R. Pilkington: *Metal Technology*, 1983, **10**, 461.

39. H.K.D.H. Bhadeshia: *Materials Science and Technology*, 1989, **5**, 131–137.

40. B.J. Cane and J.A. Williams: *International Mat. Rev.*, 1987, **32**, 241–262.

41. R. Singh and S. Banerjee: *Scripta Metallurgica et Materialia*, 1990, **24**,1093–1098.

42. S.D. Mann and B.C. Muddle: *Microstructures and Mechanical Properties of Aging Material*, P.K. Liaw, R. Viswanathan, K.L. Murty, E.P. Simonen and D. Frear eds, The Minerals, Metals and Materials Society, Warrendale, Pennsylvania, 1993, 301–308.

43. S.D. Mann and B.C. Muddle: *Microstructures and Mechanical Properties of Aging Material*, P.K. Liaw, R. Viswanathan, K.L. Murty, E.P. Simonen and D. Frear eds, The Minerals, Metals and Materials Society, Warrendale, Pennsylvania, 1993, 309–317.

44. S.D. Mann and B.C. Muddle: *Micron*, 1994, **25**, 499–503.

45. H.K.D.H. Bhadeshia: *Mathematical Modelling of Weld Phenomena 2*, H. Cerjak and H.K.D.H. Bhadeshia eds, Institute of Materials, London, 1995, 71–118.
46. J. Chance and N. Ridley: *Metallurgical Transactions A*, 1981, **21A**, 1205–1213.
47. P. Wilson: *Ph. D. Thesis*, University of Cambridge, 1991.
48. R.C. Thomson and H.K.D.H. Bhadeshia: *Materials Science and Technology*, 1994, **10**, 193–204.
49. R.C. Thomson and H.K.D.H. Bhadeshia: *Materials Science and Technology*, 1994, **10**, 205–208.
50. D. Venugopalan and J.S. Kirkaldy: *Hardenability concepts with applications to steels*, D.V. Doane and J.S. Kirkaldy eds, AIME, Warrendale, Pennsylvania, 1978, 249–268.
51. X. Du, R.C. Thomson, J. Whiteman and H.K.D.H. Bhadeshia: *Materials Science and Engineering A*, 1992, **A155**, 197–205.
52. R.C. Thomson and H.K.D.H. Bhadeshia: *Metall. Trans. A*, 1992, **23A**, 1171–1179.
53. J.D. Robson and H.K.D.H. Bhadeshia: *Creep Resistant Metallic Materials*, published by Vitkovice, Czech Republic, 1996, 83–91.
54. C.D. Lundin, S.C. Kelly, R. Menon and B.J. Kruse: *Bulletin 315 of the Welding Research Council*, United Engineering Center, New York, June 1986, 1–66.
55. J. Pilling and N. Ridley: *Metallurgical Transactions A*, 1982, 13**A**, 557–563.
56. J. Dobrzanski, P. Milinskp and J. Woszczek: *Creep Resistant Metallic Materials*, published by Vitkovice, Czech Republic, 1996, 152–163.
57. B. Lonsdale and P.E.J. Flewitt: *Materials Science and Engineering*, 1979, **39**, 217–229.
58. J.D. Parker: *Creep Resistant Metallic Materials*, published by Vitkovice, Czech Republic, 1996, 122–130.
59. B.J. Cane and J.A. Williams: *International Materials Reviews*, 1987, **32**, 241–262.
60. R.W. Evans and B. Wilshire: *Creep of Metals and Alloys*, Institute of Metals, London, 1985.
61. G.W. Rowe: *Elements of Metalworking Theory*, Edward Arnold, London, 1979, 16.
62. B.F. Dyson and D. McLean: *Metal Science*, 1972, **6**, 220.
63. B.J. Cane: *Metal Science*, 1979, **13**, 287–294.
64. B.F. Dyson and D. McLean: *Metal Science*, 1977, **10**, 37–45.
65. S.M. Beech, D.J. Gooch and A. Strang: *Third International Charles Parsons Conference on Materials Engineering in Turbines and Compressors*, 1995, 277–291.
66. B.J. Cane: *Metal Science*, 1981, **13**, 287–294.
67. C.H. Wells, D.O. Harris, S. Rao, D. Johnson and F.A. Ammirato: *Electric Power Research Institute Report CS-5208*, Palo Alto, California, 1987.

68. S.M. Beech, J.W. Selway and A.D. Batte: *Second International Conference of Creep and Fracture of Engineering Materials and Structures*, Swansea, 1984, 925–936.

69. B.J. Cane and C.J. Middleton: *Metal Science*, 1981, **15**, 295–301.

70. B.F. Dyson: *Metal Science*, 1976, **10**, 349–353.

71. M.S. Shammas: *Central Electricity Generating Board report TPRD/L/31299/ R87*, Leatherhead, Surrey, 1987, 1–59.

72. M.S. Shammas: *Refurbishment and Life Extension of Steam Plant*, Inst. Mechanical Engineering, London, 1987, 289–295.

73. D. Hull and D.E. Rimmer: *Philosophical Magazine*, 1953, **4**, 673.

74. R. Raj and M.F. Ashby, *Acta Metallurgica*, 1975, **23**, 653.

75. G.K. Walker and H.E. Evans: *Metal Science*, 1970, **4**, 155.

76. A. Strang, S.M. Beech and D.J. Gooch: *Creep Resistant Metallic Materials*, published by Vitkovice, Czech Republic, 1996, 102–121.

77. B.J. Cane: *Metal Science*, 1978, **12**, 102–108.

78. C.E. Price: *Acta Metallurgica*, 1966, **14**, 1781.

79. B. Neubauer and U. Wedel: *Advances in Life Prediction Methods*, D.A. Woodford and J.R. Whitehead eds, American Society of Mechanical Engineers, New York, U.S.A., 1983, 307–314.

80. B. Neubauer: *Second International Conference of Creep and Fracture of Engineering Materials and Structures*, Swansea, B. Wilshire and D.R.J. Owen eds, Pineridge Press, 1984.

81. R. Wu, J. Storesund and R. Sandström: *Materials Science and Technology*, 1993, **9**, 773–780.

82. C.J. Middleton: *Metal Science*, 1981, **15**, 154–167.

83. S.R. Holdsworth and D.V. Thornton: *Microstructures and Mechanical Properties of Aging Material*, P.K. Liaw, R. Viswanathan, K.L. Murty, E.P. Simonen and D. Frear eds, The Minerals, Metals & Materials Society, Pennsylvania, USA, 1993, 83–89.

84. S. Wignarajah, I. Masumoto and T. Hara: *ISIJ international*, 1990, **30**, 58–63.

85. G. Langford and M. Cohen: *Metallurgical Transactions A*, 1975, **6A**, 901–910.

86. A. Orlová, Z. Tabolová and J. Čadek: *Philosophical Magazine*, 1972, **26**, 1263–1274.

87. H. Yoshizawa, K. Morishima, M. Nakashiro, S. Kihara and H. Umaki: *Creep Resistant Metallic Materials*, published by Vitkovice, Czech Republic, 1996, 164–173.

88. Bild: *Creep Resistant Metallic Materials*, published by Vitkovice, Czech Republic, 1996.

89. V. Bina, O. Bielak, J. Hakl, F. Hnilica and J. Vrtel: *Creep Resistant Metallic Materials*, published by Vitkovice, Czech Republic, 1996, 174–182.

90. M. Nakashiro, H. Yoneyama, S. Shibata, H. Umaki and N. Izawa: *Creep Resistant Metallic Materials*, published by Vitkovice, Czech Republic, 1996, 183–192.

91. G.E. Lucas: *Metallurgical Transactions A*, 1990, **21A**, 1105–1119.
92. J. Purmenský and J. Wozniak: *Creep Resistant Metallic Materials*, published by Vitkovice, Czech Republic, 1996, 142–151.
93. M.J. Barba: *Mem. Soc. Ing. Civils*, Part I, 1880, 682.
94. J.F. Knott: *Micromechanisms of Fracture and Their Structural Significance*, Second Griffith Conference, Institute of Materials, London, 1995, 3–14.
95. C.J. McMahon Jr., and M. Cohen: *Acta Metallurgica*, 1965, **13**, 591.
96. D.A. Curry and J.F. Knott: *Metal Science*, 1978, **12**, 511.
97. G.E. Lucas, G.R. Odette, J.W. Sheckherd and M.K. Krishnadev: *Fusion Technology*, 1986, **10**, 728–733.
98. B.A. Senior: *Central Electricity Generating Board Report TPRD/L/3220/R87*, 1987.
99. B. Cane: *Advances in Turbine Materials, Design and Manufacturing*, Proceedings of the Fourth International Charles, A. Strang, W.M. Banks, R.D. Conroy and M.J. Goulette eds, 1997, 554–574.

Materials and Processes for High Temperature Surface Engineering

J. R. NICHOLLS[*] and D. S. RICKERBY[†]
* Cranfield University, Cranfield, Bedford MK43 0AL, UK
† Rolls Royce Plc, PO Box 31, Derby DE24 8BJ, UK

ABSTRACT

Surface engineering plays an important role in the operation of all high temperature power or process plant, whether it be by alloy design aimed at forming a protective thermally grown oxide or through the custom engineering of surface alloys or coatings to resist the arduous high temperature service environments.

As examples, within the modern aero gas turbine engine up to 50% of components are surface engineered, with high temperature coatings varying from the relatively simple diffusion aluminised coatings to more exotic (and expensive) thermally sprayed and physical vapour deposited coatings that are used for example to provide thermal protection.

For industrial power plant, whether diesel or gas turbine driven, engines must be multi-fuel capable, running on a range of fuels that may vary from natural gas, through bio-fuels to distillates and light crude oils. This leads to a materials challenge to design coating systems that are resistant to a range of highly corrosive environments but capable of providing the operating lives required to meet these industrial duty cycles.

In other areas of fossil fuel fired plant, materials may be required to resist highly sulphidising, low oxygen activity gaseous atmospheres in which rapid corrosion degradation can take place. Often highly alloyed, and therefore expensive materials are used for this service to reduce the rates of corrosion. Surface engineered low alloyed steels offers an alternative option provided coating processes can be identified capable of coating large components both during initial plant construction and repair.

Regardless of the application, the move towards 'designed surfaces' is driven by the often conflicting requirements to manufacture with materials that require different surface and bulk properties. This paper will review both new materials and processes used to surface engineer high temperature components and will illustrate how the coating/substrate system should be regarded as a 'composite' system at the plant design stage, thus allowing engineers the greater freedom to design components with the best bulk (substrate) materials, whilst protecting the component from the working environment by some form of surface treatment.

1   INTRODUCTION

The insatiable desire for higher performance, cleaner and more fuel economic power plant, be it an aero gas turbine, a medium speed diesel, automotive petrol engine, or a future generation multifuel capable industrial power plant, can only be met by concurrent engineering practices involving the best in design concepts, with the most advanced materials and manufacturing technologies.[1-4] For components operating at high temperatures within such plant, surface engineering is an important part of this concurrent engineering practice,[3,4] with coating systems and surface treatments increasingly being relied upon to deliver the reliability and plant performance for advanced power unit designs.

1.1 High Temperature Materials Issues in Advanced Power Plant

The materials issues for advanced steam turbines,[5] supercritical boilers,[6] heat exchangers,[7] chemical and petrochemical plant[8] and industrial gas turbines[9] are considered elsewhere in this congress. This paper will consider only those issues which reflect on the use of high temperature surface engineering in such plants.

1.1.1 Steam Turbines

Recent developments[3] have seen 'state-of-the-art' steam temperatures increase from 540°C to around 620°C. Future developments can be expected to see steam turbines operating at even higher temperatures.[5] The increase in steam temperatures from 540–620°C has been achieved through the development of new ferritic steels, while the exploitation of nickel based alloys (whose use is well established in the aero and industrial gas turbine industries) will enable operating steam temperatures up to 700°C and above to be considered.[3,5] Coating technologies may play a role in future plant design, to protect against increased oxidation and solid particle erosion at the perceived higher operating temperatures.

1.1.2 Supercritical Boilers

The design of supercritical boilers with steam conditions of 580°C and above require that metal temperatures in the final superheaters are in the range 650–680°C.[3] The development of advanced ferritic steels to 650°C, and advanced austenitics to 680°C are being researched[6] and must form part of a future materials R&D strategy.[3] At these metal temperatures steam side and fireside corrosion would play an increasingly important role in determining tube life. Beyond these temperatures and pressures, the hottest components in the boiler are expected to be constructed from advanced alloys, such as nickel based superalloys.

In these advanced plant, fireside corrosion of water walls and superheaters is seen as an increasing problem. This will restrict the use of cheaper fuels and possible co-firing with waste or biomass, unless more corrosion materials are used. Coating technologies have a role to play, however, they must be low cost and easily applied as thick overlays. Alternatively co-extruded materials maybe used. Whichever alternative is chosen, the protection systems must be capable of resisting the fireside corrosion environments generated as a result of burning poorer grade fuels, biomass and waste.

1.1.3 Aero-gas Turbines

The working environment in the aero gas turbine is extremely demanding. Turbine inlet temperatures are increasing at a rate of 15°C per year and are expected to exceed 1450°C by the early 2000s.[10] This is reflected in the metal surface temperatures of the first stage stator vanes and turbine blades, which may see continuous service of 1050°C and peak temperatures up to 1150°C (1200°C in hot spots). Thus rotating components must have the creep strength to operate at these temperatures, plus they must be able to resist the superimposed thermal loads, and have sufficient oxidation and corrosion resistance to achieve desired service lives.

Surface engineering forms an implicit part of advanced aero engine design. Over the last 30 years there has been a steady increase in the number of coated components used in the aero gas turbine engine from around 25% for the RR-Avon engine to greater than 50% in the future Trent series of Rolls Royce engines.[4] This move towards 'designer surfaces', customised surface engineering, has been driven by the often conflicting requirements in the design of new materials of achieving adequate bulk material properties and sufficient environmental protection. This is well illustrated in Fig. 1 which compares the corrosion performance of an early uncoated blade (Fig. 1a) with that coated using the pack aluminide process (Fig. 1b). The uncoated turbine blade has been removed from an engine after 2500 hours of service between airfields with low level sea approaches and relative short duration sector times and exhibits severe hot corrosion. The aluminised blade (Fig. 1b) shows virtually no attack after similar service times. Thus today, most components in the HP and IP turbine are coated to ensure that surface related problems encountered in the high pressure high temperature module of the turbine are minimised.

Fig. 1 A comparison of the corrosion resistance of an uncoated (a) and pack aluminide coated (b) blade after 2500 h flight service.[4]

Commercially available coatings include diffusion coatings (e.g. pack aluminides), modified diffusion coatings (e.g. platinum aluminides), overlay coatings and thermal barrier coatings; all of which will be included as part of this review.

Thermal barrier coatings are widely used on static components to provide thermal protection and have been used on combustors in aero-gas turbines for over 20 years.[11,12] For example, the most common combustion chamber materials consist of a nickel based component protected with a plasma sprayed ZrO_2–$8\%Y_2O_3$ thermal barrier coating. More recently, thermal barrier coatings have been introduced into service on rotating components, turbine blades, by a number of aero engine manufacturers.[11]

Factors governing the final selection of the coating system are the operational temperature, the required design life, the initial capital cost, through life cost, ease of manufacture, including both quality assurance and logistics.[4] In future generations of aero gas turbines, one will expect to see custom design surface engineering particularly on rotating components. This could include a structured thermal barrier coating, overlaying an MCrAlY or diffusion based coating, specifically designed as a bond coat, but which additionally provides environmental protection. For more severe service an additional underlay coating may be included which limits interdiffusion between the substrate and bond coat.

1.1.4   Industrial Gas Turbines

Industrial gas turbines currently feature in the most efficient variants of advanced power plant. Their versatility allows them to burn many fuel types, including natural gas, synthetic gas (produced from coal, biomass or other feedstocks) and/or liquid fuels.[3,9] Turbine inlet temperatures are currently lower than their aero-counterparts, but are following aero-gas turbine trends and are expected to exceed 1400°C by the year 2000.[10]

The materials issues in the industrial gas turbine are similar to those identified for the aero-gas turbine, but the service temperatures are lower, although the working environment may be more severe, particularly when burning other than premium fuels.[3,9] Equally, duty cycles are such that industrial turbines lives in excess of 50 000 h are expected; considerably longer than their aero space counterparts.

The perceived increased operating temperatures, coupled with the use of dirtier fuels, both more corrosive and erosive, and expected long service lives will require custom developed coatings for such utility service. These advanced coating systems must be resistant to high temperature corrosion, erosion and ideally provide thermal protection as well. The long duty cycles at high temperatures, and the need to provide corrosion resistance against a range of fuel-borne impurities can make the industrial turbine a particularly arduous material service environment.

1.1.5   Large and Medium Speed Diesel Engines

This last example of high temperature power plant considered in this paper sees a severe mechanical and environmental load. The large, medium speed diesel engine operating under utility or marine conditions often burn a variety of fuels, many of which are low grade and often high in vanadium. Figure 2 illustrates the formation of a gutter in the exhaust valve seat.[13] In this example, the hard facing alloy deposited on the valve seat must not only resist severe vanadic corrosion, but must be able to resist the stresses generated by thermal gradients and valve closure.

Thus in this duty cycle the coating must be able to resist indentation and high temperature wear[14] in addition to severe hot corrosion. It is the entrapment of combustion debris, valve seat indentation and the cracking of any surface deposits that are formed that lead to hot gas blow through and the initiation of a gutter. Metal surface temperatures are in the range 450–550°C, but rise to in excess of 700°C within the valve when a gutter forms.

1.1.6   Summary of Surface Engineering Requirements
for High Temperature Service

From the above it is evident that the selection of a suitable treatment or coating for a given application depends on a complex interplay of surface and coating and substrate related properties determined by the specific application.

The desired properties of a coating system for high temperature service have been tabulated over a decade ago by Nicholls and Hancock[15,16] and has been adopted subsequently as a framework around which coating performance can be

Fig. 2 Severe gutter formation in a diesel engine exhaust valve, the engine was operated on residual fuel oil.[13]

readily discussed.[17,18,19] An updated version of this tabulation, emphasising the design requirements of a coating system for good oxidation/corrosion resistance and indicating the location within the coating system where these desirable properties are of most benefit, is given in Table 1. Clearly, if coatings of fixed composition are used then they must offer a compromise between the often conflicting requirements, summarised in Table 1. Thus the concept of 'Designer Surfaces' and the need to design surface engineering treatments to a given application.

The complexity of interactions between a coating and both the substrate and the environment emphasises the need to treat each application individually. It is essential therefore to carry out a complete characterisation of each projected use. Obviously, this is a costly process and may inhibit the widespread use of coatings or surface treatments within high temperature plant. Some general guidance can be obtained from existing experience and from a clear understanding of the various types of failure mechanism thereby facilitating the development of models that could be used to predict the longer term behaviour of coated components.

The considerations which govern the final selection of a surface treatment or coating system include:

   (i) The operating temperature of the component
  (ii) The service environment and duty cycle
 (iii) The desired design life

In recommending the use of a coating to solve a particular problem, the engineer or materials scientist additionally must take into consideration the following:

 (iv) The long term stability with respect not only to the oxidation/corrosion environment, but also to interdiffusion with the substrate,
  (v) The resistance to thermal shock, and
 (vi) The load bearing capability of the treated component.

Obviously, loss of surface protection can result in very high rates of attack of the substrate and consequent catastrophic failure of the component (as evident in the turbine blade and diesel valve examples cited earlier). Hence the principal reason for applying surface treatment or coating to high temperature components is to ensure that the component is capable of operating efficiently throughout the design life.

Having selected a suitable materials solution one must bear in mind that the surface engineered component has to be manufactured and at an acceptable cost. This requires that:

(vii) The manufacturing route be defined, including the cost to set up the manufacturing process (capital cost),

Table 1 Requirements of coating systems for oxidation resistance.[15]

Coating property	Requirement	Coating surface	Mid coating	Coating/substrate interface
Oxidation/corrosion resistance	Low rates of scale formation	X	–	–
	Uniform surface attack	X	–	–
	A thermodynamically stable surface oxide	X	–	–
	Ductile surface scales	X	–	–
	Adherent surface scales	X	–	–
	High concentration of scale forming elements with coating to act as reserve for scale repair	X	X	–
Interface stability	Low rate of diffusion across interface at operating temperatures	–	–	X
	Limited compositional changes across interface	–	–	X
	Absence of embrittling phase formation during service	–	–	X
Good adhesion	Matched coating and substrate properties to minimise coating mismatch and stress generation at coating/substrate interface	–	X	X
	Optimum surface condition before coating	–	–	X
	Growth stresses during coating formation should be minimised	–	X	X
Mechanical strength	Coating must withstand all stress (creep, fatigue, and impact loading) that is generated at component surface during service	–	X	–
	Well matched thermal expansion coefficients between coating and substrate to minimise thermal stressing and thermal fatigue	–	X	X

(viii) The through life costs be assessed, including the feasibility for component repair.

(ix) That procedures are established for quality assurance and finally

(x) The logistics of including the coating/surface treatment into the current manufacture are assessed.

The remainder of this paper will review high temperature surface engineering procedures in industrial usage, together with new, novel methods of providing improved protection that are still in the development stage.

Hence, this review will examine:

1. The feasibility of modifying the thermally grown oxide, to give a mechanically stable, protective oxide scale,
2. The modification of a components surface using diffusion coatings (chemical vapour deposition),
3. The combination of electroplating and chemical vapour deposition to give modified diffusion coatings,
4. The deposition of corrosion resistant overlay coatings, using either physical vapour deposition (PVD), plasma spraying techniques, or electroplating,
5. The combination of chemical vapour deposition with overlay coatings to produce compositionally gradient coatings resistant to complexed corrosion environments,
6. Thermal barrier coating systems (TBC's) deposited by either plasma spraying or electron beam physical vapour deposition (EB-PVD),

and

7. Layered thermal barrier coating concepts offering reduced thermal conductivity, while maintaining good strain tolerance.

2   THE NATURE OF HIGH TEMPERATURE SURFACE PROTECTION

Structural materials for use at high temperatures are generally highly alloyed with an iron, nickel or cobalt base, for example high temperature ferritic and stainless steels, Fecralloys and the nickel or cobalt based family of superalloys. Ceramics are also used, as structural components but this is outside the scope of this review paper. These structural materials find use in numerous high temperature processes as reviewed in the introduction to this paper with operating temperatures in the range 500–1250°C.

Oxygen, sulphur, carbon, halogens (particularly chlorine), sodium (and other alkali metals) and vanadium are the most common aggressive species present in industrial combustion environments. These species give rise to various forms of corrosive environment, including sulphidation, metal dusting, type I and type II hot corrosion, and vanadic corrosion. For details on these forms of high temperature corrosion reference is made to various recent text books on the subject.[20-22] However, as most of these high temperature environments contain oxygen to a greater or lesser degree the ability to form a stable, protective, surface oxide is paramount to good high temperature materials performance in these corrosive environments. In this respect, good, high temperature corrosion resistance depends primarily on the alloy/coatings ability to form and maintain a protective oxide scale of Cr_2O_3, Al_2O_3 or SiO_2. Thus most metallic structural components are protected by a stable, thermally grown oxide scale.

2.1   The Growth of Mechanically Stable, Protective Oxides

Protection of the material, beit a structural alloy or metallic coating, from high temperature corrosion depends on the properties of the oxide scale/corrosion product formed. This should act as a diffusion barrier to prevent further attack by the atmosphere. It should be chemically stable in the environment so that the oxide barrier is not actively dissolved[20] and should limit the transport of both the reacting gases and metallic species through the oxide.[21,22] The oxide must also be mechanically stable,[22-26] that is it should not crack or spall under the mechanical and thermal loads introduced in service, as this will lead to breakdown of the protective barrier and rapid attack of the substrate.

2.1.1   Selective Oxidation – Controlled Growth of Native Oxides

Alloys containing concentrations of either Al or Cr in excess of about 10 and 16%, respectively, will in oxidising atmospheres eventually form a scale in which alumina and chromia predominates. When Al and Cr are present in combination much lower levels of aluminium, typically 5%, will result in the establishment of a protective alumina scale. It has been frequently found that if exclusive formation of these oxides can be encouraged by, for example, oxidation in the relevant H_2–H_2O or CO–CO_2 gas mixtures, there is improved resistance to attack compared with behaviour obtained by direct exposure. The conditions for the oxidation process can be chosen by reference to an Ellingham–Richardson diagram (see Refs 21 and 22) and essentially involves selecting conditions such that the H_2/H_2O ratio is sufficient to form Al_2O_3 or Cr_2O_3, but inhibits formation of the base alloy constituent. Figure 3 shows an oxide stability diagram, derived from such thermodynamic background, for oxide systems relevant to high temperature service in the temperature range 500–1200°C. To calculate the H_2/H_2O ratio to be used to preferentially form a given oxide at 1000°C, for example, one can use the following equation.

$$\log \frac{H_2}{H_2O} = -7.3 - 0.5 \log(p.O_2) \tag{1}$$

where $p.O_2$ is oxygen partial pressure.

It should be noted that while exclusive formation of Al_2O_3 is often possible using such selective oxidation processes (except where active elements such as yttrium are present) growth of Cr_2O_3 will be accompanied by the oxidation

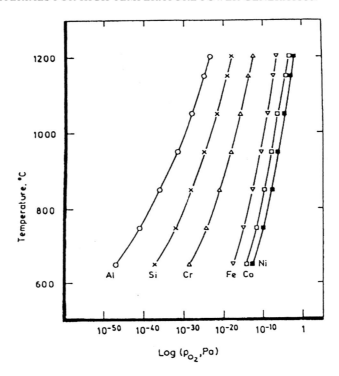

Fig. 3 Stability diagram for protective oxide scales.[19]

products of Mn, Si, and Ti. These elements are all frequently present in quantities of ~0.5% in commercial alloys and, their presence may significantly affect the protective nature of the preformed, thermally grown oxide.

When using this approach, the temperature for these selective oxidation treatments is generally chosen to be greater than the application temperatures and sufficiently high to promote relatively rapid formation of an oxide layer 1–5 μm in thickness. For example, at 900°C an alumina layer of 1 μm thickness or a chromia layer of ~3 μm is formed respectively on Fe–28Cr–4Al and Fe–28Cr.[27]

Preoxidation treatments can also be carried out in air or other oxidising atmospheres. Under these circumstances the oxide is often less adherent, particularly if the alloy does not contain additions, such as yttrium, that promote adhesion. The better adhesion achieved following selective oxidation in H_2/H_2O environments has been attributed to the doping effect of hydrogen[27,28] which is reported to enhance scale plasticity.[28] Another important factor in preoxidation is the design of the thermal cycle. It is advantageous to limit thermal cycling as much as possible to avoid stressing the oxide by differential thermal expansion effects and this effect is considered in more detail below in the section on 'Coating Performance'.

Such selective oxidation treatments have been reported to improve the adhesion of thermal barrier coatings on diffusion aluminide bond coats,[29–31] by limiting the formation of transition metal oxides at the bond coat/zirconia interface. However, this experience has not been observed in similar studies within the author's laboratory.

2.1.2   Modification of Thermally Grown Oxides

This process treatment also involves exposing the alloy or metal to the working environment, but, by altering the alloy surface composition or by the addition of trace amounts of material to the oxidising atmosphere, the naturally formed oxide becomes modified and a more protective layer can be obtained. Modifications to the oxide barrier resulting from the selective oxidation procedures outlined above can also be carried out in combination with these processes.

The nature of this type of surface treatment, precludes a general description of procedures to be used. In principle, there is available an almost infinite variety of methods/surface treatments that can be utilised. However, to illustrate this approach examples from reference 19 are cited, some of which are in the realm of 'laboratory curiosities' while others have been developed to the stage of commercial application.

Table 2 (reproduced from data provided in reference 19) gives possible processes used to modify scales formed on Fe–Cr alloys by addition of boron or silicon containing compounds to the working environment or as superficially

Table 2 Modification of thermally grown oxides.[19]

Alloy composition, wt-%	Surface treatment	Diffusion conditions	Treatment temperature, °C
Boron enrichment Fe–Cr (10–20%Cr)	–	Air + 600 ppm H_2O + B_2O_3 vapour at 600°C, 1-10 min	600
Fe–9Cr–1Mo	–	CO_2, 4 MPa, + B_2O_3 vapour at 580°C, continuous	580
Fe–2.25Cr–1Mo, Fe–9Cr–1Mo	–	Steam + B_2O_3 vapour 600°C, continuous	600
Fe–Cr alloys	Alcoholic solutions $Na_2B_4O_7$, 0.05–3M dip	Air	600–1000
Cerium enrichment Fe–20Cr–25Ni	$CeNH_4(NO_3)_4$ dip in 5w/v alcoholic solution	CO_2	750–900
Silicon enrichment Fe–9Cr–1Mo	Si enrichment by SiH_4 deposition at 450°C	H_2–H_2O (100:l)	430
Fe–20Cr–25Ni-Nb	SiO_2 solution	Steam or CO_2 2 h, then SiO_2 solution dip	800, then 850

applied oxide or borate additions to a components surface. These additions result in both a reduction of oxidation rates and improved scale adhesion.

There are numerous other additives that can be used to modify or control scale growth. A more detailed account of some of these processes is given in Refs 32–34.

2.1.3 The Mechanical Stability of Oxide Scales

Optimum resistance to oxidation results when the protective oxide layer remains adherent and uncracked, so that continued reaction is limited by solid state diffusion through an oxide layer that acts as an effective diffusion barrier as eluded to in the previous section.

Increasingly, research attention is being paid to service conditions that may prejudice the above protective oxide conditions, through mechanical failure of the scales.[23–26] Of interest is the role temperature change has on scale failure, since rapid temperature transient can generate thermal mismatch stress that lead to scale spallation.

Figure 4 illustrates a spallation map for alumina on a ferritic steel developed using Evans'[25] finite element model for wedge crack propagation. This illustrates how important the interfacial adhesion is (measured as an effect fracture energy) on the ability of an alloy to retain its protective oxide under thermal transient conditions. Recent work by Bull[35] using a scratch adhesion test and associated mathematical models has shown that the adhesive strength for alumina onto

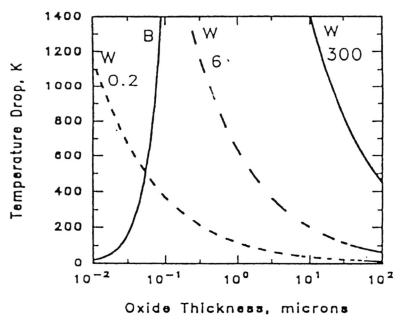

Fig. 4 Spallation map for alumina on a ferritic steel.[26]

FeCrAl alloys varies with material and takes values between 6.6 and 14.2 Jm^{-2}, i.e. values close to the middle curve in Fig. 4.

Two further significant factors, that influence the stability of this thermally grown oxide are that, the position of the wedge crack boundary in Fig. 4 (the critical temperature drop for spallation) varies with cooling rates,[25,26] dropping by approximately 150°C per order of magnitude increase in cooling rate for a 5 μm alumina scale.

Secondly, recent work by Wilber *et al.*[36] has shown that at mass gains of between 15–30 gm^{-2} (oxide thicknesses of 7.5–15 μm) for alumina scales on a range of FeCrAl based steels, the composite defect size rapidly increases as voids cluster and interact to form macro defects in the scale. A net result is the observation of a 'critical' scale thickness for the onset of spallation. An observation often reported in the literature which up until now had not been quantified.

3   CHEMICAL VAPOUR DEPOSITION (CVD) AND DIFFUSION COATING PROCESSES

Diffusion coating processes have been applied for many years to improve the environmental resistance of a base alloy by enriching the surface in Cr, Al, or Si. Pack chromising was in widespread use in the early 1950s[37,38] to increase the oxidation/corrosion resistance of low alloy steels. In the late 1960s aluminising was first used for the protection of superalloy gas turbine aerofoils.[39–41] Aluminide diffusion coatings are now routinely applied to nickel-base superalloy blades and vanes used in the hot sections of gas turbines, to enhance their resistance to high temperature oxidation and hot corrosion. Plain aluminides have provided cheap cost effective solutions to protect superalloy components within the high pressure turbine of both aero- and industrial gas turbines. As such they are probably the most widely used coatings in service within the gas

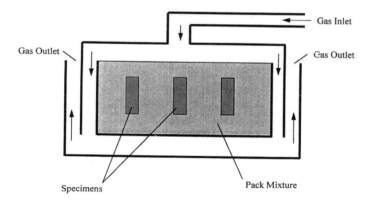

Fig. 5 Schematic diagram of a pack aluminising retort.[59]

turbine industry. There was renewed interest in siliconising and silicon modified diffusion coatings for high temperature service in the early 1970s[42–44] when novel solutions to the low temperature hot corrosion problems, associated with contaminants in industrial turbine plant burning impure fuels, were required. Siliconising had previously been dismissed as a major contender because of the strong embrittling effect on high temperature alloys and the tendency to form many low melting point metallic and oxide eutectics.[42]

These aluminising, chromising and siliconising processes result in enhanced corrosion resistance through the formation of protective thermally grown oxides of Al_2O_3, Cr_2O_3 and SiO_2 as discussed earlier. It is also possible to produce diffusion coatings containing a combination of aluminium and chromium (chrome-aluminising) or aluminium and silicon.[42–51] The performance of these modified aluminide coatings will be discussed later in this paper.

Diffusion coatings can be applied to hot gas components using several techniques including pack cementation, slurry cementation, and metallising. Comprehensive reviews of the methods of deposition of diffusion coatings are given in Refs 37, 40 and 44–51. Fluidised bed techniques have been used to deposit diffusion coatings on a laboratory scale.[52] This approach should permit the coating of large components with a close tolerance on coating thickness, as a result of the uniform temperatures and, therefore, chemical activities possible within a fluidised bed, if scale up from laboratory to commercial operation proves possible.

3.1   The Pack Cementation Process

Pack cementation is still the traditional route used to apply diffusion coatings. For example within the gas turbine industry pack cementation was introduced into aeroengine service to coat turbine components during the late 1950s. The coatings produced were based on the formation of nickel or cobalt aluminides and this class of coating is still extensively used, probably satisfying some 80–90% of the current world market, for protecting turbine blades under aero, marine and industrial turbine service.

In this process components to be coated are buried in a 'pack', contained in a sealed retort. (Figure 5, shows a schematic diagram of the retort used at 'Cranfield' to produce diffusion coatings.) The retort is heated to the desired processing temperature under either an inert gas or hydrogen atmosphere to prevent oxidation. The exact process cycle, time, and temperature are dependent on the required coating, coating thickness, and subsequent substrate heat treatment. The pack contains a donor alloy that releases solute material at a known rate and hence determines the pack activity, a halide activator that dissociates during the process cycle and acts to transport solute material from the pack to the component to be coated, and an inert oxide diluent to prevent pack sintering.

Typical pack compositions used to produce a range of metallic coatings are given in Table 3. Of these the two diffusion coating processes that are most

Table 3 Typical pack compositions and deposition temperatures for halide activated pack cementation.

Coating	Pack composition, wt-%			Deposition temperature, °C
Al	15Al	$80Al_2O_3$	$5NH_4Cl$	815–900
Cr	48Cr	$48Al_2O_3$	$4NH_4Cl$	850–1050
Ti	77Ti	$20TiO_2$	$3NH_4Cl$	–
Si	5Si	$92Al_2O_3$	$3NH_4Cl$	–

widely used are 'aluminising' and 'chromising'. The pack aluminising process will be used as an example. The formation of aluminide coatings by pack cementation have been extensively studied and details of the process and characteristics of the coating are well documented in the literature.[16,29–51,53–57]

During the aluminising process material from the pack is transferred to the component surface through the formation of intermediate volatile aluminium mono-halide gas and as such the coating process is probably more accurately described as a chemical vapour deposition process. Interdiffusion between the depositing aluminium and the substrate alloy results in the formation of the intermetallic coating, primarily NiAl or CoAl, depending on the alloy base, but containing to a degree most of the elements present in the base alloy either in solution or as dispersed phases.

The deposition rate and morphology of the coating depend on pack activity, process time and temperature. Coatings are classified as either 'low activity', when outward diffusion of nickel occurs, or 'high activity' when inward diffusion of aluminium occurs. In the latter case a surface layer of Ni_2Al_3 forms and a further heat treatment is required to convert this brittle surface layer to NiAl. This step is usually combined with the heat treatment required to recover substrate properties. Figure 6 illustrates a typical aluminide coating deposited onto a nickel based superalloy using a high activity coating process. The coating was deposited using an aluminiding pack contain 2%Al at 900°C, and was heat treated for 2 h at 1120°C, then 24 h at 845°C.

Clearly, the properties of the aluminide coating (or for that matter any diffusion coating) depend upon the process methodology, the substrate composition and the subsequent heat treatment. Typically, aluminide coatings contain in excess of 30 wt%Al and are deposited to thicknesses between 30–100 μm depending on the type of aluminide formed. They offer satisfactory performance for many aviation, industrial and marine engine applications. Under severe hot corrosion conditions, or at temperatures above 1100°C, aluminide coatings offer limited protection. Hence modified aluminide coatings have been developed offering improved corrosion resistance.

3.2   Modified Aluminide Coatings

Modified aluminides have been fabricated using one of the following techniques[53]

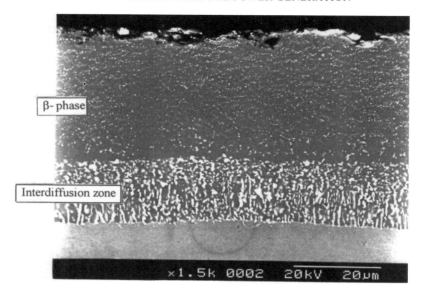

Fig. 6 A typical high activity aluminide coating deposited onto a nickel base superalloy.

(i) co-deposition of elements from the pack or slurry,
(ii) pretreatments of the superalloy before pack aluminising, for example chromising prior to pack aluminising,
(iii) deposition of a metallic layer using electroplating or PVD (physical vapour deposition) techniques. For example, a platinum-aluminide coating is formed by depositing platinum onto the superalloy prior to aluminising.

Alloying additions include Cr, Si, Ta, the rare earths and precious metals. Many of these coatings are now commercially available as detailed in Table 4.

Probably, the most significant advance in this area was the development of the platinum modified aluminide class of coating. The first commercial coating of

Table 4 Some commercially available modified aluminide coatings.

Process	Specification	Manufacturer
Pack aluminised (high activity, low activity)	PWA73	Numerous
Pack chromised	PWA70; HI12; MDC3V; RT5	Numerous
Two step pack chromised + pack aluminised	PWA62; HI32; RT17; SylCrAl	Numerous
Codeposition, chromium aluminising	HI15	Alloy surfaces
Platinum coating (electroplated) + pack aluminised	LDC2, RT22	TEW, Chromalloy
Ti/Si slurry diffusion	Elcoat 360	Elbar
Al/Si slurry diffusion	Sermaloy J, Sermaloy 1515	Sermetel

Fig. 7 Back scattered electron image of a platinum aluminide coating.[59]

this class was designated LDC-2 and was produced by electrodepositing a thin layer of platinum, followed by a pack aluminising treatment at 1100°C.[58] Since 1970 this process has been licensed to many manufacturers who produce their own variants on this basic coating design. Figure 7 illustrates the microstructure of the RT22 variant, marketed by Chromalloy. The SEM micrograph is a backscattered electron image with the platinum rich aluminide phase within the coating shown in light contrast.[59]

4   OVERLAY COATINGS PROCESSES

Diffusion coatings, by the nature of their formation, imply a strong inter-dependence on substrate composition in determining both their corrosion resistance and mechanical properties, hence, the possibility of depositing a more 'ideal' coating, with a good balance between oxidation, corrosion and ductility has stimulated much research interest. Early coatings of this type were alloys based on cobalt (CoCrAlY) containing chromium additions in the range 20–40%, aluminium additions between 12–20% and yttrium levels around 0.5% with the most successful coating being Co25Cr14Al0.5Y.[60] Recent coatings are more complex, based on the M–CrAl–X system, where M is Ni, Co, Fe or a combination of these and X is an active element, for example Y, Si, Ta, Hf etc.[54, 61–67] The composition of the M–Cr–Al system is selected to give a good balance between corrosion resistance and coating ductility, while the active element ad-

Table 5 Some typical commercial overlay coatings.

	Composition	Specification	Deposition process
CoCrAlY	Co–23Cr–12Al–0.5Y	ATD2B	EB PVD
	Co–18Cr–11Al–0.3Y	ATD5B	EB PVD
	Co–18Cr–8Al–0.5Y	LCO29	Argon shrouded plasma spray
CoNiCrAlY	Co–32Ni–21Cr–8Al–0.5Y	LCO22	Argon shrouded plasma spray
	Co–23Ni–30Cr–3Al–0.5Y	LCO37	Argon shrouded plasma spray
NiCoCrAlY	Ni–23Co–18Cr–12.5Al–0.3Y	ATD7	EB PVD
		PWA270	
NiCrAlY	Ni–20Cr–11Al–0.3Y	ATD10	EB PVD

dition(s) can enhance oxide scale adhesion and decrease oxidation rates. Current thinking suggests that a combination of active elements is beneficial in reducing coating degradation through their synergistic interaction. Overlay coatings have been deposited using a range of techniques including physical vapour deposition (PVD)[54,60–63,68–70] thermal spraying,[68,71–79] composite electroplating processes[80–92] and laser fusion.[80,84] Table 5 summarises some typical commercial overlay coatings and their methods of deposition.

The earliest production method was electron beam PVD.[60] However, because of the high capital cost in setting up a commercial EB-PVD plant, plasma spray methods have found wide acceptance, particularly the argon shrouded and vacuum plasma spray processes.[72–75] More recently high velocity oxyfuel spraying processes,[76,79] composite electroplating[80–82] and laser fusion[83,84] methods have been used to deposit overlay coating systems. However, coatings produced by EB PVD processes are still considered the commercial standard against which other process routes are compared.

4.1   Physical Vapour Deposition

Physical vapour deposition processes encompass both electron beam evaporation and sputtering, either of which may be used in conjunction with ion plating.

Commercial EB PVD coatings are produced in a vacuum environment, 10^{-2}–10^{-4} Pa, with deposition rates often exceeding 25 mm min^{-1}. Components are generally preheated in vacuum to between 800–1100°C, and are rotated within the evaporant cloud during the deposition process.[68] Rotation attempts to ensure uniform coverage as the process is primarily 'line of sight' at these pressures. High deposition temperatures result in increased surface diffusivity which reduces the density of leader defects (columnar grains with unbonded interfaces) and also permits some minimal interdiffusion between the coating and substrate during the coating process cycle, ensuring good adhesion. Hence coating spallation, a problem with many processing routes is not a problem provided the substrates are properly cleaned prior to coating. The high rate of deposition possible with this process means that a through-put of some 500 standard sized

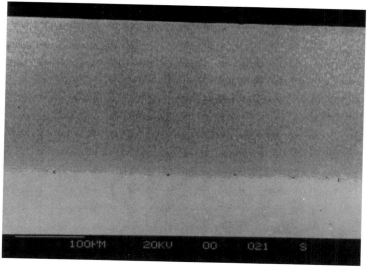

Fig. 8 Micrograph of a commercial CoCrAlY coating (ATD 5B) deposited by EB-PVD.[16]

parts is possible in an 8 hour shift,[68] from commercial equipment operated on a continuous, rather than batch, process. Post coating processing includes glass bead peaning and heat treatment which results in complete closure of any remaining leaders. Figure 8 is a micrograph of a CoCrAlY coating (ATD5B) produced by an EB-PVD processing route.

In the sputtering process, positive gas ions (usually argon) produced in a glow discharge or plasma bombard a target of coating material dislodging groups of atoms. These then enter the vapour phase and deposit onto substrates to be coated.[85] Deposition rates are much slower than the E.B. evaporation route, generally 10–20 μm/hour for a diode system or up to 50 μm/hour for a planar magnetron system. Magnetron systems (having magnetic plasma confinement adjacent to the target) usually result in more uniform deposition and can virtually eliminate substrate heating during the deposition process. Working pressures are of the order of $1-10^{-2}$ Pa depending on whether a DC glow discharge or R. F. Plasma is used to generate ion bombardment and hence offers excellent throwing power and good overall coverage of components to be coated. Because of the low process temperature, coatings are invariably heat-treated following deposition to produce the desired properties.

Figure 9, illustrates a CoCrAlY coating with platinum underlay deposited using a sputtering route onto an IN792 substrate.[63] The platinum underlay was proposed to improve coating performance under marine gas turbine conditions by limiting the diffusion of alloying additions from the substrate into the overlay coating.

Fig. 9 Micrograph of a sputter deposited CoCrAlY, plus platinum underlay, deposited on IN792.[16,63]

Both EB evaporation and sputtering may be combined with ion plating. Ion plating is essentially a PVD process in a soft vacuum $(1-10^{-1}$ Pa) with evaporant depositing onto substrates held at a high negative potential (often between 2–5 kV).[86] During ion plating the components to be coated are initially bombarded with positive ions, formed in the discharge, which remove oxides and other contaminants from the surface. When the surfaces are sufficiently clean the vapour source is energised and metal evaporant enters the discharge and is deposited onto the sample. With EB evaporation and ion plating deposition rates are typically 10–20 μm/min.

4.2 Spraying Processes

Plasma spraying[71–73,75] has the advantage of being able to deposit metals, ceramics, or a combination of these, generating homogeneous coatings with microstructures consisting of fine equiaxed grains, i.e. no columnar/leader defects. High deposition rates are possible with little significant change in composition occurring from the powder feedstock through to the coating, even when the elements in the coating have widely differing vapour pressures. The most obvious limitations of this process are that the coating process is 'line of site', requiring complexed robotic manipulation for complete coverage, and that the more reactive elements may well oxidise during the spraying process if conducted in air. Porosity problems previously reported for plasma spray coatings can largely be overcome using post coatings thermomechanical treatments. Figure 10 shows a fully processed, argon shrouded plasma sprayed LCO22 (CoNiCrAlY) overlay coating.

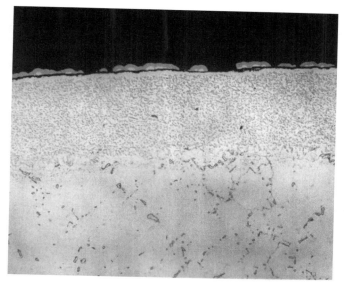

Fig. 10 Fully processed, argon shrouded plasma sprayed CoNiCrAlY overlay coating (LCO22).[19]

Carrying out the plasma spraying process in a closed chamber under reduced pressure (5–7×10^3 Pa), known as low pressure plasma spraying (LPPS) or vacuum plasma spraying (VPS), permits close control of the gaseous environment. This removes the possibility of unwanted gas-metal reactions and also permits high particle velocities giving improved adhesion and higher densities, thus overcoming many of the limitations of plasma spraying at increased capital plant cost. Figure 11, shows a typical low pressure plasma spray facility, consisting of a vacuum chamber, plasma spray torch and robotic manipulator to permit all round coating.

In the high velocity oxy-fuel (HVOF) spraying process fuel is burnt with oxygen at a high pressure and generates a high velocity exhaust jet. Fuel gases may include acetylene, kerosene, propane, propylene, hydrogen, methacetylene-propadiene (MAPP) mixtures etc. The ratio of the gas flow rates determines the temperature of the flame. The temperature of oxygen-acetylene reaches a maximum at 3170°C if mixed 1.5 to 1 (by volume) and oxygen-propylene reaches the maximum temperature of 2900°C at the ratio of 4:1.[76] The velocity of the exhaust jet in the Jet-Kote torch is about 2000 m/s.[77] The main advantage of this process is the shorter residence time in the flame (lower powder temperature) and higher kinetic energy of the particles impacting. This produces a dense coating with less degradation of the powder during spraying.

The most widely sprayed powders are carbides, although there is considerable interest in using the HVOF process to deposit MCrAlY overlay coatings. The

Fig. 11 A typical low pressure plasma spray facility.[19]

bond strength of HVOF sprayed coatings can be as high as 90 MPa with porosity less than 1%.[77] Typical coating thicknesses are in the range 100–300 μm.

4.3   Laser Processes

Lasers can be used to form coatings directly by (i) preplacing a powder onto the substrate, (ii) blowing the powder into a laser generated melt pool, or (iii) applying the clad in wire or sheet form. Alternatively, coatings produced by other routes such as plasma spraying or electroplating may be modified by surface treatment with the laser. By careful selection of laser conditions dilution of the coating by substrate melting can be minimised and large areas can be surfaced by a series of overlapping tracks. Power, beam diameter, and beam velocity are the main process variables with typical values being 1.5 kW, 5 mm, and 10 mm s^{-1}, respectively. As with plasma spraying it is necessary to apply an inert gas shroud to reduce oxidation of the coating. Powell *et al.*[83] suggest that adhesion is improved if some melting of the substrate occurs.

Problems associated with reflectivity of the surface occur particularly with use of wire and sheet, and for the preplaced powder method there are difficulties in maintaining the powder on the surface. Thus the blown powder method is favoured by Steen.[84] Porosity in coatings can occur from cavities between overlapping tracks, solidification cavities, or gas evolution, but problems can generally be overcome by careful choice of processing parameters.

Residual stress in a laser coating are tensile and occur from restrained contraction of the prior molten clad. Cracking may result, but can usually be overcome by preheating or pretensioning the substrate before coating.[84]

5   PERFORMANCE OF DIFFUSION AND OVERLAY COATINGS

5.1   *Corrosion Resistance of Diffusion and Overlay Coatings*

Hot corrosion problems (type I and type II hot corrosion, vanadic corrosion) are a direct result of salt contaminants such as Na_2SO_4, $NaCl$ and V_2O_5 which in combination produce low melting point deposits which dissolve the protective surface oxides. A number of fluxing mechanisms has been proposed to account for the different corrosion morphologies that are observed,[20-22] and this has resulted in the general classification of high temperature (type I, 800–950°C), hot corrosion low temperature (type II, 650–800°C) hot corrosion and vanadic corrosion (535–950°C). These corrosion processes can be separated into an initiation and propagation stage. During the initiation stage the corrosion rate is comparatively low as breakdown of the surface oxide occurs. However, once this has happened and repair of the oxide is no longer possible, then the propagation phase results in the rapid consumption of the alloy. Since the coating provides for the repair of the protective surface oxide scales, the initiation stage can be extended, ideally for the design life of the component. However, once coating penetration occurs, the propagation stage results in catastrophic corrosion rates.

Methods of evaluating the corrosion resistance of coatings range from simple laboratory tests through to burner rig and engine testing.[87] Table 6 gives the ranking order for a range of diffusion and overlay coatings obtained from engine tests undertaken as part of the COST programme[88] together with results from Cranfield burner rig tests.[89] Over the temperature range 700–950°C the platinum, modified aluminides performed exceptionally well. Of the other diffusion coatings, chromised and chrome-aluminised coatings generally fared worse than the conventional aluminide, although chromium rich coatings have been shown to offer improved corrosion resistance in an industrial turbine environment.[88,90]

Overlay coatings of classic design, with 18–22Cr and 8–12%Al, generally perform better at higher temperatures where oxidation is the dominant failure mode (above 900°C) reflecting the good adherence of the thin alumina scales which is promoted by the presence of active elements such as Yttrium. Generally under these high temperature oxidising conditions NiCrAlY's and NiCoCrAlY's out perform the cobalt based systems. This is illustrated schematically in Fig. 12 reproduced from a paper by Novak (1994)[91] and is similar to an early diagram by Mom in 1981.[92]

However, at low temperatures where type II hot corrosion predominates, 650–800°C, corrosion rates for the NiCrAlY and NiCoCrAlY overlay coatings can be relatively high. CoCrAlY's generally out perform NiCrAlY based systems, with the high chromium containing CoCrAlY's showing best performance.[90-92]

Methods have been investigated to improve the traditional MCrAlY coatings by use of a platinum underlayer and overlayers[63] (Platinum has been used to improve the hot corrosion resistance of superalloys).[93] Other additions such as Ti, Zr, Hf, Si and Ta have been examined.[53,61,63,87,90,94] Surface modification by

Table 6 Ranking, from best to worst, of coatings included in various engine and burner rig tests.[19, 50, 88, 89, 90, 100]

Burner rig tests	
Cranfield	
500 h at 700°C	{Platinum aluminide (RT22); HA aluminide} {PS CoNiCrAlY (LCO22); PS CoNiCrAlY + Pulse Al} {(SIP NiCrTiAl; SIP CoCrAlY (ATD2)} ...
500 h at 700°C	{SIP NiCrTiAl} {Platinum aluminide (JML1)} {PS CoNiCrAlY + pulse Al} (HA aluminide) {PS CoNiCrAlY (LCO22)} ... {EB PVD CoCrAlY (ATD6)} {chromised} {chromium aluminised}
500 h at 850°C	{Platinum aluminide (JML1)} {HA aluminide} {PS CoNiCrAlY + pulse Al} ... {SIP NiCrTiAl} PS CoCrAlY {EB PVD CoCrAlY (ATD6)} {chromised} {chromium aluminised}
NPL	
990 h at 850°C, marine	{Platinum aluminide (LDC2) (738 and B1900)} {chromised (738 and IN100) {Co23Cr3A12B3Si} ... {aluminised (738)} {platinum aluminide (LDC2) (IN100)} {aluminised (INI100)}
GE	
1000 h at 732°C	{Co + 40–48%Cr + 0.6–5%Si; Co + 40–48%Cr + 0.1%Y} {platinum aluminide} {Co + 20–30%Cr + 6–12%AlY} ...
1000 h at 871°C	{Platinum aluminide; Co + 20–30%Cr + 6–12%AlY; Co + 40–48%Cr + 0.6–5%Si; Co + 40–48%Cr + 0.1%Y} ...
ARL	
600 h at 700°C	{NiCrAl1-6Zr} {AlSiTi (Elcot 360)} ... {HA aluminide; Al–Si} {NiCrAl1-4Hf}
MTU	
1000 h at 1050°C	{LPPS CoNiCrAlY (LCO22)} {platinum aluminide (RT22)} {LA alumide (Codep)} ...
400 h at 1100°C	{LPPS CoNiCrAlY (LCO22)} {platinum aluminide (RT22)} ... {LA alumide (Codep)}
100 h at 900°C	{PVD NiCoCrAlY (PWA 270)} {LPPS NiCoCrAlY; PVD 58CrNiCoAlY} ... {LPPS 58CrNiCoAlY}
Engine tests NLR	
3000 h at 950°C maximum	{Platinum aluminide (RT22), LDC2)} {aluminide (PWA 73)} ... {chromised + aluminised}
270 h at 1040°C maximum	{Al–Si slurry coating on LA aluminide (Codep)} {gas phased aluminide on LA aluminide (Codep); pulse aluminised on LA aluminide (Codep)} {LA aluminide (Codep)} ...
RAE/Lufthansa/BB	
7000 h scheduled aircraft cycle	{Platinum aluminide (LDC2); PVD CoCrAlY (PWA 68)} {platinum aluminide (RT22)} {chromium aluminide (PWA 70/73)} {aluminide (PWA 73)} ...
500 h military cycle (1000°C mean)	{PVD CoCrAlY; PS CoCrAlY} ... {aluminide}
20 000 h industrial cycle	{chromised} {aluminide: chromium aluminide; platinum aluminide (LDC2)} ...

{} Indicates coatings of equal status listed in order of decreasing resistance to attack.
... Coating penetrated through to substrate.
PS plasma sprayed; SIP sputter ion plating; HA high activity; LA low activity; EB PVD electron beam physical vapour deposition. LPPS low pressure plasma spraying.

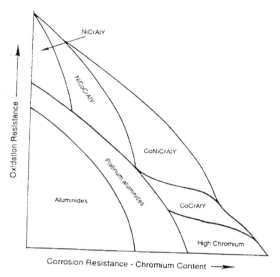

Fig. 12 Schematic diagram of the relative oxidation and corrosion performance of diffusion and overlay coatings.[91]

CVD,[95] PVD[66] or slurry cementation techniques,[66] and reprocessing of the coating surface using laser beams[53] have also been considered with varying degrees of success.

Surface modification results in the formation of a duplex coating structure and this can result in improved performance, for example, in the Cranfield burner rig tests a pulse aluminised CoNiCrAlY coating exhibits superior corrosion resistance at 750 and 850°C compared to its plasma sprayed counterpart.[89] Similarly, silicon modifications to the surface of CoCrAlY coatings[66] have also been shown to improve resistance to low temperature hot corrosion. Recently, such surface modifications have been extended further by grading not only the aluminium profile, but also that of chromium to further improve oxidation/corrosion resistance. These 'smart coating concepts[97] are discussed later in this paper.

In summary, the corrosion resistance of conventional aluminide coatings may be improved by additions of Pt, Cr and Si depending on the service environment. These coatings often exhibit superior performance compared to MCrAlY overlay coatings, under low temperature hot corrosion conditions. At higher temperatures MCrAlX coatings perform well (where X may include Y, Hf, Si or a combination of these), particularly if the environment is primarily oxidising. It should also be remembered that overlay coatings may be deposited up to 5 times as thick as a diffusion coating which can be significant in terms of overall coating lifetime.

In addition to corrosion resistance, surface coatings can have a major influence on the erosion/corrosion resistance of hot-end components. As mentioned earlier, environmental resistance depends on the formation of thin protective oxide scales and therefore the interaction between an erodent and the oxide scale that is formed on a component is of particular importance in determining the erosion/corrosion behaviour of a particular coating.

The effect of particle impingement on degradation rate depends on whether the impact event produces failure within the protective oxide scale. The higher the impact energy (larger particles, higher velocities) the more likely it is that scale failure will occur and this will result in an increase in corrosion rate[98,99] and reduce incubation times to the onset of catastropic corrosion.

5.2 Thermal Stability of Diffusion and Overlay Coatings

Structural stability of coatings is also an important factor if they are to maintain their protective qualities over extended periods of time at high temperatures. Coatings degrade not only by loss of scale forming elements to the surface, but also by interdiffusion with the substrate. This can result in additional problems such as the formation of TCP phases e.g. sigma, below the coating, which cause embrittlement of the substrate.

The thermal stability of coatings tends to decrease as the temperature increases and interdiffusion rates become higher. The rate of degradation of the coating increasing in the ratio 1:6:30 as the temperature is raised from 750:850:950°C.[15] In aluminide diffusion coatings this degradation results in the breakdown of the outer β-NiAl phase by the formation of a γ'-Ni$_3$Al network which after extended exposure time can completely penetrate through the β-phase. These γ' leaders behave as short circuit corrosion paths which results in rapid failure of the coating. The thermal stability of diffusion coatings, including pack aluminide, chromium aluminide and platinum aluminide variants were studied in the early eighties.[100,101] These studies have shown that the thermal stability of β-NiAl can be improved by the addition of platinum. The two layered PtAl$_2$ β structure degrades to single phased β with no γ' leaders even after extended exposure time, e.g. 18 000 h at 850°C.[15] This is a major factor which contributes to the improved corrosion resistance of Pt-aluminide coatings compared to conventional aluminides.

Few papers have been published on the metallurgical stability of overlay coatings. Until recently, with the interest in NiCrAlY and NiCoCrAlY alloys as bond coats, interdiffusion with the substrate has not been considered a major failure mode of overlay coatings. Nevertheless, substrate interaction effects have still been observed. For example hafnium diffusion from hafnium modified MM200 has been found to improve the hot corrosion resistance of CoCrAlY coatings.[54] Movement of other elements, e.g. tungsten and molybdenum, from the substrate, into the coating may well be detrimental to corrosion resistance. The beneficial effect of a platinum underlayer,[62] under marine conditions, may well be due to it acting as a diffusion barrier.

One of the most comprehensive studies on the stability of overlays is that of Mazars *et al.*[102] in 1986 who have examined in detail the diffusion degradation of NiCrAlY, CoCrAlY, and FeCrAlY systems. From their work they conclude that diffusion stability can be considered a life limiting factor if coatings are operated at temperatures > 1000°C for prolonged periods of time. Of the systems studied, limited interdiffusion was observed in the NiCrAlY/Ni based, NiCrAlY/Fe based, and CoCrAlY/Fe based systems and hence these systems are of particular interest for high temperature applications with the NiCrAlY/Ni base system most relevant to turbine service. Their studies would suggest that some 25 μm of β phase would be lost from a NiCrAlY/Ni-based alloy system after 1000 h at 1100°C due to interdiffusion. Whereas a CoCrAlY coating on a Ni-based alloy would be severely degraded after this exposure with β depletion extending through a typical coating (200 μm thick) after 300 h at 1100°C.

5.3   Mechanical Properties of Diffusion and Overlay Coatings

Probably the most important property of a coating for service at elevated temperature is its resistance to cracking by thermally induced stresses. A summary of the ductile-brittle transition temperatures (DBTT) for a range of diffusion and overlay coatings is shown in Fig. 13, in which it can been seen that the DBTT of the aluminides is higher than that observed for many of the overlay coatings. Since the peak tensile surface strains on the turbine blades are likely to occur at relatively low temperatures,[103] then from a ductility point of view these diffusion coatings will be inadequate or only marginally adequate for relatively high strain applications.

Unlike the aluminides, the ductile to brittle transition of overlay coatings, and hence their resistance to cracking, can be modified by varying the coating composition. It is evident from Fig. 13 that, for a given Cr and Al level, the ductility of NiCrAlY coatings is significantly better than that of CoCrAlY coatings. The ductile to brittle transition temperature is found to increase with Cr and Al level. This is not unexpected since coatings with a high volume fraction of CoAl or NiAl shows ductile to brittle transition temperatures close to those for the diffusion aluminides.

This transition from ductile behaviour to brittle behaviour, as the temperature drops, is of significance when considering the thermo-mechanical cycles seen in service. Should coatings crack on cooling, there is a danger of the crack propagating to the coating substrate interface. It may arrest there, may propagate along the interface or may run into the substrate component, thereby decreasing the components life. In an investigation of a plasma sprayed NiCoCrAlY coating on a single crystal superalloy at 650°C, a marked reduction in the fatigue life of coated components at low strain ranges was observed.[104] The mechanical behaviour of coated components is therefore of major importance but is beyond the scope of this review. Readers are recommended to read the papers by Strang and Lang,[105] Gayda *et al.*,[104] and Au *et al.*,[106] for a further appreciation of the subject.

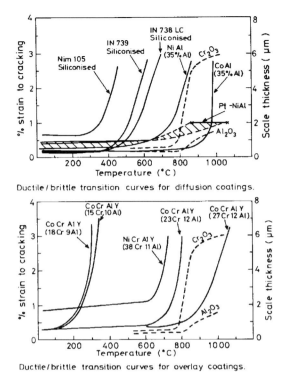

Ductile/brittle transition curves for diffusion coatings.

Ductile/brittle transition curves for overlay coatings.

Fig. 13 Ductile to brittle transition curves for diffusion (a) and overlay (b) coatings.[16]

6   CERAMIC COATINGS

In its broadest sense ceramic coatings are fundamental to high temperature corrosion protection, for the naturally grown oxides (section 2) could be considered as 'as grown' ceramic corrosion resistant barriers. However, in this paper discussion over the use of ceramic coatings as corrosion (diffusion) barriers and as thermal barriers will be restricted to those coatings deposited as overlays.

6.1   Corrosion Resistant Ceramic Coatings–Diffusion Barrier Coatings
In attempting to predict the behaviour of a ceramic coating it is useful to have some measure of its effectiveness of these coatings as a diffusion barrier and this can be obtained by considering the growth rates of the various oxides on alloys or elements (section 2). The ranking of the three oxide systems in conferring protection is $SiO_2 > Al_2O_3 > Cr_2O_3$. The effectiveness of SiO_2 as a diffusion barrier, is highlighted by the work of Bennett *et al.* who have shown that the presence of a continuous layer of amorphous silica eventually controls overall diffusion within an oxide system.[107] More recent research has been undertaken to

deposit amorphous silica and alumina coatings using vapour deposition techniques[108–110] or laser fusion methods.[111] This recent work by Ansair *et al.*[111] has demonstrated the excellent behaviour of amorphous SiO_2 in resisting attack by a variety of aggressive atmospheres. The eventual failure of these coatings was shown to result from crystallisation of the amorphous layer to α-crystobalite after prolonged exposure at high temperatures.

In addition to bulk diffusion, transport across these corrosion (diffusion) barrier coating systems is affected by microcracks and pore and by gross spalling which results from thermal stresses. Thermal stresses will occur in all applications with temperature cycling because of the large mismatch in linear expansion coefficient between the ceramic coating and the metallic substrate (Table 7).[112] For overlay coatings a simple expedient to minimise the effects of thermal expansion mismatch is to produce the coating with the substrate heated to service temperature, Thus, if the component can survive the first thermal cycle to ambient temperature as part of the coating process, there are good prospects that it would also survive subsequent cycles of similar magnitude.

6.2   Thermal Barrier Coatings

Thermal barrier coatings have been used since the early 1950s and have been effectively used to protect piston crowns and valve faces in diesel engines and flare head and primary zone sections of combustors against the effects of hot spots within turbines, giving considerably improved component lives. Because of this success much attention has been directed towards the use of thermal barriers

Table 7 Coefficients of thermal expansion (α) of common metal substrates and ceramic coatings.[112]

Substrate material	$\alpha \times 10^6$, K^{-1}	Coating material	$\alpha \times 10^6$, K^{-1}
Carbon steel	12, 14	Al_2O_3	8, 9
		Cr_2O_3	7, 7.3, 7.8, 9.6
		NiO	14, 17.1
		CoO	15
18Cr–8Ni steel	14.8, 19.0	MgO	13.9, 12.9
		$MgO–Al_2O_3$ spinel	9.1
25Cr–20Ni steel	16.5, 18.3, 20	SiO_2 crystalline 300–1200°C	2.0
		SiO_2 cystalline, RT–300°C	43.0
		SiO_2 amorphous, RT–1200°C	0.5
		Si_3N_4 (reaction bonded)	3
		Si_3 (hot pressed)	3
		Si-Al-O-N on Sialon	3.2, 3
Nickel base superalloy	9.7–12.2	SiC	4.8, 3.7–4.2
Nickel base superalloy	13.1–15.2	–	–
Nickel base superalloy	17.9–19.2	ZrO_2	8, 10
		ZrO_2 + MgO	12.2, 10

on blades and vanes[113–116] but only recently have they been used on highly stressed turbine components within commercial gas turbine engines.[114,116]

Current commercial thermal barrier coatings are duplexed, consisting of a plasma sprayed yttria or magnesia stabilised zirconia (yttria stabilised TBC's are currently the preferred system) thermal barrier, 0.2–0.4 mm thick, deposited on top of an MCrAlY bond coat. The purpose of the bond coat is to act as a key between the substrate and ceramic layer, to provide corrosion protection to the base alloy (as the thermal barrier is porous and therefore permeable to the ingress of corrodents) and to accommodate the mechanical strains generated by differences in the Youngs modulus and thermal expansion coefficients of the ceramic and substrate alloy. Figure 14 shows a commercial plasma sprayed thermal barrier coating.

A complex stress distribution is present in as sprayed partially stabilised zirconia (PSZ) thermal barrier coatings.[117] This is the result of thermal expansion mismatches and also volume shrinkage during solidification (~10%) thermal expansion anisotropy, and volume changes resulting from solid state phase transformation, and stress gradients due to temperature variations during spraying. Further, in service the coatings are designed to toughen as a result of a stress induced phase transformation. Therefore, these layers are extensively cracked and indeed it has been suggested[118] that crack opening and closing give these coatings a pseudoductility. Obviously, oxidation resistance is minimal in this instance, but if the outer layer can be sealed by some process such as laser glazing[119] improved corrosion resistance would result.

Fig. 14 A commercial air plasma sprayed thermal barrier coating.

Fig. 15 Fractograph of an EB-PVD thermal barrier coating, highlighting the columnar, strain tolerant microstructure.

Recent studies[113–116,120–123] have demonstrated that thermal barrier coatings deposited using EB.PVD have superior properties to those produced by plasma spraying, and this is thought to be due to the columnar growth morphologies of EB.PVD coatings.

Figure 15 illustrates the strain tolerant columnar microstructure that is produced using electron beam physical vapour deposition. It is this strain tolerant microstructure that has allowed TBC's to be successfully operate on highly stressed turbine components without spalling. The use of TBC's on turbine blades and airfoils offers two major benefits. Firstly, for the same gas path operating temperature, cooling requirements can be reduced. Secondly, for the same cooling air flow, the hot gas temperature can be raised significantly. These two limiting conditions give rise to lower specific fuel consumption or higher output powers respectively. In practice some intermediate design solution is to be expected.

7 FUTURE TRENDS AND NEW COATING CONCEPTS

Although coatings are available that offer adequate oxidation/corrosion resistance for many applications, improvements in coating performance can still be made. The drive to operate plant hotter, in the name of increased efficiency, coupled with the desire for increased component lives requires that new generations of coatings be developed. As operating temperatures are increased, it no longer becomes possible to achieve the desired service lives using diffusion coatings and overlay coatings therefore are the only possible development route.

For metallic systems, much work is underway to develop overlay coatings containing multiple active elements. This acknowledges that although many of the active elements (Y, Hf, Ta, Si etc) produce similar improvements in oxidation behaviour, they function by different mechanisms and hence show maximum benefit at different operating temperatures. Since high temperature components encounter a range of temperatures, clear benefits are possible from the inclusion of multiple active elements in a coating.

But even within the MCrAlX systems diffusion of elements between the substrate and coating can have a major influence on coating performance. Therefore to provide long term stability it is necessary to develop diffusion barrier coatings to minimise the interdiffusion between the coating and the substrate. Some interdiffusion is of course necessary to give good adhesion hence the diffusion barriers must be tailored to limit the movement of particular problematic elements.

Once the concept of a diffusion barrier is accepted as a method of providing good interface stability, one is no longer constrained in the design of the best overlay coating. No longer is substrate compatibility a requirement in specifying the overlay coating composition.

By removing this constraint it is possible to design overlay coatings with optimised oxidation or corrosion resistance.[69,70] These studies have shown that optimum performance is achieved using different base coating compositions. Although in general both classes of coating alloy possess high chromium and aluminium levels than in current commercial overlay coatings.

For high temperature oxidation and type I hot corrosion resistance, NiCrAlY coatings that form a stable alumina scale offer best performance. Whereas under type II hot corrosion conditions high chromium containing MCrAlY coatings exhibit least attack. Recent work between Cranfield and Birmingham University[97,124] has focused on developing a single coating that is resistant to both forms of attack. This 'smart coating' is a functionally gradient overlay coating that is produced using a combination plasma spray and diffusion techniques. It contains a high chromium enriched region, midway through the coating, that limits type II pitting attack overcoated with a β aluminide rich zone to provide high temperature oxidation resistance and resistance to type I hot corrosion. Table 8 compares the corrosion resistance of various smart overlay coatings with RT22 and Sermetel 1515 using a salt recoat test procedure. Tests were conducted at 700°C and 800°C for 500 hours. It can be seen that this smart overlay coating concept out performs RT22 under type II hot corrosion conditions and matches the performance of Sermetel 1515.

Further, research at Cranfield is looking towards the next generation of EB-PVD thermal barrier coatings for airfoil applications. Although strain tolerant,

Table 8 Mean metal recession following 500 h exposure to a salt recoat test in air/ SO$_2$ at 700 and 800°C

Coating reference	Original coating thickness (µm)	Coating loss (700°C)	Coating loss (800°C)	Comment
RT22	82	41	(132)	Reference coatings
Sermetel 1515	78	27	5	
SmC 153	446	48	19	SMARTCOAT
SmC 155	350	26	25	variants
SmC 155H	280	28	56	

( ) corrosion penetration through the coating into the IN738 substrate.

current EB-PVD thermal barrier coatings have thermal conductivities significantly less than their plasma sprayed counterparts (1.8 W/mK compared to 0.8 W/mK).[125] Work is underway to produce EB-PVD coating microstructures that match the thermal conductivity of the plasma sprayed coating, while still retaining good thermal shock resistance (strain tolerance) and erosion resistance. Reduced thermal conductivities have been achieved by introducing layers into each column of the EB-PVD columnar microstructure. The layers act to reflect photon thermal transport and scatter phonon thermal transport so reducing the coatings overall thermal conductivity.[125] Thermal conductivities of the order of 1.2 W/mK can be achieved using these layered structures, approaching those of plasma sprayed TBC's.

Extending these concepts further, one can envisage, the 'ideal' hot gas path coating. It would consist of a nanostructured, multilayered, strain tolerant thermal barrier coating, deposited by EB-PVD onto a bond coat which exhibited good corrosion resistance but more importantly matched thermal expansion coefficients. Within this bond coat, composition can be graded to provide the required corrosion resistance, coating mechanical properties and a diffusion barrier at the interface with the substrate.

This conceptual coating is possibly only a few years away, for research is underway in many of the fundamentals necessary to produce such '*functionally graded coatings*'.

REFERENCES

1. 'The Power Generating Industry', *Technology Foresight Programme report*, Institute of Materials, 1994.
2. S.C. Miller: 'Design and Manufacture – Getting it Together', *Fellowship of Engineering Lecture*, University Newcastle upon Tyne, April 1990.
3. 'Requirements for Materials Research and Development for Coal Fired Power Plant: into the 21st Century', Institute of Materials, Sept. 1997.

4. D.S. Rickerby and H.C. Low: 'Towards Designers Surfaces in the Aero Gas Turbine', *4th European Propulsion Forum*, Bath, June 1993.

5. D. Thornton: 'Materials for Advanced High Efficiency Steam Turbines', this publication,

6. A. Fleming: 'Materials Developments for Supercritical Boilers', this publication,

7. F. Starr: 'Advanced Materials for Heat Exchangers', this publication,

8. D. Wray: 'High Temperature Materials for Chemical and Petrochemical Plant', this publication,

9. D. Allen: 'High Temperature Alloys for Advanced Industrial Gas Turbines', this publication,

10. P. Chiesa *et al.*: *ASME Int. Gas Turbine and Aeroengine Congress*, Cincinatti, May 1993.

11. *Coatings for High Temperature Structural Materials*, NMAB, NRC, National Academy Press, Washington DC, 1996.

12. D.F. Betteridge: 'Surface Engineering in the Aero Industry: Past, Present and Future', *Surface Engineering and Heat Treatments*, P.H. Morton ed., Institute of Materials, 1991.

13. J.R. Nicholls and D.A. Triner: *Diesel Engine Combustion Chamber Materials for Heavy Fuel Operation*, Institute of Marine Engineers, London 1990, 121–130.

14. M.G. Kingston-Jones, J.R. Thomas and A.S. Radcliffe: *ibid*, 1990, 15–28.

15. P. Hancock and J.R. Nicholls: *Coating for heat engines*, (Workshop Proc.), R.L. Clarke *et al.* eds, Washington, DC, US Department of Energy, 1984, 31–58.

16. J.R. Nicholls and P. Hancock: *Ind. Corros.*, July 1987, **5**(4), 8–18.

17. T.N. Rhys-Jones and D.F. Bettridge: Proc. Conf. on *High temperature materials and processing techniques for structural applications*, Paris, France, Sept. 1987, ASM.

18. T.N. Rhys-Jones: *Mater. Sci. Technol.*, 1988, **4**(5), 421–430.

19. S.R.J. Saunders and J.R. Nicholls: 'Coatings and surface treatments for high temperature oxidation resistance', *Mater. Sci. and Technol.*, 1989, **5**, 780–798.

20. C.S. Giggins and F.S. Pettit: 'Hot Corrosion Degradation of Metals and Alloys - A Unified Theory', *PWA-Report FR-11545*, 1979.

21. N. Birks and G.H. Meier: *Introduction to High Temperature Oxidation of Metals*, Edward Arnold Publishers, London, 1983.

22. P. Kofstad: *High Temperature Corrosion*, Elsevier, 1988.

23. M. Shutze: *Oxid. Met.*, 1985, **24**, 199–232.

24. *Materials at High Temperatures*, **12**(2–3), 'Special Issue on the Mechanical Properties of Protective Oxide Scales'

25. H.E. Evans: *International Materials Reviews*, 1995, **40**, 1–40.

26. J.R. Nicholls, H.E. Evans and S.R.J. Saunders: *Materials at High Temperatures*, 1997, **14**, 5–13.

27. F.H. Stott, F.M.F. Chong, and C.A. Stirling: *High temperature corrosion in energy systems*', Conf. Proc., M.F. Rothman ed., Warrendale, PA AIME 1985, 253–267.
28. P. Kofstad: *High Temperature Corrosion*, Elsevier Applied Science, London 1988, 120.
29. W. Lih, E. Chang, B.C. Wu and C.H. Chao: *Materials Science & Engineering A124*, 1990, 215.
30. B.C. Wu, C.H. Chao, B.C.H. Chao and M.L. Tsai: *Oxidation of Metals*, 1992, **38**, 99.
31. J.H. Sun, E. Chang, B.C. Wu, C.H. Chao and A. Peng: *Oxidation of Metals*, 1992, **40**, 229.
32. S.R.J. Saunders: *Met. Technol.*, 1984, **11**, 465–473.
33. J. Little: *Corrosion*, 1989.
34. P.Y. Hou and J. Stringer: *J. Electrochem. Soc.*, 1987, **134**(7), 1836–1849.
35. S.J. Bull: *Oxidation of Metals*, 1997.
36. J.P. Wilber, J.R. Nicholls and M.J. Bennett: *Microscopy of Oxidation 3*, Institute of Materials, 1997, 207–220.
37. R.L. Samuel and N.A. Lockington: *Met. Treat. Drop Forging*, 1951, **18**, 354–359, 407–415, 440–444, 495–502 and 506.
38. A.H. Sully and E.A. Brandes: *Chromium*, 2edn. Chap. 7; London, Butterworths, 1967.
39. G.W. Goward, D.H. Boone, and C.S. Giggins: *Trans. ASM*, 1967, **60**, 228–241.
40. G.W. Goward and D.H. Boone: *Oxid. Met.*, 1971, **3**, 475–495.
41. S.J. Grisaffe: *The Superalloys*, C.T. Sims and W.C. Hagel eds, New York, John Wiley, 1972, 341.
42. P. Felix and E. Erdos: *Werkst. Korro.*, 1972, **23**, 626–636.
43. F. Fitzer and J. Schlichting: *High temperature corrosion*, R.A. Rapp ed., Houston, TX, NACE, 1983, 604–614.
44. R. Buuer, H.W. Grunling, and K. Schneider: in Proc. 1st Conf. on '*Advanced materials for alternate fuel capable directly fired heat engines*, J.W. Firbanks and J. Stringer eds, Palo Alto, CA, Electric Power Research Institute (EPRI), 1989, 505.
45. E. Fitzer and H.J. Maurer: *Materials and coatings to resist high temperature corrosion*, D.R. Holmes and A. Rahmel eds, London, Applied Science, 1978, 253–269.
46. R. Pichoir: *Materials and coatings to resist high temperature corrosion*, D.R. Holmes and A. Rahmel eds, London Applied Science, 1978, 271–291.
47. R. Pichoir: *High temperature alloys for gas turbines*, D. Coutsouradis *et al.* eds, London, Applied Science, 1978, 191–208.
48. B.K. Gupta and L.L. Seigle: *Thin Solid Films*, 1980, **73**, 365–371.
49. R. Pichoir and J.E. Restall: *Coatings for heat engines* (Workshop Proc.), R.L. Clarke *et al.* eds, Washington, DC, US Department of Energy, 1984, 257–281.

50. J.E. Restall, M. Malik, and L. Singheiser: *High temperature alloys for gas turbines and other applications 1986*, Conf. Proc., W. Betz *et al.* eds, Dordrecht, D. Reidel Publishing Co, 1986, 357–404.

51. R. Mevrel, C. Duret, and R. Pichoir: *Mater. Sci. Technol.*, 1986, **2**(3), 201–206.

52. W.T. Wu, W.L. Wang, and T.F. Li: *High temperature corrosion*, R.A. Rapp ed., Houston, TX, NACE, 1983, 598–603.

53. C. Duret, A. Davin, G. Marrijnissen, and R. Pichoir: *High temperature alloys for gas turbines*, Conf. Proc. R. Brunetaud *et al.* eds, Dordrecht, D. Reidel Publishing Co, 1982, 53–87.

54. G.W. Goward: *High temperature corrosion*, Conf. Proc. NACE-6, R.A. Rapp ed., Houston, TX, NACE, 1983, 553–560.

55. G.W. Goward and L.W. Cannon: 'Pack cementation coatings for superalloys, history, theory and practice', *ASME Paper 87-GT-50*, New York, American Society Mechanical Engineers, 1988.

56. S.R. Levine and R.M. Caves: *J. Electrochem. Soc.*, 1973, **120**(8), C232.

57. R. Bianco and R.A. Rapp: *J. Electrochem. Soc.*, 1993, **140**(4), 1181–1191.

58. G. Lehnert and H. Meinhardt: *Surface Treatment*, 1972, **1**, 72.

59. G. Fisher: 'The optimisation of bond coat oxides for improved thermal barrier coating adhesion', *PhD Thesis*, Cranfield University, 1997.

60. F.T. Talboom, R.C. Elam and L.W. Wilson: 'Evaluation of Advanced Superalloy Protection Systems', *Report CR7813*, NASA, Houston TX, 1970.

61. J.P. Coad and R.A. Dugdale: Proc. Conf. on *Ion plating and allied techniques*, London, CEP Consultants, May 1979, 186–196.

62. R.L. Clarke: Proc. 4th Conf. on *Gas turbine materials is a marine environment*, Annapolis, MD, Naval Sea Systems Command, June 1979, 189–219.

63. J.T. Prater, J.W. Patten, D.D. Hayes, and R.W. Moss: Proc. 2nd Conf. on *Advanced materials for alternate fuel capable heat engines*, J.W. Firbanks and J. Stringer eds, *Report No. 2639SR*, 7/29-7/43; Palo Alto, CA, EPRI, 1981.

64. K.N. Stafford, D.P. Whittle, P.J. Hunt, and A.K. Misra: Proc. 4th Conf. on *Gas turbine materials in a marine environment*, Annapolis, MD, Naval Sea Systems Command, June 1979, 221–252.

65. J.A. Goebel, R.J. Hecht, and J.R. Vargas: Proc. 4th Conf. on *Gas turbine materials in a marine environment*, Annapolis, MD, Naval Sea Systems Command, June 1979, 635–653.

66. J.A. Goebel, C.S. Giggins, M. Krasij, and J. Stringer: Proc. 2nd Conf. on *Advanced materials for alternate fuel capable heat engines*, J.W. Firbanks and J. Stringer eds, Report no. 2639SR, 7/1; Palo Alto, CA, EPRI, 1981.

67. R.G. Corey, R.H. Barkalow, A.S. Khan, and R.J. Hecht: Proc. 2nd Conf. on *Advanced materials for alternate fuel capable heat engines*, J.W. Firbanks and J. Stringer eds, Report No. 2639SR, 7/17; Palo Alto, CA, EPRI, 1981.

68. D.H. Boone: *Materials coating techniques*, Conf. Proc. LS-106, Chap. 8; Neuilly-sure-Seine, AGARD, 1980.

69. J.R. Nicholls, P. Hancock, and L.H. Al-Yasiri: *Mater. Sci. Technol.*, 1989, **5**, 799–805.

70. J.R. Nicholls, K.J. Lawson, L.H. Al-Yasiri and P. Hancock: to be published in *Corrosion Science*, 1993.

71. R.C. Tucker, T.A. Taylor, and M.H. Weatherby: Proc. 3rd Conf. on *Gas turbine materials in a marine environment*, Bath, UK Ministry of Defence, Session 7, paper 2 Sept. 1976.

72. A.R. Nicoll: *Coatings and surface treatment for corrosion and wear resistance*, K.N. Strafford *et al.* eds, Chichester, Ellis Horwood, 1984, 180.

73. T.A. Taylor, M.P. Overs, B.J. Gill, R.C. Tucker: *J. Vac. Sci. Tech.*, 1985, **3**(6), 2526–2531.

74. J.H. Wood, P.W. Schilke, M.F. Collins: *ASME Paper No. 85-GT-9*, 1985.

75. J.E. Restall and M.I. Wood: *Mater. Sci. Tech.*, 1986, **2**(3) 225–231.

76. K. Niederberger and B. Schiffer: *Eigenschaften verschidener Gas und deren Einfluss* (1990), beim thermnischen spirtzen. Thermische Spirtzkonferenez, Essen, Germayn, 29–31 August 1990, 1–5.

77. H. Kreye: 'High velocity flame spraying processes and coating characteristics', *Second Plasma Technik*, Symposium, Lucerne, Switzerland, 5–7 June 1991, 39–47.

78. R.W. Kaufol, A.J. Rotolico, J. Nerz and B.A. Kushner: 'Deposition of Coatings Using a New High Velocity Combustion Spray Gun', *Thermal Spray Research and Applications*, T.F. Bernecki ed., Materials Park, Ohio: ASM International 1990, 561–569.

79. L. Russo and M. Dorfman: 'High-Temperature Oxidation of MCrAlY Coatings Produced by HVOF', *Proceedings of the International Thermal Spray Conference*, A. Ohmori ed., Japan: High Temperature Society of Japan, 1179–1194.

80. E.C. Kedward: *Metallurgia*, 1969, **79**, 225–228.

81. J. Forster, B.P. Cameron, J.A. Carews: *Trans. Inst. Metal Finish*, 1985, **63**, 115–119.

82. F.J. Honey, E.C. Kedward and V. Wride: *J. Vac. Sci. Tech.*, 1986.

83. J. Powell, P.S. Henry, and W.M. Steen: Preprints Conf on *Surface engineering with lasers*, London, May, The Institute of Metals, paper 17, 1985.

84. W.M. Steen: in Proc. Conf. on *Applied laser tooling*, G.D.D. Soares and M. Perez-Amor eds, Dordrecht, Martinue Nijhoff, 1987, 131–211.

85. R.F. Bunshah: Proc. Conf. *Materials coating techniques*, Neuilly-sur-Seine, AGARD, 1980.

86. D.G. Teer: Proc. conf *Materials coating techniques*, Neuilly-sur-Seine, AGARD, 1980.

87. S.J. Saunders and J.R. Nicholls: *Thin Solid Films*, 1984, **119**, 247–269.

88. M. Malik, R. Morbioli and P. Huber: *High temperature alloys for gas turbines*, Conf. Proc., R. Brunetaud *et al.* eds, Dordrecht, D. Reidel Publishing Co. 1982, 87–98.

89. J.R. Nicholls, D.J. Stephenson, P. Hancock, M.I. Wood and J.E. Restall: Proc. Workshop on *Gas turbine materials in a marine environment*, Bath, UK, Ministry of Defence, Session 2, Paper 7, Nov. 1984.

90. K.L. Luthra and O.H. LeBlanc: *Mater. Sci. Eng.*, 1987, **88**(1), 329–336.

91. R.C. Novak: 1994, cited in reference 11.

92. A.J. Mom: NLR Report, *MP 81003U*, Amsterdam 1981.

93. D.R. Coupland, C.W. Hall and I.R. McGill: *Platinum Metal Review*, 1982, **26**(4), 146–157.

94. N.S. Bornstein and J. Smeggil: *Corrosion of Metals Processed by Directed Energy Beams*, Met. Soc. AIME, 1982, 147–158.

95. J.E. Restall and C. Hayman: *Coatings for heat engines* (Workshop Proc.), R.L. Clarke *et al.* eds, Washington, DC, US Department of Energy, 1984, 347–357.

96. N.S. Bornstein and J. Smeggil: *Corrosion of Metals processed by Directed Energy Beams*, Met. Soc. AIME, 1982, 147–158.

97. J.R. Nicholls, S. Neseyif, M. Taylor and H.E. Evans: 'Developments of smart overlay coatings', *LINK Surface Eng. PVD Workshop*, TWI, Cambridge, May 1997.

98. P. Hancock, J.R. Nicholls and D.J. Stephenson: *Surface and Coatings Tech.*, 1987, **32**, 285–304.

99. D.J. Stephenson: *Corrosion Sci.*, 1989, **29**, 647–656.

100. S.P. Cooper and A. Strang: *High temperature alloys for gas turbines* Conf. Proc., R. Brunetaud *et al.* eds, Dordrecht, D. Reidel Publishing Co., 1982, 249–260.

101. E. Lang and E. Bullock: 'European concerted action, COST 50 - Materials for gas turbines', *EUR Report 8242 EN*, 1982.

102. P. Mazars, D. Maresse, and C. Lopvet: *High temperature alloys for gas turbines-1986*, V. Betz et al. eds, Dordrecht, D. Reidel Publishing Co., 1986, 1183–1192.

103. R.J.E. Glenny: *High temperature materials in gas turbines*, P.R. Sahm and M.O. Speidel eds, Amsterdam, Elsevier, 1974, 257.

104. J. Gayda, T.P. Gabb, R.V. Miner and G.R. Halford: *Proc. TMS-AIME Annual Symp. 1987*, P.K. Liaw and T. Nicholas eds, TMS, 1987, 217–223.

105. A. Strang and E. Lang: '*High Temperature Materials for Gas Turbines*, R. Brunetaud *et al.* eds, D. Reidel Publishing Company, Dordrecht, Holland, 1982, 469–506.

106. P. Au, R.V. Dainty and P.C. Patnaik: Proc. Conf. *Surface modification technologies II*, T.S. Sudarshan and D.G. Bhat eds, TMS-AIME, 1990, 729–748.

107. M.J. Bennett, J.A. Desport, and P.A. Labun: *Oxid. Met.*, 1984, **22**(5/6), 291–306.

108. M.J. Bennett, A.T. Tuson, C.F. Knights and C.F. Ayres: *Mater. Sci. Technol.*, 1989, **5**, 841–852.

109. M.J. Bennett, C.F. Knights, C.F. Ayres, A.T. Tuson, J.A. Desport, D.S. Rickerby, S.R.J. Saunders and K.S. Coley: *Mater. Sci. Eng.*, 1991, **A139**, 91–102.

110. R.W.J. Morssinkof, T. Fransen, M.D. Hensinkveld and P.J. Gellings: *Mater. Sci. Eng.*, 1989, **A121**, 449–445.

111. A.A. Ansari, S.R.J. Saunders, M.J. Bennett, A.T. Tuson, C.F. Ayres, and W.H. Steen: *Mater. Sci. Eng.*, 1987, **88**, 135–142.

112. P. Hancock: *Mater. Sci. Eng.*, 1987, **88**, 303–311.

113. D.S. Duvall and D.L. Ruckle: *ASME paper 82-GT-327*, 1982.

114. J.W. Fairbanks and R.J. Hecht: *Mater. Sci. Eng.*, 1987, **88**, 321–330.

115. T.E. Strangman: 'Development and performance of physical vapour deposition thermal barrier coatings systems', Paper presented at the 1987 Proceedings of the *Workshop on Coatings for Advanced Heat Engines*, Castine, Maine, July 27–30, Washington DC: US Department of Energy, 1987.

116. S. Bose and J. Demasi-Marcin: 'Thermal barrier coating experience in gas turbine engines at Pratt and Whitney', *Thermal Barrier Coatings Workshop*, NASA-CP-3312. Cleveland, Ohio, National Aeronautics and Space Administration Lewis Research Center, 1995, 63–77.

117. R. Burgel and I. Kvernes: High temperature alloys for gas turbines and other applications 1986 (Conf. Proc.), W. Betz *et al.* eds, Dordrecht, D. Reidel Publishing Co, 1986, 327–356.

118. A. Bennett: *Mater. Sci. Technol.* 1986, **2**(3), 257–261.

119. F.H. Stott: *Mater Sci. Technol.*, 1988, **4**(5), 431–438.

120. K.D. Sheffler and D.K. Gupta: *Trans. ASME. J. Eng. for Gas Turbines and Power*, 1988, **110**, 605–609.

121. F.C. Toriz, A.B. Thakker and S.K. Gupta: *Surface Coating Tech.*, 1989, **39/40**, 161–172.

122. *Thermal Barrier Coatings Workshop*, National Aeronautics and Space Administration, Lewis Research Centre, Cleveland Ohio, NASA-CP-3312, 1995.

123. *Thermal Barrier Coatings*, Meeting of AGARD Structures and Materials Panel, Aalborg, Denmark, Advisory Group for Aerospace Research and Development, Report AGARD-R-823, Oct. 1997.

124. J.R. Nicholls: 'Smart Coatings - A Bright Future', *Materials World*, 1996, **4**(1), 19–21.

125. J.R. Nicholls, K.J. Lawson, D.S. Rickerby and P. Morrell: 'Advanced Processing of TBC's for Reduced Thermal Conductivity', Presented at AGARD Meeting, Denmark 14–16 Oct. 1997, *AGARD Report No. 823*, April 1998.

High Temperature Alloys for Advanced Industrial Gas Turbines

B. J. PIEARCEY

ABB ALSTOM Power Technology Centre, Whetstone, Leicester, UK

ABSTRACT

Gas Turbines have found widespread use in industrial and marine applications for the provision of motive power and the generation of electricity. Their development has followed that of aero engines to a large extent but their scope has widened in a number of ways. Aero engines are designed with a more severe weight penalty than industrial gas turbines whereas the latter may have to use poorer quality fuels, are often subject to prolonged industrial and marine atmospheric corrosion and are expected to operate for periods in excess of 100 000 hours. The lower importance of power/weight ratio, in many applications, has enabled manufacturers to use cast iron and steel rather than titanium alloys for major components which has resulted in the unit cost of power to the customer being much less than that for aero engines.

This paper provides an overview of requirements, selection, and future developments of materials for industrial gas turbines. In the first part, the basic types of industrial gas turbines are described, together with the efficiency requirements and design procedure and other factors which govern the selection of materials. Next, the property requirements are summarised and related to particular components, including casings, combustion chambers and exhaust systems, blades and discs. Future developments in industrial gas turbine design and materials and process development are evaluated. In the final part of the paper, opportunities for Materials Engineers at Universities, Research Institutions and Industry, Manufacturers of Material Product and The Institute of Materials are defined which will improve the reliability of gas turbine engines and reduce costs for the industry.

THE INDUSTRIAL GAS TURBINE–BASIC CONCEPTS

Industrial gas turbines generally employ a single shaft gas generator which is comprised of a compressor, a combustor and a turbine. The gas generator is directly coupled via a drive shaft with a power turbine for power generation purposes or used to drive a separate power turbine for mechanical drive applications. The latter are sometimes referred to as 'twin shaft' but are more correctly described as 'twin spool' to distinguish them from the modern aero engine which may employ two or three concentric shafts to control the rotational speed of the fan disc, low pressure and high pressure compressor independently. Many applications provide both electrical and mechanical power and, in addition, use is made of the waste heat to provide environmental and process heating. Various designs are in service where a drive shaft is coupled to the compressor rather than or in addition to that connected to the power turbine.

Output is usually determined by the physical size of the engine ranging from machines less than one metre long rated at 2.5 MW to machines in excess of ten metres long rated at over 400 MW. Many of the systems used for power generation are classed as 'aero derivative' since the gas generators are based on the use of existing designs of aero engine. A matching power turbine is attached to provide electrical power via a generator and/or mechanical power via a gear box. In combined cycle power systems the exhaust gas from the engine is used to raise steam which is then used to power a steam turbine.

Efficiency Considerations

A recent review of the opportunities for improving the efficiency of gas turbines[1] describes how the controlling parameters affect the thermodynamic cycle and how these improvements might be realised by a combination of materials selection and component design. A key development has been the improvement in compressor performance to achieve higher stage loading, overall pressure ratio and system efficiency. Pressure ratios of over 20:1 are now common enabling firing temperatures of over 1400°C to be realised. Further improvement in efficiency can be realised by improved design of gas path seals, reduction of friction in bearings and the insulation of the hot gas path. The consequence for materials selection is that work lost as heat may raise the temperature of the compressed air in some engines to over 500°C at the high pressure end. Both compressor blades and discs attain the air temperature and consideration must be given to the long time temperature capability of each stage of the compressor. The higher compressor delivery temperature together with improvements in combustion technology provides high pressure gas at temperatures limited only by the materials available to withstand creep and cyclic strain for thousands of cycles. These conditions affect the materials used for combustors and transition ducts as well as the nozzle guide vanes and turbine blades in the turbine section. Whilst the use of cooled components may enable a particular material to be used at a higher temperature, the use of cooling does result in a loss in efficiency because it is obtained by the bleed-off of compressor air.

Factors Affecting Materials Selection

The choice of any material raises a number of issues in addition to the feasibility of component manufacture. Accurate, traceable materials data are required at several stages in the design and manufacture of gas turbines and, in recent years attention has turned to computerising related activities[2] including, for example, product data management. For example, for a turbine blade:

• materials selection during conceptual design uses simple material properties such as creep strength at a particular set of conditions, cost/availability, previous experience in related engines, the need for coatings or surface treatment, etc.

- the purchase of material or a component according to a material specification against which material suppliers' test certificates are checked and statistical process controls updated.
- 1-dimensional analysis (i.e. along the blade axis) to assess creep performance under expected conditions leading to the initial 3D finite element model describing the blade geometry.
- thermal, stress and dynamic analyses leading to temperature/stress profiles.
- consideration of component life from a detailed knowledge of creep and fatigue behaviour.
- knowledge of the effects of corrosion and exposure at temperature.

Each of these tasks requires a variety of physical, thermal, mechanical and environmental material properties presented in various formats, usually in the form of computerised data files for input into design, analysis or lifing codes. In an industrial gas turbine manufacturing environment, this provides the initial impetus for computerising material property data. This apparently simple task is complicated by two chief factors. Firstly, material property data are of little value if the material pedigree and metadata concerning the origin and conditions applying to the test data are absent or poorly described. Secondly, the exact definition of property requirements may vary with different design practices or when anisotropic materials are used.

Conceptual design of a turbine blade simply requires rupture strength data at a limited number of target engine conditions; whereas the generation of blade geometry from 1-D creep calculations requires mathematical descriptions or tabulated creep rupture strengths over a range of temperatures and duration. Generally, thermal analyses indicate stresses exceeding the elastic limit in relatively cool parts of the blade but with hotter regions, such as blade trailing edge, substantial creep may occur. To determine creep-relaxed stresses, and therefore when component failure will occur, lifing equations describing the overall creep strain versus time are applied as subroutines within finite element analyses. Each stage of the design and analysis process requires different data in different formats, all nominally generated from creep tests, and which the Database must be capable of providing. Each candidate material must, therefore, be characterised in terms of its physical and mechanical properties and the effect of exposure for long times and/or many cycles at elevated temperature has to be evaluated by means of creep rupture and fatigue testing. Such test programmes typically cost in excess of £100k for each material in addition to the cost of engine testing of components. A well characterised material, such as those investigated by international consortia, may eventually cost as much as £1 million.

The selection of a material for a particular component cannot be considered in isolation since the component may be joined to another. A small difference in the coefficient of expansion may create unacceptable stresses within the joint, making the choice unacceptable. Mechanical or welded joints are made between many components, each of which may operate over a range of temperature. The

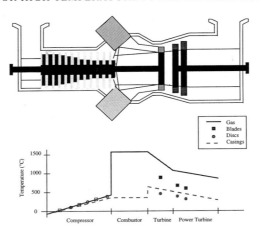

Fig. 1 Typical industrial gas turbine layout and temperature profile.

layout of the major components in a typical engine is illustrated in Fig. 1 to-gether with a temperature profile of the gas stream and major components.

Environmental Consideration

Whilst a full thermal and elastic analysis will provide a good approximation of the temperature and stress profiles within each component, consideration must also be given to the effects on the local environment of a gas turbine engine from both the input and output points of view. The inlet air may be contaminated with local atmospheric industrial pollution which may contain sulphur and chlorine, salt and particulate materials such as sand. Erosion and stress corrosion of compressor blades may result and consideration must be given to the use of protective coatings. The fuel used may also be contaminated with sand and salt water which will be introduced into the gas stream via the combustion chambers. The combination of salt and sulphur leads to rapid hot corrosion in some nickel base alloys in the range 750 to 850°C and protective coatings are essential for high strength alloys particularly those with low chromium contents.

On the output side, environmental legislation is demanding that gas turbine emissions contain ever decreasing quantities of nitrous oxides and carbon di-oxide. Where the exhaust gases are used to provide heating or to raise steam in combined cycle operations, consideration must be given to corrosion of boiler tubes and gas path lining materials resulting from the salt and sulphur content.

Cost and availability

The cost of the gas turbine engine is very dependent on the cost of materials selected which, in turn, depend on their composition and method of manu-facture. However, even though the capital cost of a natural gas fired power station is substantially less in £/MW than its equivalent coal fired station, the choice of gas turbine and its design require that both the engine manufacturer and the eventual owner determine a compromise between original capital cost, reliability and fuel efficiency. Higher compression ratios with higher combustion temperatures require more expensive alloys to provide the same reliability but the additional component cost is repaid in terms of fuel efficiency.

The tremendous growth in production in both aerospace and land based gas turbines has created supply problems for some gas turbine components, particularly for large nickel base alloy discs. Alternative methods of manufacture such as powdermet and 'spray casting' are being tested but, so far, have not proved so economical for the same reliability.

HIGH TEMPERATURE ALLOYS FOR INDUSTRIAL GAS TURBINES

The challenge for materials engineers is to select materials for all the components in the engine capable of surviving for a predetermined life, usually of the order of 100 000 hours and 10 000 cycles; a cycle in industrial terms being a start-up to shutdown. The number of cycles in each application may vary and the life of a particular engine may be determined either by hours of endurance or the number of cycles.

As each section of the engine is designed the designer calls upon a database of approved candidate materials for each component. An approved material implies that sufficient knowledge and property data is available to enable the design intent to be met in the environment defined. This information includes material product specification, sources of supply, corrosion resistance and physical and mechanical properties. As the demand for ever more efficient engines grows, it usually follows that the temperature of the component increases; materials selection, in the first analysis is more often determined by the availability of the elevated temperature properties of the existing material rather than the automatic substitution of another material. Any material has a temperature limit beyond which its strength is negligible; clearly, an alternative material must be chosen if this temperature is approached in any part of the component when every means of limiting the temperature have been investigated.

The Characterisation of Materials

Design engineers need to know the operating temperature, stress and strain of each part of a component and their variation within each operating cycle. These values are determined by thermal and elastic analyses in the first instance, an iterative procedure since the result depends on the material selected. This stage is usually followed by a 'lifing' analysis where the time and cycle dependent properties are investigated to determine whether each component will withstand the environment applied for the declared life of the component. An abbreviated summary of the material properties required is shown in Table 1.[3]

The material suppliers usually provide much of the physical and mechanical property data but a large number of materials have been put into use with limited

Table 1 Material properties typically required for gas turbine components.

Physical properties	Tensile properties

• Thermal conductivity
• Specific heat
• Density
• Coefficient of thermal expansion (CTE)
• Elastic compliances
	• Stress–strain curves
	• Ultimate strength
	   Proof stress at 0.1 to 5% strain,
	   % Elongation & % reduction of area
	   Static Modulus of elasticity

Fatigue data
(Load controlled)
• HCF, stress to produce failure up to
   10^8 cycles vs. temperature
• LCF, stress to produce failure up to
   5.10^4 cycles vs. temperature, axial data at
   various Kt values and R-ratios
	Creep rupture data
	(Iso-load, Iso-temperature).
	• Stress to produce rupture at target life
	   vs. temperature.
	• Stress to produce finite creep strain at target life,
	   Values 0.01 to 5% strain up to 10^5 hrs. vs.
	   temperature.

Fatigue data (Strain controlled) • Strain range to produce failure up to 5.10^4 cycles vs. temperature.	Supplementary data • Goodman diagrams • Cyclic stress–strain curves vs. temperature • Oxidation and corrosion rates • Melting point • Hardness and impact • Susceptibility to oxygen or hydrogen embrittlement

and, at best, short time lifing data because of the tremendous expense of testing. Such data is closely guarded by engine manufacturers because of the common misunderstanding that even limited data provides a competitive advantage. This situation will persist until some means is found of sharing the cost of such test programmes and sufficient data can be generated about each material to provide for a satisfactory statistical analysis.

'High temperature alloy' is a relative term meaning different temperatures for each class of material. Because of the constraints of cost and reliability, industrial gas turbines are manufactured using high temperature cast irons and steels rather than titanium alloys for external casings and compressors but the most advanced aerospace alloys are used for combustors and turbine components.

Examples of the materials used in industrial gas turbines may usefully be described by classification by engine component as follows.

Structural Materials

The gas turbine may be considered as a structure which co-ordinates the functions of an air compressor, a combustor, a high temperature turbine, an exhaust system, shafts and bearings. The structural parts include casings to enclose each section, combustion chambers and ducting to deliver the high pressure and high temperature gas to the turbine, rings to hold compressor stator blades and turbine nozzle guide vanes and the exhaust ducting to carry the products of combustion to heat exchangers and the external atmosphere.

Casings

Casings have a dual function, as well as their essential role as a structural part, they may also provide containment in the event of a blade failure, a role which will limit the extent of damage to external structures and personnel. The lower half of such a casing is shown in Fig. 2 which illustrates a fully assembled rotor comprising an 18 stage compressor and 3 stage turbine being lowered onto its bearings. The casing is approximately 10 metres in length.

The choice of material is primarily determined by cost, so long as the material can withstand the applied stress and temperature for the life of the component. Creep rupture data is used to provide a selection of 'candidate' materials. The rupture strength of some of the materials commonly used in industrial gas turbines is shown in Fig. 3, the stress to produce rupture in 1000 hours is shown for illustration, longer time data being required for component 'lifing'.

Clearly, where the overall weight of each component does not affect the life of the engine or its size can be accommodated in the structure, low stresses can be achieved by thick sections and relatively high temperature capabilities realised by some relatively 'ordinary' materials. However, care must be taken not to introduce stress due to thermal gradients. Cast iron, typically Grade '420/12 + Mo', is used extensively for the whole casing even though its operating temperature may exceed 400°C in some locations.

In the hot section of the gas turbine, the structural parts are used to attach the nozzle guide vanes which are in direct contact with the gas stream. These parts are manufactured from materials with a higher temperature capability than alloy steel and cast iron, for example, the creep resistant stainless steels FV448 and FV507 and the nickel base alloy Inconel alloy 718. All these materials are used in

Fig. 2 A turbine rotor being lowered into a cast iron casing during the assembly of a 250 MW gas turbine.

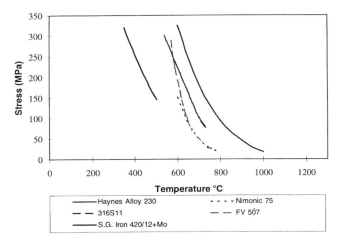

Fig. 3 Comparison of the creep rupture strength of some structural materials.

the forms of bar, forgings and castings in many types of structural part and fastener. In the hottest regions, components are manufactured from alloys which have useful strength in the temperature range 600 to 900°C, typically nickel-base alloys such as the Nimonic and Inconel series, e.g. Nimonic 75.

Combustion Chambers

Nimonic 75 bar, sheet and plate has been used for many years to construct combustion chambers and transition ducts for gas turbines engines. The material can be easily formed and welded and has good ductility, a characteristic which is found to provide good resistance to thermal fatigue. As the combustion temperature is increased, Nimonic 75 is being replaced with Haynes alloy 230, an alloy having a temperature capability 100°C higher.

No metallic material has been developed to withstand the temperatures currently achieved in the combustor and extensive use is now made of thermal barrier coatings to reduce the surface temperature of the metallic component. Yttria stabilised Zirconia, plasma sprayed over an 'MCrAlY' bond coat, is commonly used to provide an additional 50°C temperature gradient between the gas flame and the combustor wall. However, the use of water injection, designed to meet environmental restrictions regarding the output of nitrous oxides, may result in premature spalling of the ceramic oxide.

Exhaust Systems

The exhaust system is primarily designed to transport the products of combustion to the atmosphere via recovery units which are used for recuperation or the raising of steam and process and/or environmental heating. Having expanded during the delivery of energy to the turbine and power turbine the volume flow of the combustion products is greatly increased and the temperature of the inner skin is reduced to less than 700°C, sometimes with the assistance of external water spray. The stress and temperature can be accommodated by alloys like Nimonic 75 up to 700°C, 12% Cr ferritic and 18% Cr 8% Ni austenitic stainless steels backed up by structural members made of structural steels are used up to 600°C.

Blade Materials

Compressor blades, turbine blades and power turbine blades are all designed to withstand a complex stress and temperature range which has resulted in the development of a wide range of material products. These components, in both aerospace and land based applications, are particularly sensitive to density because of the centrifugal stress created by the rotation of the turbine. As the turbine varies in speed during start up and shut down, blades are subject to natural frequency vibration resulting in an additional high frequency load. Furthermore, as the temperature of the gas path increases, the components become subject to temperature gradients which may add to or reduce the applied stress, the combination of cyclic stress and temperature leading to combined creep and fatigue damage, further compounding the problem of stress analysis and the selection of material.

Whereas Whittle managed to produce a whole turbine using the 1% Cr steam turbine alloy, designers are now confronted with a wide range of steels, titanium and nickel based alloys which are selected to meet the design criteria selected for a particular engine and power output. The temperature capability of a material may be judged by creep rupture testing and comparison of the stress to produce rupture in a specified time. The popular Larson–Miller parameter is often used for interpolation of creep rupture data but gross errors can result if it is used for extrapolation; the creep strength of many materials varies as a function of time. A comparison of candidate blade materials is shown in Fig. 4 where the high

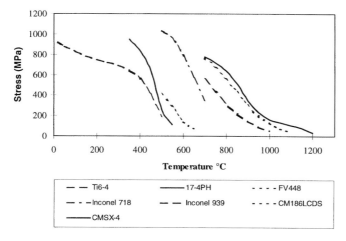

Fig. 4 Comparison of the creep rupture strength of rotor blade materials.

strength wrought alloys used for compressor blades are compared with conventionally cast, directionally solidified and single crystal turbine blade alloys.

Wrought nickel base alloys are still used for turbine blades in modern engines at temperatures up to 700°C. They have higher fatigue strength than cast alloys and are often used as unshrouded blades in turbines. However, they have been replaced with cast alloys in the hotter sections because of their inferior creep strength. The other advantage of the investment casting process is that it can be used to provide components with complex cooling passages and enable the temperature of the turbine blade to be maintained within the creep capability of the material even in an environment approaching the melting point of the alloy.

Disc Materials

Each disc and blade assembly is designed as a whole because the disc has to carry the whole weight of the blades. The centrifugal force acting on the blades is transmitted to the disc, usually via a 'dovetail' fixing in the compressor and a 'fir tree' root fixing in the turbine to develop a radial stress. In addition, the forces acting on the disc, which has to be restrained from bursting, create hoop stresses similar in magnitude to the radial stress so that the overall stress system in any part of the disc is bi-axial. This simple analysis explains why it is often difficult to predict the properties of disc materials since most testing is uniaxial. In addition to the stresses created by centrifugal force, the turbine disc is subject to severe thermal gradients resulting in complex tri-axial and cyclic stresses acting on the component exacerbated in the high stress concentrations in the locations of the blade root fixing and the disc bore. If this scenario was not already too complex for analysis, gas turbines are subject to overspeeding increasing the centrifugal stress by as much as 20% if the driven load is suddenly rejected. This 'overspeed criteria' must be accommodated in the first instance and so a minimum value of the ultimate tensile stress can be specified at the operating temperature. The ultimate tensile strength of several disc materials is shown in Fig. 5.

The outstanding strength of Inconel 718, even at temperatures over 500°C explains why the alloy has found such widespread use in industrial gas turbine engines. In addition to its strength the alloy has excellent corrosion resistance although its properties may be affected by stress activated grain boundary oxidation (SAGBO)[4] at temperatures around 600°C.

FUTURE DEVELOPMENTS

The selection of alloys for advanced industrial gas turbines will continue to be constrained by the factors of cost, reliability and meeting the design intent. The continuous requirement to improve efficiency means that materials will be required with improved temperature capability. It must be remembered, however, that material products for gas turbines must be fully characterised before they can be considered for use as a gas turbine component, and then engine tested to satisfy customer and insurance requirements.

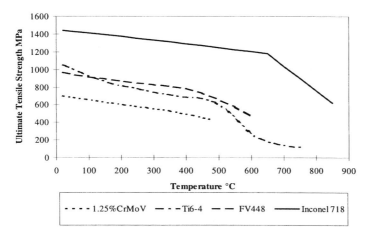

Fig. 5 Comparison of the ultimate tensile strength of some gas turbine disc materials.

The Importance of Strain Control

Frequent reference is made of the effect of stress when considering materials selection; whether it is the ultimate stress, the stress to produce rupture in a finite time or the cyclic stress range to produce failure in a finite number of cycles. In the last five years, two technologies have become widely available, finite element analysis and robust testing machines with advanced control systems which are able to exert a predetermined strain on a test piece at high loads and specified frequencies. In many components the stress at a particular location in a component is controlled by the applied strain, which may result from either mechanical loading or thermal gradients. The value of strain can now be more accurately predicted by finite element analysis and tests devised to determine the life of the material product subjected to a controlled strain range. This technology has enabled the materials engineer to distinguish between successful materials which are strong and strain tolerant and other strong materials which appear to fail prematurely. Conversely, where materials with otherwise desirable properties but poor strain tolerance become available, design engineers are now finding ways to limit the strain requirement of the component and incorporate the materials in their designs. Typical examples of such materials include intermetallic compounds and ceramics.

New Materials and Processes

No picture of the current status of high temperature alloys for industrial gas turbines would be complete without a look at the future. In the first instance, the terms 'alloy' and 'material' must be replaced with 'material product' since the properties of a material cannot be defined without a knowledge of its form, processing history and its surface condition. The scope of materials selection must include non-metallic materials in both monolithic and composite forms. In fact, it

Table 2 Material products being considered for advanced industrial gas turbines.

Component	Material product	Advantages	Disadvantages
Structural parts	Alloy steel castings	Sounder castings	Higher cost
	High temperature austenitic cast irons	Low cost alloy useful at $-40>T>400°C$	Porosity high coefficient of thermal expansion
Combustors	Thick thermal barrier coatings on metals	Reduced metal temperature	Unsuitable for water injection
	Ceramic matrix composites	More compatible with thick thermal barriers	Need oxidation protection
Exhaust systems	Ceramics	Low cost insulation	Low strain tolerance
	Coated structural steels	Lower cost than stainless steel	Low temperature capability
Blades	Defect tolerant single crystal superalloy	Lower cost and tolerant of small secondary grains	Defect is source of failure, location must be controlled and detectable
	Titanium aluminide	Low density	Low strain tolerance
Discs	Dual alloy composites	Variable properties to suit local environment	Difficult and expensive to manufacture

is simpler to consider any component in everyday use as a composite material product since most are found to be inhomogeneous and anisotropic and make use of surface coatings and other treatments which modify their mechanical, physical and chemical properties. Material products being considered or under development and testing for advance industrial gas turbines are summarised in Table 2.

Novel Component Design

There is usually a limit to the extent to which a change in material product can solve the problem of premature failure or inadequate life of a component. The solution to such problems and the overall improvement in gas turbine efficiency is often found in novel design or the application of coatings and/or other surface treatment. New designs can often be implemented by the use of existing material products. Examples include,

Design Feature/Process	Improvement
the use of aerospace materials	established confidence in meeting the design criteria
improved machining and polishing	better surface finish and larger corner radii reduce stress concentration resulting in longer life
improved combustion to reduce hot spots	lower maximum temperature requirement
the steam cooling of blades	improved heat transfer and lower blade temperature
air & water spray cooling of exhaust and recuperation of the extracted heat	lower cost material but lagging and cooling system required
removal of impurities in existing alloys and coatings	longer life improved ductility
manufacturing processes which reduce segregation and anisotropy	more consistent properties and reduced statistical deviation from norm.

OPPORTUNITIES FOR MATERIALS ENGINEERS, MANUFACTURERS AND THE INSTITUTE

The industrial gas turbine industry has entered a period of substantial growth as the World's energy requirement increases to meet the needs of industry and the general expectation of an improved standard of living. Current application of gas turbines involve a wide variety of fuels from super-clean natural gas to low calorific value bio-waste and the gasification of poor quality coal and oil shale.

The UK Power Engineering Sector directly employs around 160 000 people and has a turnover of around £14 billion, approximately £4 billion greater than the aerospace industry. Including the electrical supply industry this represents approximately 7.3% of the Gross Domestic Product, making it one of the major contributors to the wealth creation of the UK. In order to ensure that this sector is able to continue to trade effectively in world markets, the UK Technology Forsight initiative, through the Energy Panel, has been instrumental in establishing the future R&D requirement in a variety of technical fields. In particular, the Energy Panels' Clean Coal Taskforce and The Institute of Materials[5] have produced a report that proposes a single UK Strategy for materials R&D for power generation systems; it recommends,

1. A rolling five year co-ordinated programme of materials R&D be initiated.
2. A power generation materials committee reporting into an overall industrially led Power Generation Committee.
3. A significant re-focusing of investment by industry and government should take place.
4. Support for power generation materials research in universities should be more industry driven.
5. The total estimated size of the required materials programmes should be in excess of £20 million per annum.

The scope of materials knowledge required to ensure the optimisation of efficiency and reliability of any gas turbine installation expands by the day and the demand for Materials Engineers and the innovation of new material products is never satisfied. Research and development must, however, take into account the particular requirements of the Design Engineer and the characterisation of both established and novel material products must be demonstrated over the temperature, stress, strain and life required to qualify as candidate materials for a gas turbine component. The test data acquired must be stored in machine-readable form for further analysis by materials engineers who are responsible for the derivation of design data and the publication and review of material specifications of a quality which ensures the product's consistency.

For many years, material scientists have attempted to explain the temperature and strain dependence of the mechanical properties of materials and the diversity of the resulting constitutive equations attests to the lack of sufficient test data over the range of dependent variables to provide adequate statistical analysis. They would have a better chance of achieving this end if a 'Global Database' of test data were created. Generic models of materials performance, probably for each material class, would provide the Design Engineer with data of known confidence, a factor which will determine the reliability of any engine.

Manufacturers of material products will have more success in marketing their products if they can provide the user with access to a database of physical and mechanical properties which demonstrates the superiority or equivalence of their product at a competitive price. Collaboration between manufacturers and users in the accumulation of test data will allow the derivation of reliable design data with known confidence limits thereby decreasing their product liability requirement where component failure is the result of poor design or misuse.

The need for databases to store the properties of material products is now well established and many organisations have been successful in developing their own systems. However, no standard has yet been agreed which defines a material product precisely nor a standard to define a list of attributes which uniquely defines all the properties we commonly use throughout the materials world. It is recommended that The Institute of Materials should take the responsibility for such definitions and require that all data published in its associated publications, usually in the form of figures, is submitted for entry to its own materials database with appropriate citations for its authors. Such a database would allow the publication of a single but important datum which might otherwise never be published. The advisory role of the Institute would be strengthened by the resource and provide an opportunity for additional revenue. If we start now, the equivalent of a hundred data complications similar to that produced by Woolman and Mottram[6] would be available to design engineers within the first decade of the millennium.

SUMMARY AND CONCLUSIONS

High temperature alloys for industrial gas turbines include common engineering alloys such as cast irons and steels as well as those developed specifically for aero engine gas turbines. The range of material products includes conventionally processed bar, forgings and sand castings in addition to those produced by more sophisticated processes such as consumable electrode melted ingot and forging and precision castings to provide complex cooling passages in conventionally cast, directionally solidified and single crystal forms.

The gas turbine designer needs to have access to 'candidate' materials for each component application. Each material product must be characterised by a specification which will ensure consistency of product and machine readable data to carry out thermal, stress and 'lifting' calculations by computerised methods.

The daunting task of predicting gas turbine product reliability will be made easier by the establishment of a global database of the material properties of alloys over a temperature range of minus 50°C to the melting point of the material. Materials Engineers, Manufactures of material products and The Institute of Materials have the opportunity to create such an environment.

ACKNOWLEDGEMENTS

The author is indebted to his colleagues for their comments and assistance, especially Derek Allen, Peter Barnard, Chris Bullough, Richard Laycock, Nick Otter and Eric Royle; and to ABB ALSTOM Power. for permission to publish this paper.

REFERENCES

1. C.H. Buck: 'The Future of the Industrial Gas Turbine', *Advances in Turbine Materials, Design and Manufacturing, Proceedings of the Fourth International Charles Parsons Turbine Conference*, Institute of Materials, 1997.
2. C.K. Bullough and B.J. Piearcey: 'Optimising Materials Assessment for Industrial Gas Turbines Using Computerised Methods', *Proceedings of the third International Charles Parsons Turbine Conference*, Institute of Materials, 1995.
3. B.J. Piearcey and C.K. Bullough: 'An On-line Database for Gas Turbine Materials', *Paper 97-GT-167* presented at the ASME Conference, June 1997.
4. W. Carpenter, B.S.J. Kang and K.M. Chang: 'SAGBO Mechanism on High Temperature Cracking Behaviour of Ni-base Superalloys', *Superalloys 718, 625, 706 and Various Derivatives*, E.A. Loria, The Minerals & Metals Society, 1997.
5. D.H. Allen, S. Beech, L. Buchanan, J. Oakey and R. Vanstone: 'Requirements for Materials R&D for Coal-Fuelled Power Plant: Into the 21st Century', Institute of Materials, to be published.
6. J. Woolman and R.A. Mottram: *The Mechanical Properties of the British Standard En Steels*, BISRA, Pergammon Press, 1969.

Ceramics in Aero Gas Turbines – an Engineer's View

J. DOMINY*, W.J. EVANS[†] and S.L. DUFF[†]
* *Rolls-Royce plc, Derby, UK*
† *IRC in High Performance Materials, University of Wales, Swansea, UK*

INTRODUCTION

For many years it has been predicted that various forms of ceramic material would play a significant and important role in the development of the gas turbine engine for aerospace, marine and land-based applications. Indeed, much of the composites work that was in progress in the 1950s and early 60s was aimed at early ceramic matrix composite (CMC) materials. The champions of ceramics have prophesied an optimistic outlook since the very earliest days of the gas turbine engine. Yet there is still no civil aircraft engine using ceramics in anything like a structural application and with one or two exceptions, ceramics fare little better in military engines.

The question addressed by this paper is why ceramic materials have still to gain a significant foothold in the gas turbine engine, and what can be done to expedite their arrival and allow designers to exploit the properties that they undoubtedly posses. There are lessons and parallels that can be learned from the introduction of other 'new' materials, such as carbon fibre reinforced polymer matrix composites.

THE BENEFITS OF CERAMICS

In spite of the views sometimes expressed, ceramics have no right to a place in the engine merely because they are perceived as 'advanced' or because research effort, be it academic or industrial, is directed towards them. In today's industry, any new material must earn its place in the engine, demonstrating both technical capability and, most particularly, cost effectiveness. The hard fact is that there is currently no 'hole' in the capability of the gas turbine engine, no component that cannot be made in an existing material. So, why should we continue with ceramics?

Ceramics offer two properties of particular interest to the designer. The first is a high temperature capability. A number of hot components in the engine must be cooled in order that they operate within the temperature constraints. In these cases, ceramic allows the cooling to be reduced or eliminated. Since the cooling requires air drawn from the core of the engine, the reduction in flow is a direct benefit to the fuel consumption of the engine. The ability to simplify or eliminate the cooling system has the added bonus of reduced weight and cost. Each of these is a significant advantage and a major issue in the sale of engines. A further advantage is the low density of ceramics. Most of the components that appear to be likely candidates for ceramic applications use one of the nickel alloys, all of which have high densities. The use of ceramics in these parts offers a direct weight advantage. While the designer of the civil engine will be interested in weight, a realistic projection of the likely application of ceramics means that the total volume will not make much of an impact on the whole engine. However, his military colleague will be much more interested in low density, since the projected application of ceramics in the engine will be towards the rear of the engine. In a combat aircraft, this has major implications on the centre of gravity and through this a synergetic effect on the whole aircraft. This makes ceramics a much more attractive proposition.

CERAMICS – THE PAST

If ceramics offer the advantages of low density and high temperature compared with even the most advanced metals, both of which are such essential requirements in the gas turbine, then why have ceramics not made the long predicted impact on engine design. To understand this we must first address a number of problems that currently inhibit their use:

Toughness

The basic criticism levelled at ceramics is their lack of ductility. While the lay expectation that ceramics inevitably behave like domestic tableware is pessimistic, nevertheless, even engineering ceramics tend to be brittle. Unlike some other modern brittle materials, such as polymer composites, which exhibit significant elastic behaviour before an eventual brittle failure, ceramics have a very low strain to failure. As a consequence, it is not possible to design a structure that is tolerant to damage or that will have a 'graceful' failure. Thus, while it is quite possible to produce components that will survive normal running, it is extremely difficult to design them to tolerate the effects of the real world. Figure 1 illustrates an experimental turbine assembly with monolithic silicon nitride blades. The inset shows the 'after' illustration demonstrating the effect of a combination of a low strain to failure and brittle fracture. It is unfortunate that nature has played this trick on us, since in other respects, modern monolithic systems offer a stiff, strong, reliable material with a low density which is able to run at high temperatures.

Toughness is the ability of a material to resist flaws propagating to failure. This can only be realised through an ability to absorb energy released at the crack tip. In metals this is achieved by the plasticity in the region ahead of the crack or defect. Some years ago, it was anticipated that CMC materials would solve the toughness problem by offering a form of 'pseudo-plasticity' (Fig. 2). While this has been achieved to an extent, the resultant material has a range of new problems which must also be overcome. Toughness, in the context of

Fig. 1 Before and after views of a test on a monolithic ceramic turbine wheel illustrating the problem of engineering with intolerant materials.

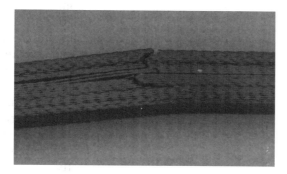

Fig. 2 Pseudo-plasticity demonstrated by delamination in a CMC bend specimen.

ceramics, is a strange term, which can have many meanings. There is a well-known demonstration of 'toughness' in carbon–carbon (a brittle material) where a nail is driven through a plate of the material thereby demonstrating its toughness and resilience. This is only possible because of the voidage within the material which allows the brittle internal structure to collapse within itself as the nail passes through. Is this toughness? We would suggest not, although it does demonstrate a degree of tolerance to some forms of damage.

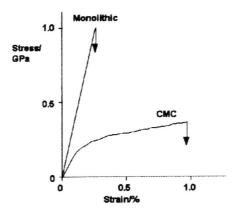

Fig. 3 Comparative behaviour of a high performance monolithic ceramic and a typical ceramic matrix composite based on silicon carbide fibres.

In spite of this, in the aero engine industry most of the research emphasis is now on CMC materials such as SiC/SiC and SiC/Alumina, at the expense of the monolithic ceramics. While it is questionable that real and worthwhile toughness can be achieved in a ceramic, there is no doubt that continuous fibre reinforced ceramic composites do offer a greater resistance to damage than the monolithics (Fig. 3).

Reliability

While CMCs show better damage tolerance than monolithics, this can be at the expense of the reliability of the material. The allowable design parameters for any material are derived from the mean value of the data and the standard deviation. Greater or lesser degrees of sophistication may be applied to the analysis (such as Weibull) but the basis remains the same. Thus, any material that has a relatively large standard deviation will suffer a high penalty in relation to the mean value of the data. CMC materials exhibit more variation in their behaviour than many monolithics. It is perhaps not surprising that CMCs suffer in this way considering their hetrogenious nature.

Strength

In comparison with a metal, CMCs offer quite high values of stiffness (Young's modulus). In specific terms, where density is introduced (stiffness/density), then CMCs become highly competitive at temperatures above 1000°C, particularly compared to the high-density super-alloys with which they compete. This is not, unfortunately, true of strength. Figure 4 shows the comparison of the specific strength of a CMC with high temperature metals, typically turbine blade materials. The graph illustrates that while the CMC does retain a degree of specific strength to higher temperatures than the metals, the strength is relatively low. In

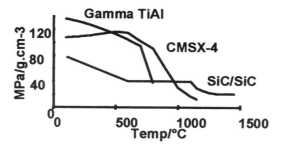

Fig. 4 Comparison of the specific strength of a CMC and competing metal alloys.

absolute terms, the usable strength is very low compared with that of metals. If we are to use CMCs, then it will be necessary to evolve designs that exploit this combination of high stiffness and low strength.

Once again, consider a parallel with carbon/epoxy composites. In that case nature has been kind to us and provided a material that has more useable strength than stiffness. This means that a polymer composite component will generally reach a limiting deformation before the material fails. This is a particularly useful property in a material system since, in most engineering design, the ultimate criteria is usually deformation rather than absolute strength. The consequence is that while there are still scientific arguments regarding the nature of failure and failure criteria in polymer composites, the engineering community has used the material because in practise it is rarely necessary to have a complete understanding of the mechanics of the failures involved. The point to note, however, is that almost sixty years since the introduction of what we would today understand as structural composites, the arguments over the nature of failure still rage. There are, however, few enough parts that are designed against ultimate strength criteria that it is usually possible to establish an acceptable empirical failure criterion for the component as a whole.

In the case of CMCs this is not a luxury that we enjoy. Because of the high stiffness and low strengths exhibited by these materials, the dominant design criteria will almost certainly be a strength criterion. CMCs, therefore, will only gain widespread and general application in the aerospace industry when we gain a high level of understanding of the failure modes and mechanisms within the material to the extent that strength can be realistically predicted. Bearing in mind the complexity of CMC as a material system, we would suggest that this is a substantial piece of work. It will require the co-operation of interested parties in academia and industry if a useful methodology is to be offered in a reasonable time scale.

Temperature

The main attraction of CMCs is their ability to withstand high temperatures for long periods of time. In reality, the very high temperatures are rarely achieved.

For component lives in thousands and tens of thousands of hours, the realistic CMCs offer only 1200°C with possible excursions to 1400°C. At these temperatures the material will still require cooling in parts of the turbine, and since it is possible to produce more sophisticated cooling features in metals, CMCs begin to lose their advantage. Nevertheless, there is still a 'window' in which ceramic materials offer advantages, but it is not as wide as might be expected. Carbon–Carbon, in principle, is the answer to the wish list due to its stability at extremely high temperatures. Unfortunately, the material, like any carbon, is a distant relation to coal and will not survive an oxidising environment. Substantial work has been carried out on protective coatings, but while these are adequate for some spacecraft applications, they do not come close to achieving the durability necessary for aircraft propulsion systems, which require component lives of several thousand hours (Fig. 5). These materials have, however, proved very effective in non-oxidising environments, such as rocket motors and brake discs.

THE PRESENT

Although the above may appear to paint a bleak picture, there is light at the end of the tunnel. CMC materials will find their way into the engine in niche applications where their advantages can be exploited while their disadvantages are

Fig. 5 Before and after views of a test on a carbon–carbon component showing the oxidation damage due to a break-down of the protective coating.

Fig. 6 The turbine tip seal segment which was recently successfully tested in a Rolls-Royce Trent engine. The lower photograph shows a number of the segments fitted in the assembly.

suppressed. An example of this is the turbine seal segment which has been successfully demonstrated in a Rolls-Royce Trent engine, and which is the first engine run of this component (Fig. 6). This segment forms part of a sealing ring surrounding the rotating turbine blade tips. The requirement of the material is contradictory. On the one hand it must be robust enough to operate at very high temperatures, while on the other it must be soft enough to abrade when the blades rub without damaging the tips. In comparison with conventional metal designs, CMC is able to withstand the temperatures without the need for cooling, has a much lower weight and is better able to withstand erosion. The abradable coating – a material called Hassmap which has been developed at Rolls-Royce – is better able to withstand blade tips rub while resisting oxidation and maintaining the optimum clearance. The mechanical and thermal loads on the seal segment are relatively low and within the capability of the material system.

Another application in development is the exhaust vane shown in Fig. 7. Once again, the density and thermal stability of CMC can be exploited without the mechanical properties of the material becoming an issue. The future of CMCs as materials for aerofoil stators in a turbine system is reasonably bright. However, it would be a very brave prophet indeed who predicted their use as rotating turbine blades, since the very high stresses due to the centrifugal loading put the necessary material properties well beyond the realistic properties of CMCs.

Fig. 7 CMC exhaust vanes (various material systems and suppliers).

THE FUTURE

The components discussed above all meet the conditions that the applied loads are relatively low and the components are relatively small. These are important for two reasons.

The first is that, as we have already seen, the CMC community has yet to get to grips with the mechanical behaviour of the material, at least in engineering terms. It may well be that, in reality, the problems have been overstated here and that, if creep related failures dominate, then an engineering database can be compiled relatively economically. As we stand today the nature of practical failure modes in real components which must run with reliability approaching absolute, is not sufficiently well understood to allow industry to proceed with structural CMC components.

The second is cost. Today the emphasis in the aero engine industry, as in the whole of the air transport sector, is cost; both cost of ownership and first cost to the manufacturer. All of the current material systems that are commercially available are extremely expensive, even compared with the super-alloys with which they compete. If CMC materials are to make a significant impact on the engine, then systems and processes that are inherently lower cost must be brought to fruition. Any new material must fight to earn its place in the modern gas turbine and it is increasingly difficult to justify the use of materials that lead to an increase in the manufacturing cost of the engine. Until these issues can be resolved, it will be difficult to bring CMCs to general application.

CONCLUSION

CMC materials have loitered at the edge of gas turbine engine technology since their introduction in the 1940s, but have never been invited to the party. The reasons for this are essentially the difficulty of defining their engineering behaviour, their failure to live up to their potential and their high and uncompetitive cost.

In spite of this, there are a number of CMC components under development for niche applications at Rolls-Royce and its competitors. In these components the advantages of CMCs can be exploited in areas where their disadvantages are not exposed. Pressures to introduce these materials arise largely from difficulty in accommodating the continual rise in peak engine temperature with 'conventional' super-alloys.

However, until a well understood, reliable and economical system becomes available, CMCs will be niche players. If these limitations are to be overcome, then the community must work together to develop materials that are properly characterised, in a way that can be assimilated by industry and which they can afford to use. The current piece-meal approach to CMCs is unlikely to achieve this objective.

This is a team game. To succeed it will be necessary to effectively co-ordinate the skills and expertise that reside with the user, the material supplier and in academia. There must be a focus that allows them to work together to introduce cost effective ceramic components into industry.

ACKNOWLEDGEMENTS

The authors wish to thank Rolls-Royce plc for permission to publish and present this paper. They also thank their colleagues at both Swansea and Derby for contributing data. This paper represents the personal opinions of the authors and is not necessarily the view of Rolls-Royce plc.

BIBLIOGRAPHY

S.F. Duffy, J.L. Palko and J.P. Gyekenesi: 'Structural Reliability analysis of laminated CMC components', *J. Engineering for Gas Turbines and Power*, 1993, **v115**, 103–108.

J.C. McNulty and F.W. Zole: 'Application of weakest-link fracture statistics to fibre reinforced ceramic matrix composites', *J. Am. Ceram. Soc.*, 1997, **80**(6), 1535–1543.

C.J. Gilbert, R.H. Dowskaldt and R.O. Ritchie: 'Behaviour of cyclic fatigue cracks in monolithic silicon carbide', *J. Am. Ceram. Soc.*, 1995, **78**(9), 2291–3000.

A. Chulya, J.P. Gyekenesi and R. Bhatt: 'Mechanical behaviour of Fibre reinforced SiC/RBSN ceramic matrix composites: Theory and experiment', *J. Engineering for Gas Turbines and Power*, 1993, **v115**, 91–102.

S.L. Duff: 'Investigation of fatigued CMC test pieces', *M. Res. thesis*, Department of Materials Engineering, University of Wales, Swansea, 1997.

Subject Index